Nanoscale Science and Technology

NATO ASI Series

Advanced Science Institutes Series

A Series presenting the results of activities sponsored by the NATO Science Committee, which aims at the dissemination of advanced scientific and technological knowledge, with a view to strengthening links between scientific communities.

The Series is published by an international board of publishers in conjunction with the NATO Scientific Affairs Division

A	**Life Sciences**	Plenum Publishing Corporation
B	**Physics**	London and New York
C	**Mathematical and Physical Sciences**	Kluwer Academic Publishers
D	**Behavioural and Social Sciences**	Dordrecht, Boston and London
E	**Applied Sciences**	
F	**Computer and Systems Sciences**	Springer-Verlag
G	**Ecological Sciences**	Berlin, Heidelberg, New York, London,
H	**Cell Biology**	Paris and Tokyo
I	**Global Environmental Change**	

PARTNERSHIP SUB-SERIES

1.	**Disarmament Technologies**	Kluwer Academic Publishers
2.	**Environment**	Springer-Verlag / Kluwer Academic Publishers
3.	**High Technology**	Kluwer Academic Publishers
4.	**Science and Technology Policy**	Kluwer Academic Publishers
5.	**Computer Networking**	Kluwer Academic Publishers

The Partnership Sub-Series incorporates activities undertaken in collaboration with NATO's Cooperation Partners, the countries of the CIS and Central and Eastern Europe, in Priority Areas of concern to those countries.

NATO-PCO-DATA BASE

The electronic index to the NATO ASI Series provides full bibliographical references (with keywords and/or abstracts) to more than 50000 contributions from international scientists published in all sections of the NATO ASI Series.
Access to the NATO-PCO-DATA BASE is possible in two ways:

– via online FILE 128 (NATO-PCO-DATA BASE) hosted by ESRIN,
Via Galileo Galilei, I-00044 Frascati, Italy.

– via CD-ROM "NATO-PCO-DATA BASE" with user-friendly retrieval software in English, French and German (© WTV GmbH and DATAWARE Technologies Inc. 1989).

The CD-ROM can be ordered through any member of the Board of Publishers or through NATO-PCO, Overijse, Belgium.

Series E: Applied Sciences - Vol. 348

Nanoscale Science and Technology

edited by

N. García
Laboratorio de Física de Sistemas Pequeños,
Consejo Superior de Investigaciones Científicas,
Madrid, Spain

M. Nieto-Vesperinas
Instituto de Ciencia de Materiales de Madrid,
Consejo Superior de Investigaciones Científicas,
Madrid, Spain

and

H. Rohrer
IBM Research Division,
Zurich Research Laboratory,
Rueschlikon, Switzerland

Springer Science+Business Media, B.V.

Proceedings of the NATO Advanced Research Workshop on
Nanoscale Science and Technology
Toledo, Spain
May 11-16, 1997

A C.I.P. Catalogue record for this book is available from the Library of Congress.

ISBN 978-94-010-6109-4   ISBN 978-94-011-5024-8 (eBook)
DOI 10.1007/978-94-011-5024-8

Printed on acid-free paper

This book contains the proceedings of a NATO Advanced Study Institute held within the programme of activities of the NATO Special Programme on *Nanoscale Science* as part of the activities of the NATO Science Committee.

Other books previously published as a result of the activities of the Special Programme are:

NASTASI, M., PARKING, D.M. and GLEITER, H. (eds.), *Mechanical Properties and Deformation Behavior of Materials Having Ultra-Fine Microstructures*. (E233) 1993 ISBN 0-7923-2195-2

VU THIEN BINH, GARCIA, N. and DRANSFELD, K. (eds.), *Nanosources and Manipulation of Atoms under High Fields and Temperatures: Applications*. (E235) 1993 ISBN 0-7923-2266-5

LEBURTON, J.-P., PASCUAL, J. and SOTOMAYOR TORRES, C. (eds.), *Phonons in Semiconductor Nanostructures*. (E236) 1993 ISBN 0-7923-2277-0

AVOURIS, P. (ed.), *Atomic and Nanometer-Scale Modification of Materials: Fundamentals and Applications*. (E239) 1993 ISBN 0-7923-2334-3

BLÖCHL, P. E., JOACHIM, C. and FISHER, A. J. (eds.), *Computations for the Nano-Scale*. (E240) 1993 ISBN 0-7923-2360-2

POHL, D. W. and COURJON, D. (eds.), *Near Field Optics*. (E242) 1993 ISBN 0-7923-2394-7

SALEMINK, H. W. M. and PASHLEY, M. D. (eds.), *Semiconductor Interfaces at the Sub-Nanometer Scale*. (E243) 1993 ISBN 0-7923-2397-1

BENSAHEL, D. C., CANHAM, L. T. and OSSICINI, S. (eds.), *Optical Properties of Low Dimensional Silicon Structures*. (E244) 1993 ISBN 0-7923-2446-3

HERNANDO, A. (ed.), *Nanomagnetism* (E247) 1993. ISBN 0-7923-2485-4

LOCKWOOD, D.J. and PINCZUK, A. (eds.), *Optical Phenomena in Semiconductor Structures of Reduced Dimensions* (E248) 1993. ISBN 0-7923-2512-5

GENTILI, M., GIOVANNELLA, C. and SELCI, S. (eds.), *Nanolithography: A Borderland Between STM, EB, IB, and X-Ray Lithographies* (E264) 1994. ISBN 0-7923-2794-2

GÜNTHERODT, H.-J., ANSELMETTI, D. and MEYER, E. (eds.), *Forces in Scanning Probe Methods* (E286) 1995. ISBN 0-7923-3406-X

GEWIRTH, A.A. and SIEGENTHALER, H. (eds.), *Nanoscale Probes of the Solid/Liquid Interface* (E288) 1995. ISBN 0-7923-3454-X

CERDEIRA, H.A., KRAMER, B. and SCHÖN, G. (eds.), *Quantum Dynamics of Submicron Structures* (E291) 1995. ISBN 0-7923-3469-8

WELLAND, M.E. and GIMZEWSKI, J.K. (eds.), *Ultimate Limits of Fabrication and Measurement* (E292) 1995. ISBN 0-7923-3504-X

EBERL, K., PETROFF, P.M. and DEMEESTER, P. (eds.), *Low Dimensional Structures Prepared by Epitaxial Growth or Regrowth on Patterned Substrates* (E298) 1995. ISBN 0-7923-3679-8

MARTI, O. and MÖLLER, R. (eds.), *Photons and Local Probes* (E300) 1995. ISBN 0-7923-3709-3

GUNTHER, L. and BARBARA, B. (eds.), *Quantum Tunneling of Magnetization - QTM '94* (E301) 1995. ISBN 0-7923-3775-1

PERSSON, B.N.J. and TOSATTI, E. (eds.), *Physics of Sliding Friction* (E311) 1996. ISBN 0-7923-3935-5

MARTIN, T.P. (ed.), *Large Clusters of Atoms and Molecules* (E313) 1996. ISBN 0-7923-3937-1

DUCLOY, M. and BLOCH, D. (eds.), *Quantum Optics of Confined Systems* (E314). 1996. ISBN 0-7923-3974-6

ANDREONI, W. (ed.), *The Chemical Physics of Fullereness 10 (and 5) Years Later. The Far-Reaching Impact of the Discovery of* C_{60} (E316). 1996. ISBN 0-7923-4000-0

NIETO-VESPERINAS, M. and GARCIA, N. (Eds.): *Optics at the Nanometer Scale: Imaging and Storing with Photonic Near Fields* (E319). 1996. ISBN 0-7923-4020-5

RARITY, J. and WEISBUCH, C. (Eds.): *Microcavities and Photonic Bandgaps: Physics and Applications* (E324). 1996. ISBN 0-7923-4170-8

LURYI, S., XU, J. and ZASLAVSKY, A. (Eds.): *Future Trends in Microelectronics: Reflections on the Road to Nanotechnology* (E323). 1996. ISBN 0-7923-4169-4

JAUHO, A. and BUZANEVA, E.V. (Eds.): *Frontiers in Nanoscale Science for Micron/Submicron Devices* (E328). 1996. ISBN 0-7923-4301-8

ROSEI, R. (Ed.): *Chemical, Structural and Electronic Analysis of Heterogeneous Surfaces on Nanometer Scale* (E333). 1997. ISBN 0-7923-4489-8

JOACHIM, C. (Ed.): *Atomic and Molecular Wires* (E341). 1997. ISBN 0-7923-4628-9

SOHN, L.L., KOUWENHOVEN, L.P., SCHÖN, G. (Eds.): *Mesoscopic Electron Transport.* (E345). 1997. ISBN 0-7923-4737-4

CONTENTS

Nanowires

Forces

Nanomagnetism

Mesoscopic Systems

viii

Photonics

Clusters

Solid-Liquid Interfaces

Special Issues: Microscopy and Particles

PREFACE

The present volume is the report of the concluding Forum of the NATO Special Programme on Nanoscale Science, held in Toledo, Spain, May 11-16, 1997, covering the major activities of the program and intended "to provide a better understanding and clearer perspective of recent developments in Nanoscale Science and Technology". The Programme began its activities in 1991. It aimed at interchanging views and experience relating to properties and processes on aggregates of matter with nanometer sizes, to measuring techniques, and to fabrication of nanometer scale structures and at fostering international collaboration.

The Special Programme brought together workers in nanoscience and technology across normal disciplinary boundaries. During the Programme's life-span (1991-1996) the NATO Advisory Panel solicited and approved proposals for specialised meetings and courses (30 Advanced Research Workshops and 7 Advanced Study Institutes) and made recommendations for awards to collaborative research projects (14 Collaborative Research Grants and 4 Linkage Grants) involving participation of scientists from laboratories in Cooperation Partner countries. Such activities have enhanced awareness of science and technology on the nanometer scale, and have initiated and strengthened cooperation between universities and public and private research institutions. They have also inspired many young students to come to this new exciting research direction which is now a topical field of high priority worldwide in education, research and industry.

Nanoscience is the science of dealing with nano-entities. This applies to understanding and selectively modifying properties, to manipulating, positioning and machining nano-objects, as well as to developing new concepts for treating nano-entities, especially in large numbers. One of the salient characteristics of nanoscience is that in the spirit of solid state technology as well as of mesoscopic physics, an object retains its individuality; it appears and is treated as an individual instead as member of a statistical ensemble.

The most obvious and at present economically rewarding notion of nanoscience and nanotechnology - nanoscience and nanotechnology have to work in even closer symbiosis than was the case on the micrometer level - is seen in the continuation of miniaturization from micrometer dimensions, the standard of today's microtechnology to nanometer dimensions with a precision in fabrication and machining down to the atomic scale. The main driving force behind the rapidly progressing miniaturization is the data processing industry. By and large, miniaturization in the past four decades has progressed evenly and exponentially, e.g. both for atoms per bit of information and for heat dissipation per logic operation, at the rate two to three orders of magnitude per decade. Currently there is no reason to believe that miniatuization will not progress at a similar rate in the near future, even though the technological challenges are increasingly demanding. The challenge in the coming decade will be to find methods suitable for the mass production of Gbit chips from present-day elements that can already be miniaturized sufficiently and assembled in small quantities. Afterwards, new types of elements will be required.

On the other hand, nanostructures offer new possibilities and new opportunities for science and technology distinctly different from micro and certainly from macro. Properties change qualitatively and new laws govern processes below some critical length scales or when quantum effects become dominant or, quite generally, at the transition from condensed matter to molecular behavior. Speed, intensities, and quantities increase by many orders of magnitudes, and nm to pm precision in measuring, sensing and actuation result in unprecedented sensitivity levels.The merit of orders of magnitude difference lies in the practicle aspects, e.g. mm or nm at the speed of sound makes a lot, at the speed of light not much of a difference when dealing with time scales well above psec. Ultra fast mechanics, thermal relaxation and diffusion will offer valid complements to electronics.

The fascination of nanoscience and nanotechnology beyond nanoelectronics is the prospect of using them to build sensors, actuators, and machines of similar dimension and complexity as biological organs, in other words to work on the same scale as living nature does. Living nature is but nanoscience and nanotechnology. Distributed, in situ mechanical, chemical, and electrical processing are intimately interweaved on the smallest scale. Also the energy required for processing is stored locally and also chemically transported. Everything does many things, sensors are not only sensors but at the same time complex processors, so are actuators.

The two basic technical prerequisits for nanoscience and technology are, beside the continued improvement in conventional microscopies and processing, the recent developments in molecular beam epitaxy and chemical vapor deposition and similar process techniques and the emergence of local probe methods. The first have produced man-made, artificially structured semiconductors with nearly arbitrary spatial control of composition and doping allowing to design at will the band structure and energy gap and thus to tailor the transport and optical properties of these structures. Simultaneously the characteristic dimensions of the systems under investigation have shrunk to the nanometer range. This is the case of superlattices with periods between 10 and 100 Å and of systems displaying quantum phenomena arising from reduced dimensionality such as quantum dots and wires. These systems are extremely interesting to the physicist as artificial objects approaching ideal quantum mechanical models. Thus the superlattices provide a unique opportunity to explore the electric field induced quantization, the Wannier-Stark ladder and the probable associated Bloch oscillations. They could also give rise to radically new electronic devices holding great promise for increased switching speed storage density and the integration of optical and electronic functions.

The local probe methods made possible routine imaging of a variety of properties down to atomic scale resolution in environments ranging from ultrahigh vacuum to electrolytes. They, however, are increasingly employed as tools for manipulation and modification down to the atomic level and as local sensors with unprecedented sensitivity. The local probes are the fingertips for the nanoworld.

The large variety of contributions in this volume gives a flavour of the fantastic prospects of nanoscience and nanotechnology, be it in nanoelectronics, be it beyond.

We would like to thank the NATO Scientific Committee as well as the Panel for Nanoscale Science and his director (J.A. Raussell-Colom) for supporting this Forum. Dr. A. Correia helped a great deal in the organization of the event. In addition, we also acknowledge the authorities of the Universidad de Castilla-La Mancha for providing us a beautiful site for the meeting.

H. Rohrer, N. García and M. Nieto-Vesperinas
January 1st, 1998

LOW VOLTAGE NONLINEAR CONDUCTANCE OF GOLD NANOWIRES: ROOM TEMPERATURE COULOMB BLOCKADE EFFECT?

J.L. COSTA-KRÄMER[1], N. GARCIA[1], M. JONSON[2],
I. V. KRIVE[2,3], H. OLIN[4], P.A SERENA[1],
AND R.I. SHEKHTER[2]

[1]*Laboratorio de Física de Sistemas Pequeños y Nanotecnología
Consejo Superior de Investigaciones Científicas
Serrano 144, 28006-Madrid, Spain*

[2]*Department of Applied Physics and* [4]*Department of Physics
Chalmers University of Technology and Göteborg University
S-412 96 Göteborg, Sweden*

[3]*B. I. Verkin Institute for Low Temperature Physics
& Engineering
National Academy of Science of Ukraine
4 Lenin Ave., 310164 Kharkov, Ukraine*

ABSTRACT. We present and analyze theoretically measurements of the room temperature $I - V$ characteristics of gold nanowires whose zero current conductance is quantized in units of $2e^2/h$. A faster than linear increase of current was observed at low voltages beginning from $V_c \simeq 0.1V$. We analyze the nonlinear behavior using a Luttinger liquid approach and show that it can be understood in terms of a dynamic Coulomb blockade effect.

1. Introduction

The conductance quantization (CQ) at room temperature in metallic nanowires formed by breaking a contact between two metallic electrodes appears to be a very robust phenomenon. Theoretically predicted at first as a step-like or oscillatory behavior of conductance in Scanning Tunneling Microscope (STM) calculations [1] it has been observed in wires made of different metals and produced by both STM techniques [2], by simply breaking the contact between touching macroscopic wires [3] and by using mechanically

1

N. García et al. (eds.), Nanoscale Science and Technology, 1–10.
© 1998 *Kluwer Academic Publishers.*

controllable break junctions [4]. Although the accuracy of the quantization in metallic nanowires is typically lower than for quantum contacts formed in gated semiconductor heterostructures [5] (where a consistent explanation of the phenomenon exists [5, 6]), the mere fact that plateaus appear at quantized values of conductance – $G \sim n \, (2e^2/h)$, $n = 1, 2, \ldots$ shows that electron transport through the nanowire is ballistic and that the current is carried in quantized modes (conduction channels) when its smallest width is of the order of the Fermi wavelength (about 0.5 nm).

In this work we present exprimental results for the room temperature $I - V$ characteristics of gold (Au) nanowires whose zero current conductance is quantized. The nonlinear behavior of the current over a finite voltage interval was analyzed in terms of a dynamic Coulomb blockade of the conducting mode(s) most sensitive to backscattering. The experiments support this picture according to which the current carried by these modes is quenched at low bias voltages due to a strong renormalization of transmission probability caused by electron-electron interactions and the slow rate of charge relaxation in quasi-1D systems. Appealing to theory we propose the blockade to be lifted at higher voltages, hence explaining the observed nonlinearity.

2. Experimental intensity-voltage curves

Two different methods for producing nanowires were used in the work reported here. First, macroscopic Au wires were put in contact in ultra high vacuum after they had been cleaned by heating as described in detail in Ref. [7]. In this process nanowires tend to form where the macroscopic wires make contact enabling $I - V$ curves to be measured as they are pulled apart. We focus on the *nonlinear* character of these curves; typical results are shown in Figs. 1 and 2. The following remarkable properties can be noted: (i) the characteristic crossover voltage $V_c \sim 0.1$ V to the nonlinear regime is sample-independent and anomalously small in comparison with the atomic energy scale ~ 1 eV set by the spacing of quantized mode energies in the contact; (ii) the nonlinear character of the $I - V$ curves does not depend on the conductance in the zero voltage limit and is therefore independent of the number of conduction channels; (iii) the nonlinear contribution to the current (i.e., what remains after a linear contribution has been subtracted off) is positive and can be fitted to a power law, V^q, where $q \sim 2 - 4$ (cf. below); (iv) the current has a tendency to go back to having a linear voltage dependence at the voltages when the conductance has increased by approximately one quantum.

In a different set of measurements at ambient pressure and room temperature the STM technique was used [2]. The mechanical stability achieved

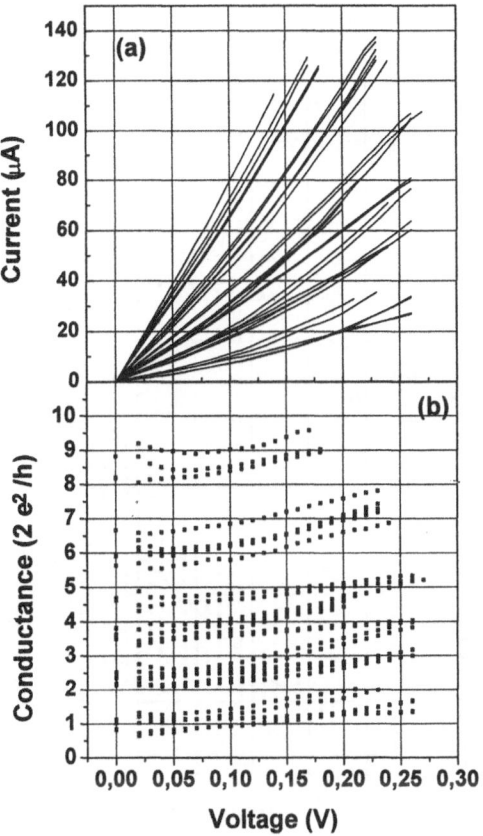

Figure 1. (a) Typical current-voltage (*I-V*) curves corresponding to different quantized conductance plateaus for nanowires formed between two macroscopic wires. The current increases faster than linear with voltage although the opposite is expected in a single-particle picture. (b) Characteristic results for the conductance $G(V)$; the measured low voltage values cluster around multiples of the conductance quantum $2e^2/h$. A nonlinear increase of about one quantum unit starts at a characteristic voltage $V_c \sim 0.1\ V$ which is anomalously small compared to relevant quantized mode energies.

with this technique was not as good as with the other; only a small fraction of the formed nanowires was stable over a long enough time (>10 seconds) to make it possible to measure the $I - V$ curves. The general trend of the results shown in Fig. 3 is the same as for those obtained with the two macroscopic wires: The current increases linearly with voltage till about 0.1 V. At higher voltages the nonlinear contribution to the current increases as a power law, V^q, with q in the range 1.8-3.0 (compared to 2.5-3.8 for the first method).

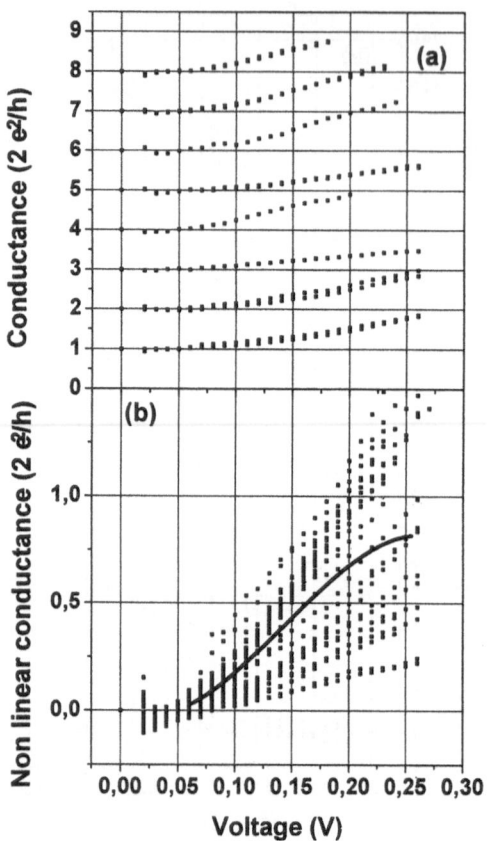

Figure 2. Conductance $G(V)$ as in Fig. 1b except that the $V = 0$ values has been (a) fitted to an integer times $2e^2/h$ and (b) subtracted off. On average (thick line) the conductance increases by almost one quantum unit.

3. Theoretical model

The first observation one needs to explain is the unusually small (when compared to the electron energies) voltage $V_c \sim 0.1$ V, which determines the onset of nonlinearity. This fact rules out any elastic scattering mechanism as well as the field effect in electron transport through the nanowire as a possible mechanism behind the observed nonlinearity. Besides, it is difficult to understand the universal character of the nonlinear behavior (especially that is does not depend on the number of channels) in a picture where the switching on of extra modes is due to field suppression of atomic-size barriers in the wire. One could also speculate that quasiparticle excitations (e.g. phonons) cause the nonlinearity through inelastic electron energy relaxation processes. However, any inelastic process should increase

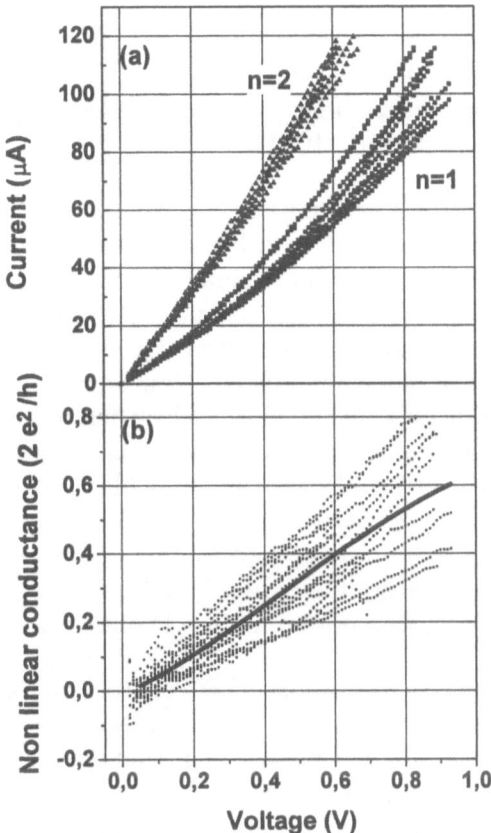

Figure 3. (a) Typical *I-V* curves corresponding to the first two quantized conductance plateaus for nanowires formed by an STM tip retracting from a substrate. The faster than linear rise in current is similar to what is shown in Fig. 1a. (b) Conductance $G(V)$ as in Fig. 1b except that the $V = 0$ values has been subtracted off. On average (thick line) the conductance of these STM-type nanowires increases slower than in the case of the nanowires formed by two macroscopic wires in contact.

the rate of relaxation with the increase of voltage thus leading to negative contribution to nonlinear current in contradiction with our measurements.

All of the above force us to suggest that many body effects (dynamic Coulomb blockade) determine the nonlinear character of charge transport through a long ($L \gg W$, W is the wire diameter) quasi 1D nanowire. Within this picture the crossover voltage to the nonlinear conductance regime can be naturally identified with the Coulomb blockade energy $\varepsilon_L = e^2/L$ ($L \gg \lambda_F$ is the length of the nanowire).

It is well known that for an "infinitely" long one-dimensional (1D) *ballistic* (this term means that the electron transport in a wire is reflectionless) wire interactions renormalize the conductance quantum to a value which depends on the interaction strength [8, 9]). However, the conductance of a *finite* ballistic quantum wire attached to leads with noninteracting electrons is quantized in integer units of the conductance quantum regardless of the interactions in the wire [10, 11].

The presence of backscattering changes the picture radically. In effectively infinite quantum wire even arbitrary small backscattering suffices to block DC currents at vanishingly small voltages and temperatures due to strong enhancement of backscattering by electron-electron interaction [12, 13]. In a real situation the quantum wire is always finite and hence the renormalization of barriers stops at a low energy scale of order e^2/L. The low voltage transport of interacting electrons in a multichannel wire becomes selective; the modes most sensitive to backscattering (see the corresponding criteria below) could be completely blocked by the Luttinger liquid renormalization effects (essentially a Coulomb blockade of transport). Other modes even in the case of strong interaction are still fully transmitting. The blockade is gradually lifted by bias voltages beginning from $V > V_c \sim e/L$ resulting in a power-law $I - V$ characteristics.

To make the above considerations more concrete we will model a nanocontact formed in CQ experiments as an atomic-size constriction (characterized by a set of transmission coefficients for quantized modes) attached to a long $(L \gg W)$ nanowire of width W. In an adiabatic model of a nanoconstriction it is reasonable to assume that T_0 is *exponentially* close to zero or unity for all channels but one (or possibly a few). This model is based essencially on the supposition that while elongation the nanocontact changes its diameter only in small region which determines the number of channels.

Since a single mode case comprises already all the physics discussed, we consider first a 1D model of interacting spinless electrons scattered by a local potential barrier and postpone comments on spin- and multichannel effects till the end. For our purpose it is important to be able to describe a situation where the bare transmission probability can be close to — but not necessarily equal to — unity so that the conductance increases by essentially one quantum when the Coulomb blockade effects are suppressed (for example by a high bias voltage). This opening up of a new conduction channel is precisely what we propose may explain the nonlinearities in the measured $I - V$ curves. For arbitrary barriers the model can be treated analytically only for weakly interacting electrons when the effect of interaction can be reduced to a barrier renormalization produced by the formation of Friedel oscillations of the electron density near the barrier. The renormalized transmission coefficient for an infinitely long wire is [14]

$$T^R(\varepsilon) = T_0 \left(\frac{\varepsilon}{D_0}\right)^{2\gamma} \left[R_0 + T_0 \left(\frac{\varepsilon}{D_0}\right)^{2\gamma}\right]^{-1}. \tag{1}$$

Here $T_0(R_0)$ is the bare transmission (reflection) coefficient ($T_0 + R_0 = 1$), ε is measured from the Fermi energy ε_F, $D_0 \simeq \hbar v_F/W$ is the high energy cutoff, and γ is a dimensionless parameter (of order $e^2/\hbar v_F$) which is determined by the strength of the electron-electron interaction [14]. Strictly speaking the parameter γ should be small for this approach to be valid, but it is believed that Eq. (1) still holds qualitatively in the range $\gamma \sim 1$ [?]. We assume that the bare transmission amplitude, T_0, is energy independent near the Fermi level. Then the whole energy dependence of $T^R(\varepsilon)$ stems from the universal effects of barrier renormalization.

The current at zero temperature can be obtained from (1) and the Landauer formula as

$$I(V) = \frac{e}{h} \int_0^{eV} d\varepsilon\, T^R(\varepsilon) \equiv \frac{e}{h} F_\gamma \left(\frac{eV}{D_0} \left[\frac{T_o}{R_0}\right]^{\frac{1}{2\gamma}}\right). \tag{2}$$

Here $F_\gamma(x)$ is a function that has to be obtained numerically except in special cases. It is illustrative to consider one such case, $\gamma = 1$, for which the conductance — defined as the ratio between current and voltage — is

$$G(V) \equiv \frac{I(V)}{V} = \frac{e^2}{h} \left[1 - \frac{\arctan(x)}{x}\right], \quad x = \frac{eV}{D_0}\sqrt{\frac{T_0}{R_0}} \tag{3}$$

For a finite wire of length L renormalization of the transmission amplitude should be stopped at energies $\varepsilon \sim \Delta = 2\pi\hbar v_F/L$. Hence Eq. (2) is valid for $V > V_c \simeq \Delta over e$, which implies that Eq. (3) represents the asymptotic behavior for $V \gg V_c$ (in the case of strong interactions the low energy cutoff V_c depends on the plasmon velocity s and is $(\Delta/e)(s/v_F)^2 \sim e/L$). At lower voltages the transmission coefficient is energy independent and hence the conductance (3) is constant

$$G(V) \sim \frac{e^2}{h} \frac{\frac{T_0}{R_0}\left(\frac{W}{L}\right)^{2\gamma}}{1 + \frac{T_0}{R_0}\left(\frac{W}{L}\right)^{2\gamma}}, \quad V < V_c, \tag{4}$$

and small $(T_0/R_0)(W/L)^{2\gamma} \ll 1$ (Cf. Fig. 4). Note that for $\gamma \sim 1$ Eq. (4) holds even for T_0 close to unity due to a strong enhancement of backscattering amplitude.

The power law $I - V$ result (4) is a quite general one for Luttinger liquid-like theories (it can also be derived in the tunnel Hamiltonian model, $T_0 \ll 1$, for the whole range of interaction strengths). In essence it is the

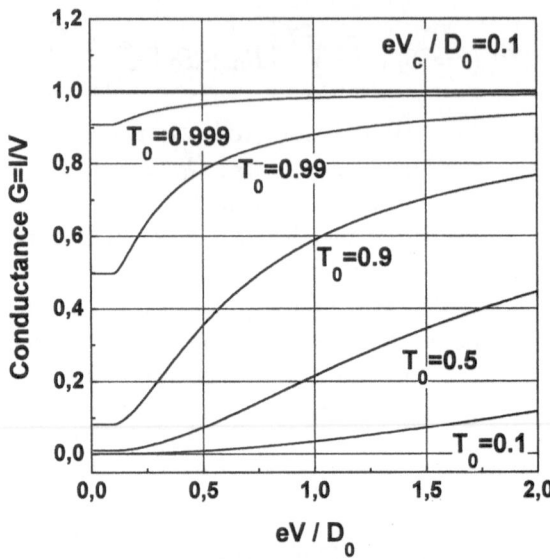

Figure 4. Conductance $G(V)$ calculated from the Tomonaga-Luttinger liquid model, using Eq. (1) for the renormalized transmission $T^R(\epsilon)$ for $\epsilon > eV_c$ and $T^R(eV_c)$ for $\epsilon < eV_c$. This represents the contribution from a conductance channel partly reflected $(T_0 < 1)$ from a potential barrier renormalized by electron-electron interactions.

most significant feature of a Luttinger liquid (together with the analogous temperature behavior).

Simple numerical estimations show that the above formulae allow us to explain at least qualitatively the $I - V$ characteristica measured. Since the *current* grows as approximately the third power of the voltage we conclude that the exponent (2γ) in Eq. (4) should be approximately equal to 2, (hence $\gamma \simeq 1$).

Plots of the nonlinear conductance $G(V)$ calculated from Eqs. (3) and (4) are shown in Fig. 4. Even for a very nearly ballistic mode with $T_0 = 0.9$ one can see that there is a significant (although not complete) restoration of transmittivity for voltages in a range relevant for our experiments. This behavior of the calculated conductance is close to what is observed experimentally in Fig. 3b. In a general situation more than one suppressed mode could be contributing to the nonlinear current. In such a case the total conductance enhancement might be close to or perhaps even larger than the conductance quantum.

Although the above formulae pertain to the case of spinless fermions, the inclusion of spin only results in a redefinition of the exponent γ — which in our approach is regarded as an input parameter — and the appearance of a factor 2 in front of the conductance.

Another question is whether the mode, which is reflected by the renormalized barrier, affects the propagation of the transmitting modes. In a weak coupling limit one can show that if the unrenormalized backscattering amplitude of a Coulomb blocked mode is small and if the intermode level spacing is larger than the relevant low-energy scale, the charge oscillations in a particular channel will only affect its own "band" electrons strongly. To put it differently - transmitting modes will propagate freely even in an interacting electron system irrespective of possibly strong renormalization of backscattering in the other modes. It is possible to extend this claim to the strongly interacting case [16].

From the details of the experiment, such as how much the wire has to be pulled between conductance steps, one can estimate the length of the nanocontact to be about 6 - 10 nm and its diameter to be roughly 2 nm [7]. Using a Fermi velocity $v_F \simeq 10^8$ cm/s one then obtains the following values for the parameters that enter the theory: $\gamma \simeq (e^2/\pi\hbar v_F)\ln(L/W) \simeq 1$; $eV_c \simeq \hbar v_F/L \simeq e^2/L \simeq 0.1$ eV; $V_{\text{sat}}/V_c \simeq L/W \simeq$ 3-5 in surprisingly good agreement with Figs. 1-3 ($V_{\text{sat}} \equiv D_0/e$ is the voltage for which $G(V)$ saturates). Moreover, the contribution of the suppressed modes at $V = 0$ is $\delta G \simeq G_0(W/L)^{2\gamma}$ or 5-10 % of a conductance quantum, which could explain the usually observed [7] fluctuations of the conductance around the quantized values.

4. Concluding remarks

Summing up it is surprising how well a Luttinger liquid approach is able to account for the measured nonlinear $I - V$ characteristics of gold nanowires. In spite of the higly simplified theoretical model (1D model, adiabatic approximation etc.) the implication is that it captures such a large part of the physics of the real nanowires that Luttinger liquid-like effects could be a reasonable explanation for the nonlinear conductance measured in nanocontacts. If so metallic nanowires are among the few systems where the fundamental properties of Luttinger liquid-like states of strongly correlated electrons have been observed. Luttinger liquid-like effects analogous to those discussed here could also be expected to play a role for the temperature dependence of the linear conductance and for the ac transport properties of nanowires. While temperature measurements at the relevant low energy scale (0.1 eV) seems unrealistic, high frequency experiments might provide additional proof in favour of interaction influenced charge transport in nanowires.

Acknowledgement

This work was supported by the EU ESPRIT Project Nanowires, by the Swedish KVA, the Swedish NFR, and by the Spanish DGICyT.

References

1. N. García, presentation at STM Workshop, ICTP, Trieste, 1987 (unpublished)
2. J. I. Pascual et al., Phys. Rev. Lett. **71**, 1852 (1993); J. I. Pascual et al., Science **267**, 1793 (1995); L. Olesen et al., Phys. Rev. Lett. **72**, 2251 (1994); J. L. Costa-Krämer, N. García, and H. Olin, Phys. Rev. B **55**, 12910 (1997).
3. J. L. Costa-Krämer, N. García, P. García-Mochales, and P. A. Serena, Surf. Sci. **342**, L1144 (1995).
4. J. M. Krans, J. M. van Ruitenbeek, V. V, Fisun, I. K. Yanson, and L. J. de Jongh, Nature **375**, 767 (1992).
5. B. J. van Wees et al., Phys. Rev. Lett. **60**, 848 (1988); D. A. Wharham et al., J. Phys. C **21**, L209 (1988).
6. L. I. Glazman et al., JETP Lett. **48**, 238 (1988); G. Kirczenow, Solid State Commun. **68**, 715 (1988); N. Garcia and L. Escapa, Appl. Phys. Lett. **54**, 1418 (1989); A. Safer and A. D. Stone, Phys. Rev. Lett. **62**, 300 (1989); E. G. Haanappel and D. van der Marel, Phys. Rev. B **39**, 5484 (1989).
7. J. L. Costa-Krämer, N. García, P. García-Mochales, P. A. Serena, M. I. Marques, and A. Correia, Phys. Rev. B **55**, 5416 (1997).
8. W. Apel and T. M. Rice, Phys. Rev. B **26**, 7063 (1982).
9. C. L. Kane and M. P. A. Fisher, Phys. Rev. B **46**, 15233 (1992).
10. D. L. Maslov and M. Stone, Phys. Rev. B **52**, R5539 (1995); V. V. Ponomorenko, Phys. Rev. B **52**, R8666 (1995); I. Safi and H. J. Shulz, Phys. Rev. B **52**, R17040 (1995).
11. K. A. Matveev and L. I. Glazman, Physica B **189**, 266 (1993).
12. C. L. Kane and M. P. A. Fisher, Phys. Rev. Lett. **68**, 1220 (1992).
13. L. I. Glazman, I. M. Ruzin, and B. I. Shklovskii, Phys. Rev. B **45**, 8454 (1992).
14. D. Yue, L. I. Glazman, and K. A. Matveev, Phys. Rev. B **49**, 1966 (1994).
15. For $\gamma = 1$ Eq. (1) was proven to be an exact solution of the Luttinger liquid model for the particular value 1/2 of the phenomenological coupling parameter g. See C.L. Kane and M.P.A. Fisher, Phys. Rev. Lett. **76**, 3192 (1996) and Ref. 9.
16. For details see: M. Jonson, I. V. Krive, and R. I. Shekhter, (unpublished).

SIMPLE MODEL FOR FORCE FLUCTUATIONS IN NANOWIRES

H. OLIN, S. BLOM, M. JONSON AND R. SHEKHTER
Department of Physics, Chalmers University of Technology and Göteborg University
SE-412 96 Göteborg, Sweden

AND

J.L. COSTA-KRÄMER, N. GARCIA AND P.A. SERENA
Laboratorio de Fisica de Sistemas Pequeños y Nanotecnologia,
Consejo Superior de Investigaciones Cientificas (CSIC)
Serrano 144, E-28006 Madrid, Spain

Abstract. When two metal electrodes are separated, a nanometer sized wire (nanowire) is formed just before the contact breaks. The electrical conduction measured during this retraction process shows signs of quantized conductance in units of $G_0 = 2e^2/h$. Recent experiments show that the force acting on the wire during separation fluctuate, which is interpreted as due to atomic rearrangements. In this report we use a simple free electron model and show that the electronic contribution to the force fluctuations are comparable to the experimentally found values, about 2 nN.

1. Introduction

The electrical conductance through a narrow constriction with a diameter of the order of the electron wavelength is quantized in units of $G_0 = 2e^2/h$ [1, 2]. Such conductance quantization is observed at low temperatures in semiconductor devices containing a two-dimensional electron gas (2DEG) [3, 4]. Similar effects are possible [5, 6] at room temperature in metallic wires with a diameter of the order of 1 nm (nanowires) and are observed using scanning tunnelling microscopy (STM) [7-13], mechanically controlled break junctions (MCBJ) [14, 15], or as recently shown [16, 17] just by using plain macroscopic wires. These techniques use the same basic principle: by pressing two metal pieces together a metallic contact is formed which can

N. García et al. (eds.), Nanoscale Science and Technology, 11–17.

be stretched to a nanowire by the subsequent separation of the electrodes. The conductance in such a system is found to decrease in abrupt steps with a height of about $2e^2/h$, just before the contact breaks.

In a recent pioneering experiment by Rubio, Agrait, and Vieira [18], following earlier attempts [19-22] , the force and the conductance were simultaneously measured during the formation and rupture of a gold nanowire. They show that the stepwise variation of the conductance is always accompanied by a jump of the force. One interpretation [10, 18, 23-25] is that the structural transformations of the nanowire, involving elastic and yielding stages, cause the stepwise variation of the conductance.

In this report we are studying the electronic contribution to the observed force fluctuations using a simple free electron approach neglecting all atomic structures of the wire: a jellium model (see also three other recent reports [26-28]). In metals the electronic contribution to the binding energy is significant (metallic binding) and one might suspect that the quantized electronic energy levels in the nanowire would be reflected in the binding energy. When a conductance mode closes it should produce a sharp change in the electronic binding energy. The quantized energy levels are of the order of eV and the wire elongation of the order of nm giving a change in force of the order of nN, the same as observed in the experiments. Considering this, we develop in this report a simple free electron model. The calculations show force fluctuations of the same size as in the experiments whenever a mode is closed or opened.

2. Model

We use a free electron model neglecting all atomic structure in the wire, a jellium model. Further, we use a cylindrical nanowire of length L and diameter d (see figure 1) Under the assumed ideal plastic deformation, the volume V of the wire will be constant during elongation. We are interested in the tensile forces acting on the wire during elongation. In general the force is the derivative of the energy with respect to distance, however, our system is open and we should instead consider the thermodynamical potential, Ω. The Fermi energy, E_F, in metals is much higher than the thermal energy and we can approximate the chemical potential by E_F and the thermodynamical potential is found to be $\Omega = E - E_F N$, where E is the energy and N the number of electrons. The force is then $F = -d\Omega/dL$. Using the adiabatic approximation [29], which is a natural starting point for these nanowires, the transverse motion of the electrons gives rise to quantized modes n of energy,

$$E_n = E_{jl} = \frac{\hbar^2}{2mR^2}\beta_{jl}^2 = \frac{\hbar^2\pi}{2mV}L\beta_{jl}^2 \qquad (1)$$

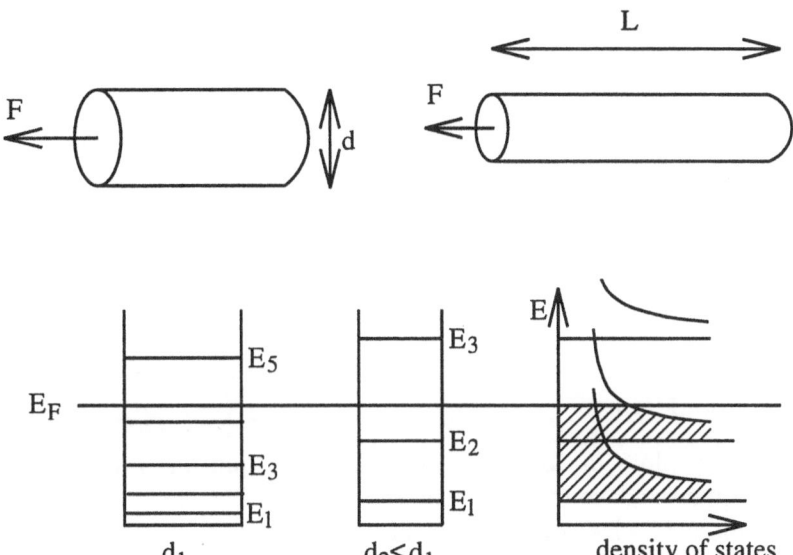

Figure 1. Model of the nanowire. The pulling force F is acting on a nanowire of length L and width d. When the wire is elongated more and more transverse quantized modes are pushed above the Fermi level and closed. The corresponding one-dimensional density of states are filled up to the Fermi level.

where β_{jl} are roots to Besselfunctions, i.e. $\beta_{jl} = 2.4048, 3.8317, \ldots$, and the degeneracy is two-fold unless j equals zero. The third equality in equation 1 is valid for a wire of constant volume. A mode is considered to be open if $E_F > E_n$. The number of electrons in the wire is

$$N = \sum_n N_n = \sum_n \int_{En}^{E_F} LD(\varepsilon - E_n)d\varepsilon = \sum_n 2L\sqrt{\frac{2m}{\pi^2\hbar^2}}\sqrt{E_F - E_n} \quad (2)$$

where $D(\varepsilon)$ is the one-dimensional density of states:

$$D(\varepsilon) = \sqrt{\frac{2m}{\pi^2\hbar^2}}\frac{1}{\sqrt{\varepsilon}} \quad (3)$$

The total electronic energy of the wire is the integral of the energy times the density of states up to the Fermi energy and summed over all open modes,

$$E = \sum_n \int_{En}^{E_F} L\varepsilon D(\varepsilon - E_n)d\varepsilon \quad (4)$$

$$= \sum_n \left\{ \frac{2}{3}L\sqrt{\frac{2m}{\pi^2\hbar^2}}(E_F - E_n)^{3/2} + E_n N_n \right\} \quad (5)$$

The thermodynamical potential is then

$$\Omega = E - E_F N = -\sum_n \frac{4}{3} L \sqrt{\frac{2m}{\pi^2 \hbar^2}} \left(E_F - E_n \right)^{3/2} \tag{6}$$

and the derivative gives the force,

$$F = -\frac{d\Omega}{dL} = \sum_n \sqrt{\frac{2m}{\pi^2 \hbar^2}} \left\{ \frac{4}{3} \left(E_F - E_n \right)^{3/2} - 2 \left(E_F - E_n \right)^{1/2} E_n \right\} \tag{7}$$

3. Force calculations

Figure 2 shows plots of the force according to equation 7 and the corresponding quantized conductance of the wire during elongation, for two different choices of wire-volume. The number of modes that contribute to the conductance is taken from equation 1. Whenever a mode closes the conductance jumps one quantum unit and an abrupt change in the force appears. The peak-to-peak amplitude of the force fluctuations between two modes is about 2 nN.

4. Discussion

The force from our calculation shown in figure 2 agrees both qualitatively as well as quantitatively with experiments [18].

Force fluctuations are also seen in molecular dynamics simulations [10, 21, 23, 25] and the jumps in conductance are interpreted as due to atomic rearrangements. However, because of the experimental like conditions in these simulations, it is difficult to separate the different contributions to the binding energy. Our interpretation is more or less the reversed: the electronic contribution to the binding energy is so large that the change of the quantized energy levels in the wire, with a corresponding quantized conductance, causes the force fluctuations. These force fluctuations might then give rise to atomic rearrangement but not necessarily. Although this is a bit like the story about the hen and the egg, our simple model shows that the electronic contribution must be considered seriously because it constitutes a significant part of the metallic binding energy in these nanowires.

One electronic contribution to the binding energy which is neglected in the present model is the Coulomb interaction. In metallic binding the electrostatic energy could be of the same order as the kinetic one and a natural extension of the present model would be to include the electrostatic energy which would change the electronic energy. The force fluctuations would, however, still be present.

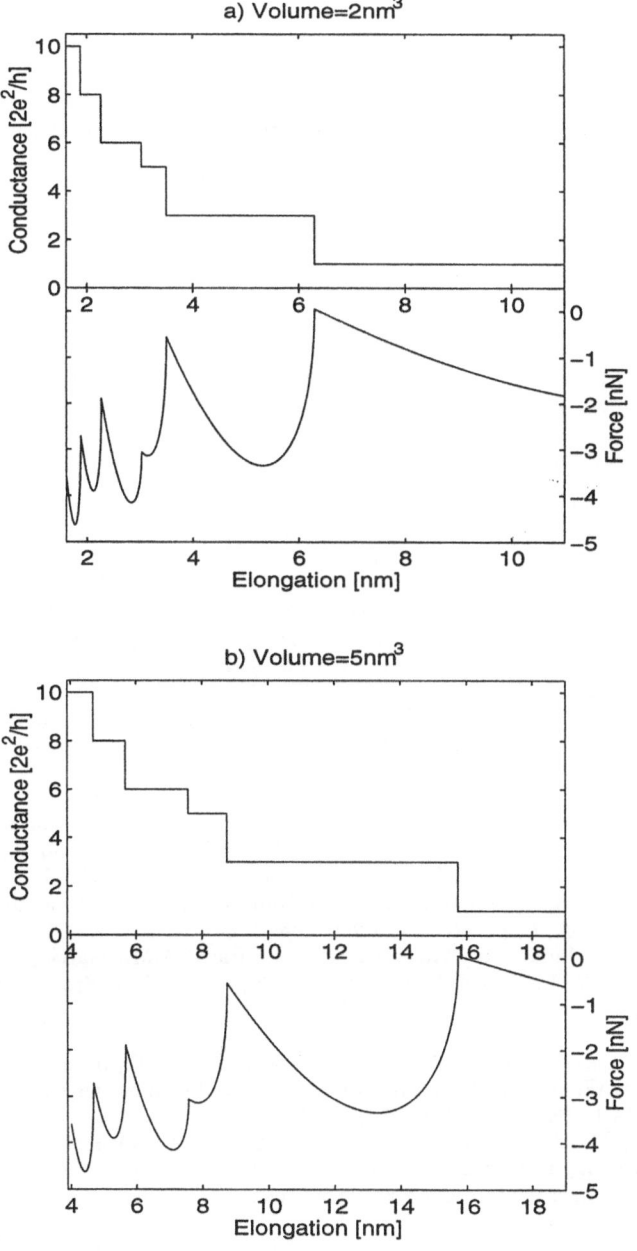

Figure 2. Calculated conductance and force as a function of elongation for gold nanowires with constant volume a) 2 nm^3 and b) 5 nm^3. Whenever a mode closes the conductance jumps one quantum unit and an abrupt change in the force appear. The peak-to-peak amplitude of the force fluctuations between two modes is about 2 nN.

5. Conclusion

We have shown, using a simple free electron model, that the electronic contribution to the force fluctuations is comparable to the experimentally found values. This could be of importance to understand the mechanism of formation of metal nanowires as well as in the wider context of nanomechanics.

Acknowledgement

This work was supported by the European ESPRIT project Nanowires, the Spanish DGCIT and CICyT, and the Swedish NFR and TFR agencies.

References

1. Landauer, R. (1989) Conductance determined by transmission: probes and quantised constriction resistance, *J. Phys. Condens. Matter*, 1, 8099-8110.
2. Beenakker, C.W.J. and van Houten, H. (1991) Quantum transport in semiconductor nanostructures, *Solid State Physics*, 44, 1-228.
3. Wharam, D.A., Thornton, T.J., Newbury, R., Pepper, M., Ahmed, H., Frost, J.E.F., Hasko, D.G., Peacock, D.C., Ritchie, D.A. and Jones, G.A.C. (1988) One-dimensional transport and the quantisation of the ballistic resistance, *J. Phys. C*, 21, L209-L214.
4. van Wees, B.J., van Houten, H., Beenakker, C.W.J., Williamson, J.G., van der Marel, D. and Foxton, C.T. (1988) Quantized conductance of point contacts in a two-dimensional electron gas, *Phys. Rev. Lett.*, 60, 848-850.
5. Garcia, N. (1987) STM Workshop, (ICTP Trieste) unpublished.
6. Garcia, N. and Escapa, L. (1989) Is the observed quantized conductance on small contacts due to coherent ballistic transport?, *Appl. Phys. Lett.*, 54, 1418.
7. Pascual, J.I., Mendez, J., Gómez-Herrero, J., Baro, A.M., Garcia, N. and Binh, V.T. (1993) Quantum Contact in Gold Nanostructures by Scanning-Tunneling-Microscopy, *Phys. Rev. Lett.*, 71, 1852-1855.
8. Pascual, J.I., Mendez, J., Gmez-Herrero, J., Baro, A.M., Garcia, N., Landman, U., Luedtke, W.D., Bogachek, E.N. and Cheng, H.P. (1995) Properties of Metallic Nanowires - From Conductance Quantization to Localization, *Science*, 267, 1793-1795.
9. Olesen, L., Lægsgaard, E., Stensgaard, I., Besenbacher, F., Schiøtz, J., Stoltze, P., Jacobsen, K.W. and Nørskov, J.K. (1994) Quantized Conductance in an Atom-Sized Point-Contact, *Phys. Rev. Lett.*, 72, 2251.
10. Brandbyge, M., Schiøtz, J., Sørensen, M.R., Stoltze, P., Jacobsen, K.W., Nørskov, J.K., Olesen, L., Lægsgaard, E., Stensgaard, I. and Besenbacher, F. (1995) Quantized conductance in atom-sized wires between two metals, *Phys. Rev. B*, 52, 8499-8514.
11. Agrait, N., Rodrigo, J.G. and Vieira, S. (1993) Conductance Steps and Quantization in Atomic-Size Contacts, *Phys. Rev. B*, 47, 12345-12348.
12. Costa-Krämer, J.L., Garcia, N. and Olin, H. (1997) Conductance quantization histograms of gold nanowires at 4 K, *Phys. Rev. B*, 55, 12910.
13. Costa-Krämer, J.L., Garcia, N. and Olin, H. (1997) Conductance quantization in bismuth nanowires at 4K, *Phys. Rev. Lett.*, 78, 4990.
14. Muller, C.J., van Ruitenbeek, J.M. and de Jongh, L.J. (1992) Conductance and Supercurrent Discontinuities in Atomic-Scale Metallic Constrictions of Variable Width,

Phys. Rev. Lett., **69**, 140-143.

15. Krans, J.M., van Ruitenbeek, J.M., Fisun, V.V., Yanson, I.K., and de Jongh, L.J. (1995) The Signature of Conductance Quantization in Metallic Point Contacts, *Nature*, **375**, 767-769.

16. Costa-Krämer, J.L., Garcia, N., Garcia-Mochales, P. and Serena, P.A. (1995) Nanowire formation in macroscopic metallic contacts: quantum mechanical conductance tapping a table top, *Surf. Sci. Lett*, **342**, L1144-L1149.

17. Garcia, N. and Costa-Krämer, J.L. (1996) Quantum-Level Phenomena in Nanowires, *Europhysics News*, **27**, 89-91.

18. Rubio, G., Agrait, N. and Vieira, S. (1996) Atomic-sized metallic contacts: Mechanical properties and electronic transport, *Phys. Rev. Lett.*, **76**, 2302-2305.

19. Agrait, N., Rubio, G. and Vieira, S. (1995) Plastic deformation of nanometer-scale gold connective necks, *Phys. Rev. Lett.*, **74**, 3995-3998.

20. Stalder, A. and Dürig, U. (1996) Study of plastic flow in ultrasmall Au contacts, *J. Vac. Sci. Techn. B*, **14**, 1259-1263.

21. Landman, U., Luedtke, W.D., Burnham, N.A. and Colton, R.J. (1990) Atomistic Mechanisms and Dynamics of Adhesion, Nanoindentation, and Fracture, *Science*, **248**, 454-461.

22. Stalder, A. and Dürig, U. (1996) Study of yielding mechanics in nanometer-sized Au contacts, *Appl. Phys. Lett:*, **68**, 637-639.

23. Landman, U., Luedtke, W.D., Salisbury, B.E. and Whetten, R.L. (1996) Reversible Manipulations of Room Temperature Mechanical and Quantum Transport Properties in Nanowire Junctions, *Phys. Rev. Lett.*, **77**, 1362-1365.

24. Torres, J.A. and Saenz, J.J. (1996) Conductance and mechanical properties of atomic-size metallic contacts: A simple model, *Phys. Rev. Lett.*, **77**, 2245-2248.

25. Todorov, T.N. and Sutton, A.P. (1996) Force and conductance jumps in atomic-scale metallic contacts, *Phys. Rev. B*, **54**, 14234-14237.

26. Stafford, C.A., Baeriswyl, D. and Brki, J. (1997) Jellium model of metallic nanocohesion, UF-IPT 1997/03-001.

27. van Ruitenbeek, J.M., Devoret, M.H., Esteve, D. and Urbina, C. (1997) Conductance quantization in metals: the influence of subband formation on the relative stability of specific contact diameters, *manuscript*.

28. Yannouleas, C. and Landman, U. (1997) On mesoscopic forces and quantized conductance in model metallic nanowires, *J. Phys. Chem. A*, **101**, 4780-4786.

29. Glazman, L.I., Lesovik, G.B., Khmel'nitskii, D.E. and Shekhter, R.I. (1988) Reflectionless quantum transport and fundamental ballistic-resistance steps in microscopic constrictions, *JETP Lett.*, **48**, 238-241.

Real time control of nanowire formation

L. Samuelson, S.-B. Carlsson, T. Junno, Hongqi Xu and L. Montelius
Solid State Physics and Nanometer Structure Consortium, Lund University, Box 118, S-221 00 Lund, Sweden

ABSTRACT

We have produced metallic nanowires by manipulating Au nanodiscs with the tip of an atomic force microscope in between electron beam defined Au electrodes. During the process we continuously monitored the electrical characteristics of the wire formation, and observed steps in the conductance in units of integer values of the conductance quantum $G_0 = 2e^2/h$. We also let the fabricated nanowires self-develop, and found wires stable at a certain conductance plateau $N \cdot G_0$ for 30 minutes at room temperature.

INTRODUCTION

Quantized conductance has been observed, at low temperatures in semiconductor samples [1], and at elevated temperatures in metallic nanowires. For the metallic systems different experimental methods has been used, including break junction [2] and scanning tunneling microscope techniques [3] as well as macroscopic wires [4]. Here we present a technique to form the nanowires in which we use pre-fabricated metallic nanodiscs that are manipulated by an atomic force microscope (AFM) into the gap between two electrodes. During this manipulation we monitor the position of the objects with an AFM in non-contact mode and measure continuously the electrical characteristics. The studies were performed at room temperature in ambient air. The measurements show that it is possible to form ohmic connections between the nanodiscs and electrodes. Furthermore, conductance quantization could be observed on making and breaking the ohmic connections, and these conductance plateaus could be stable for long times, exceeding 30 minutes.

EXPERIMENT

The structures were fabricated on Si substrates with a layer of 300 nm SiO_2 on top. Bonding pads of Ti and Au were evaporated through a metal mask, and then the electrodes were defined by electron beam (e-beam) lithography, followed by evaporation of 25 Å of Ti and 250 Å of gold and lift-off. The 300 A thick Au

N. García et al. (eds.), Nanoscale Science and Technology, 19–22.
© 1998 *Kluwer Academic Publishers.*

nanodiscs with a diameter of 30-100 nm were defined in a second step of e-beam lithography, metal evaporation and lift-off. The sample was wire-bonded and transferred to an AFM, operating under ambient air conditions. The experimental setup consists of a Topometrix Explorer AFM with a modified sample holder, allowing electrical connections to a sample mounted on a standard DIL chip carrier. The non-contact mode of AFM operation, in combination with electron-beam deposited tips, was used for imaging. To manipulate the nanodiscs we used the algorithm described in Ref. 5. In short, the discs were repositioned on the substrate surface by pushing them mechanically with the AFM tip, with the feedback-loop turned off. The oscillations of the tip were not turned off. However, during the pushing action the actual movement of the tip perpendicular to the surface was strongly dampened due to the mechanical interaction between the tip and sample. Two types of electrical measurements could be performed. Fast measurements (milliseconds) with a current amplifier and a digitizing oscilloscope, and slow (seconds or longer) with a standard DC-measurement equipment.

RESULTS

A constant bias, 1 mV to 200 mV, was applied between the electrodes. While continuously monitoring the current, one or more Au discs were selected and moved into the gap [Fig. 1]. If the nanodisc was pushed all the way into the gap it formed an ohmic connection between the electrodes with a resistance less then 100 Ω. Making and breaking electrical contact could be repeated many times with the same nanodisc by pushing it in and out of the gap. When measuring the conductance as a function of time as in Fig 2, when pushing the nanodisc in or out of the gap, one can observe steps in the conductance in the unit of the conductance quantum $G_0 = 2e^2/h$. Here e is the electron charge and h is Planck's constant. We could not observe any differences in the behavior of the wires in the bias range 1 mV to 200 mV. For biases larger than 250 mV the wires sometimes instantly changed behavior or broke. By moving the nanodisc at a lower rate and still continuously monitoring the conductance it is possible to push the nanodisc into the gap until a certain conductance value is reached. The AFM tip is then retracted and the system is left to develop of itself. In Fig 3. it takes more than one hour for the system to stabilize and then it stays on conductance plateau 2 for more than 30 minutes. It is yet unclear if the developing of the wire, after the tip has been retracted, is due to electromigration or nanodisc-dynamics.

SUMMARY

We have shown that, by manipulating e-beam Au nanodiscs in between two previously defined Au electrodes and continuously measuring the conductance, it is possible to have real-time control of nanowire formation, which to our knowledge is unachievable by other techniques. This system has shown steps in the conductance corresponding to the conductance quantum G_0 and wires formed in this way can be stable on a conductance plateau for more than 30 minutes.

REFERENCES

1. B. J. van Wees *et al.*, *Phys. Rev. Lett* **60**, 848 (1988)

2. C. J. Muller *et al.*, *Phys. Rev. Lett.* **69**, 140 (1992)

3. J. I. Pascual *et al.*, *Phys. Rev. Lett.* **71**, 1852 (1993)

4. J.-L. Costa-Krämer *et al.*, *Phys. Rev. B* **55**, 5416 (1997)

5. T. Junno *et al.*, *Appl. Phys. Lett.* **66**, 3627 (1995)

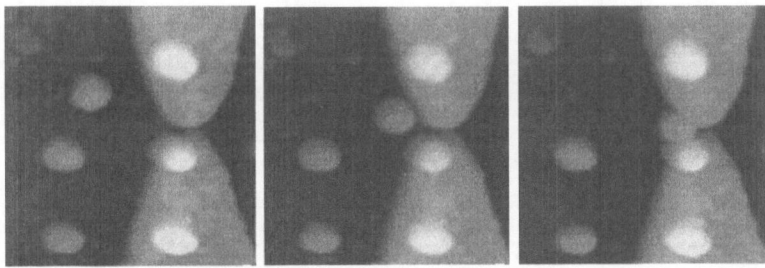

Figure 1 Three AFM images (740 nm x 740 nm) recorded during the manipulation of an Au disc into a gap between two Au electrodes.

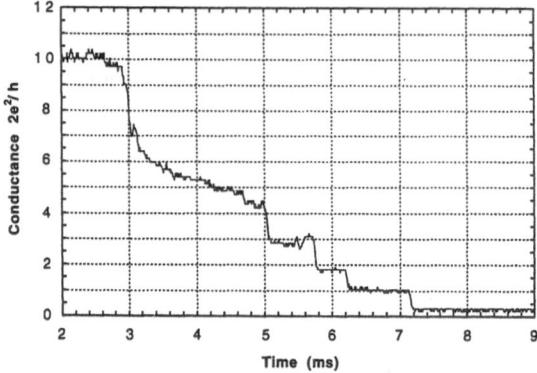

Figure 2 Dynamic behavior of the conductance as an Au nanodisc is being pushed out of contact with the electrodes. The applied bias voltage was 2 mV and a serial resistance of 400 Ω has been compensated for in the plot. Temperature 300 K.

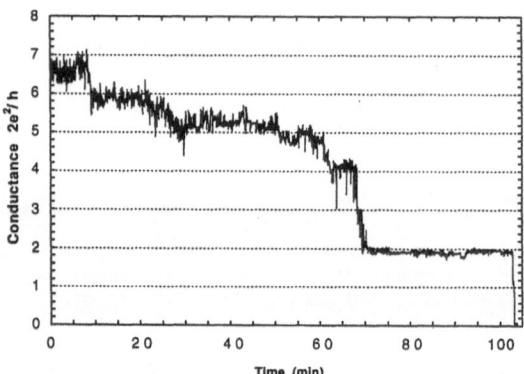

Figure 3 Long term conductance measurement of a nanowire after the AFM tip has been retracted from the sample surface. Bias voltage 50 mV, serial resistance 100 Ω and temperature 300 K.

FORCES IN SCANNING PROBE MICROSCOPY

E. Meyer, H. J. Hug, R. Lüthi, B. Stiefel and H.-J. Güntherodt
Institut für Physik, Universität Basel, Klingelbergstrasse 82,
4056 Basel, Switzerland

ABSTRACT. In the last 10 years forces in scanning probe microscopy (SPM) have been an interesting topic with continuing progress. The recently achieved true atomic resolution in dynamic force microscopy in ultrahigh vacuum (UHV) allows a comparison of scanning force microscopy (SFM) with scanning tunneling microscopy (STM) on Si(111)7x7 surfaces. The range of atomic resolution by SFM coincides with the stable tunneling range of STM. Wide area images of a quality comparable to STM have been obtained with a cleaned Si tip on stepped areas of the Si(111)7x7 and on surfaces of ionic crystals. The Magnetic Force Microscope yields new information on the magnetic domain structure of thin magnetic layers and on single flux lines and their pinning in high temperature superconductors. New visions to extend today's SFM devices to create new sensors and detectors (nano-age mechanics) are discussed.

1. Introduction

The state-of-the-art of *Forces in Scanning Probe Methods* has been reviewed in the proceedings of a NATO ASI in 1994 [1]. The highlights in the area of Scanning Force Microscopy (SPM) since 1995 are the true atomic resolution in UHV on silicon and insulators, the progress in Magnetic Force Microscopy on magnetism and superconductivity and the nano-age mechanics devices, which are focus of this review.

In addition there is progress in Friction Force Microscopy, which has been covered in the proceedings of some recent NATO ASI and NATO ARW [2]. Examples from the area of nano-tribology are:

1. Friction measurements in combination with continuum elasticity models, such as the Maugis-Dugdale model, give quantitative informations about the contact area and shear strengths of nanometer-sized contacts. The shear strengths are found to be close to the theoretical limit due to the absence of dislocations at these small scales.

2. The study of atomic-scale stick slip in combination with theoretical analysis gives insight into the fundamentals of friction and dissipation mechanisms.

3. The functionalization of probing tips gives the possibility to get material-specific contrast.

N. García et al. (eds.), Nanoscale Science and Technology, 23–39.

One of the challenges in SPM is to get a microscope with chemical sensitivity. Ultimately, chemical elements are to be analyzed on the atomic scale. One strand is the Magnetic Resonance Microscopy, originally suggested by Sidles and Rugar, where electron and nuclear spin resonance are detected with mechanical sensors. At present, a resolution in the micron range is feasable. Hopefully, individual spins can be resolved with high Q-cantilevers. Other strands are near field optical microscopy, Kelvin force microscopy and optical absorption microscopy, which yield material specific contrast in the nanometer range. Another strand, which is followed by Spence at Arizona State University is the combination of STM and mass spectroscopy, where atoms are picked up with the STM and transferred to a time-of-flight mass spectrometer. At present, we do not know which strand will lead to success. However, we are confident that SPM will become a surface science technique, that not only yields topographic information, but chemical information, local mechanical and electrical properties, the *laboratory on a tip*.

2. True atomic resolution

Since the invention of atomic force microscopy by Binnig et al. in 1986 [3], the quest for atomic resolution was a central part of research. The first results were promising: 30 Å lateral resolution on aluminum oxide [3, 1] in the first attempt. One year later, the first atomic resolution was presented by Binnig et al. on highly oriented pyrolitic graphite (HOPG) [4]. Soon afterwards, atomic resolution on a layered insulator, highly oriented boron nitride, was presented by Albrecht et al. [5]. It was Pethica, who pointed out, that the resolution might not be true atomic resolution, but due to the shear of flakes, which were attached to the probing tip [6]. Due to the commensurability of the flake and the surface, a constructive interference can occur between these finite-area surfaces, also called Moirée-effect. Atomic resolution on non-layered materials, such as LiF(001) [7], AgBr(001) [8], showed that this type of atomic resolution does not depend on the existence of flakes. Thus, the pure Moirée-effect was not generally applicable. Estimations of the contact area are based on continuum elasticity models and experimental values of the probing tip radius and the normal force (including long-range attractive forces). Typical values for the probing tip radii are 10-50nm and normal forces in ambient pressure of 1-100nN [9]. Then, the contact radius is estimated to be 1-10nm, consisting of ten to a few thousand atoms. These estimations were confirmed with normal and lateral stiffness measurements [10, 11, 12]. The observation of atomic features in contact mode was limited to the imaging of the lattice. Single point defects were not observed so far. On NaF(001), imaged in ultrahigh vacuum, a resolution of about 1nm could be observed at a step site [13]. A crucial test surface is the Si(111)7x7 surface, because of its complex surface structure. However, attempts to observe the Si(111)7x7 were rather difficult. Strong interactions between probing tip and sample (chemical reaction between probing tip and sample) were observed. Only with PTFE-coated tips, Howald et al. were able to image the Si(111)7x7 surface without damage [14]. A resolution of about 1nm was estimated from the observation of corner holes. However, individual atoms were not observed. The observation of C_{60}-molecules on NaCl(001) by Lüthi et al. is another example where a lateral resolution of about 1nm could be achieved in ultrahigh vacuum [15].

It was Ohnesorge and Binnig, who were able to get the first true atomic resolution on calcite. They immersed the surface under water, which reduced attractive forces, such as capillary forces and van der Waals forces, down to the order of 10pN [16]. Atoms were observed on the terrace and at the step edges. Thus, one could conclude that atomic resolution is achieved in liquids due to the reduced attractive forces. However, atomic resolution in air was not achieved, because of the presence of capillary forces and van der Waals forces. In vacuum, van der Waals are always attractive. Estimations of Goodman and Garcia gave values of 1-10nN, which is too large to get a single point contact [17]. The attractive forces will cause a pressure of the order of GPa in the contact zone. Thus, the tip or sample will be deformed elastically or plastically in order to increase the contact area and to reduce the pressure. Only in liquids, van der Waals forces can become small or even repulsive [18] and a single atom contact appears to be possible in contact mode.

In contrast to the contact-mode AFM, where atomic-lattice imaging was achieved rather early, non-contact AFM, also called dynamic force microscopy was not promising at the beginning. The first attempts by Martin et al. [19], McClelland et al. [20] and Nonnenmacher et al.[20a] gave resolutions in the 10nm range. The contrast was related to van der Waals forces, which were expected to be smeared out on the atomic scale. It was Giessibl, who could first achieve atomic resolution on the Si(111)7x7 with the nc-AFM [21]. Rather high amplitudes of about 300 Å were used for the cantilever oscillation and the cantilever resonance was detected with FM-detection [22]. The microscope was operated in the constant frequency mode. The presented image showed part of the unit cell. Thus, true atomic resolution of this complex surface structure was achieved. However, the operation conditions were found to be unstable, which was related to tip changes. Günthner presented true atomic resolution on Si(111)7x7, where an average tunneling current [23], $\overline{I_t}$, was used as input for the the feed-back loop (dynamic lever STM-mode). The unstable behaviour in the constant frequency mode (nc-AFM) was related to variations of frequency shift on the atomic scale not allowing stable frequency shift imaging. Lüthi et al. presented frequency shift vs. distance curves [24], where three regimes are observed:

Regime I: Long-range attractive forces, such as van der Waals and electrostatic forces, cause frequency shifts at distances as far as a few hundred Ångstroms.

Regime II: Short-range forces cause frequency shifts over a distance of approximately a nanometer. A minimum of the frequency shift is observed. True atomic resolution can be achieved with these short-range chemical forces.

Regime III: Repulsive contact forces become dominant. The conditions of the tip/sample geometry are not stable and plastic deformation is observed. This regime corresponds to the more conventional contact mode.

Using dynamic lever STM-mode, Lüthi et al. could acquire frequency shift images across a step site. Larger frequency shifts (more attractive) were observed on the lower terrace. The transition regime between upper and lower terrace was found to be atomically sharp. The explanation of these frequency shift differences between upper and lower terrace (FREDUL) were related to long-range forces [25]. On the lower terrace, the average interaction volume is bigger compared to

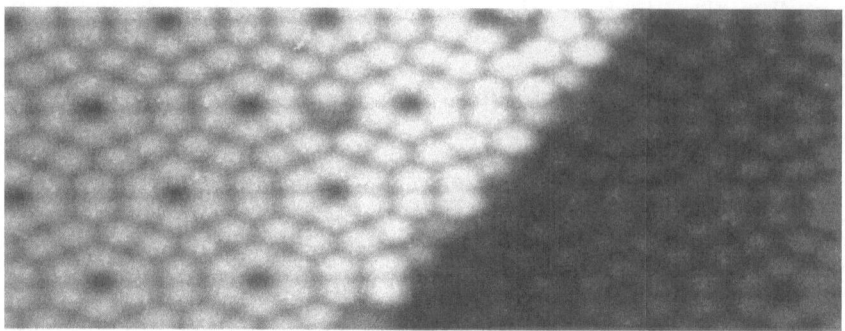

Figure 1: High resolution resonance frequency shift image of the Si(111)7x7 surface reconstruction recorded in the vicinity of a monatomic step. A time averaged tunneling current of 25pA was used as feedback input.

the upper terrace. It becomes obvious, that it is rather difficult to run the feedback in the constant frequency shift mode, because of the local changes across step sites. However, on small areas, without step sites, true atomic resolution could be achieved in the constant frequency mode (nc-AFM-mode) [26]. With the appropriate choice of a small frequency shift on the lower terrace, it is also possible to run nc-AFM across a step site without touching on the upper terrace.

In contrast to the dynamic lever STM-mode, the frequency shift in nc-AFM-mode above the adatoms was found to be stronger than above the corner holes. The observation, which is made in nc-AFM mode, is in agreement with molecular dynamics simulations of Perez et al. [27]. Thus, the strongest attraction is above the adatom sites, where the silicon tip forms a weak chemical bond with the sample. Similar observations were also made by Erlandsson et al., where the contrast of nc-AFM was related to the local reactivity of the sample [28]. Erlandsson et al. observed also a difference between the adatoms close to corner holes compared to center adatoms, where the nc-AFM was operated with slope detection (fixed excitation frequency and varying amplitude) and a tungsten tip was used. The difference between nc-AFM mode (most attractive force above adatoms) and dynamic lever STM-mode (most attractive force above corner holes), observed by Lüthi et al. [24] and Guggisberg et al. [25] is not so trivial to be explained. First, STM might be responsible due to changes of interaction distances (STM depends on electronic effects and does not keep the distance constant). Recent experiments have shown that this effect is not the primary reason. However, it is found that STM works in the branch of the frequency shift vs. distance curve with negative slope (repulsive part), whereas nc-AFM can be operated in the part with the positive slope (attractive part). Thus, nc-AFM with true atomic resolution is operated at even lower forces than typical STM operation.

Other applications of nc-AFM on semiconductors were presented by Sugawara et al. [29] on InP(001), where point defects were observed. The first application of nc-AFM on an insulator was presented by Bammerlin et al. [30]. The unreconstructed surface of NaCl(001) was observed. Only one ionic species was visible. From theoretical studies, it was concluded that the more attractive force is above the Na^+. The probing tip is a silicon tip (argon sputtered), which is assumed to be negatively charged (dangling bond). Then, the most attractive force is above

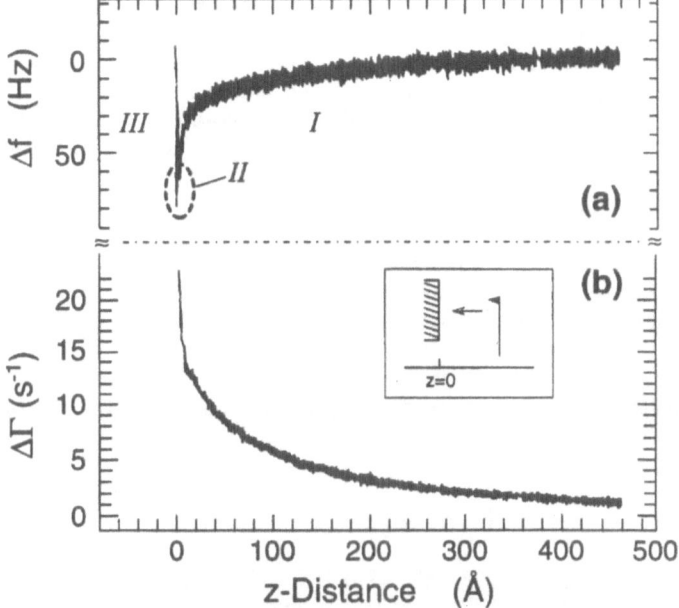

Figure 2: Distance dependence of frequency shift (a) and damping rate (b) of a silicon probing tip above NaCl(001). Long-range forces, such as van der Waals and electrostatic force determine regime I. Chemical forces are dominant in regime II, where true atomic resolution can be achieved. Repulsive contact forces take over in regime II, where plastic deformation of both probing tip and sample can occur.

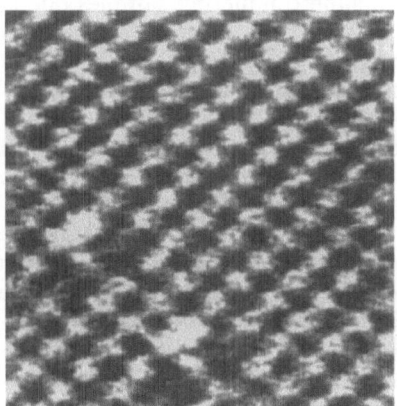

Figure 3: Non-contact force microscopy on NaCl(001). Two point defects are observable. The spacing between the periodically arranged bright spots corresponds to the distance between equally charged ions. From theoretical calculations, it is concluded that the contrast originates from the sodium ions.

the positive sodium ions. In addition, a pair of defects was observed, exhibiting stronger attraction than the rest of the imaged surface. The pair of defects was found to be stable on the time scale of minutes. After 80 minutes, the defects were found to jump one atomic spacing, which is either related to thermal motion or due to the action of the probing tip. The high stability of the defects at room temperature, the asymmetric shape and the occurence in pairs, makes it plausible that the defects are OH^--centers. Recently, Landman et al. presented molecular dynamics simulations, where the NaCl-surface was predicted to play an important role in the catalytic reaction with water [31].

With the achievement of true atomic resolution, nc-AFM has made a real break-through. Insulators, semiconductors and metals are accessible to the application of nc-AFM. The comparison of STM and AFM on conductive surfaces might be important for the understanding of STM-operation. The surface physics of insulators is still at the beginning. Possible fields of studies are: Unknown surface reconstruction of insulators, the physics and chemistry of color centers, catalytic reactions and thin oxide films of semiconductors.

3. Magnetic force microscopy

Magnetic Force Microscopy [32, 33, 34] has become a valuable tool for measuring the stray field of complicated ferromagnetic samples [35, 36, 37, 38, 39, 40] and superconductors [41, 42, 43]. The instrument consists of a micron-scale ferromagnetic tip attached to a flexible cantilever which scans close to the surface of the sample. The stray field emanating from the sample generates a force on the magnetic force microscope tip. The ability of the instrument to measure both the magnetic stray field of the sample and sample topography [44, 45, 46] allows a correlation of the measured stray field with particular surface structures. Thus it is possible to study the pinning of domain walls in ferromagnetic samples and the pinning of vortices in superconducting samples due to a structural defect observable at the surface. Other applications are the use of the stray field of the tip [47, 48, 49] for modifying the magnetic state of the sample [42, 50, 51] or for determining the sensitivity and response of magnetic heads to localized fields [52, 53, 54].

There is growing interest to move from qualitative imaging to quantitative analysis of sample properties using MFM. In general it is not possible to calculate a magnetization distribution from MFM data. In the special case of perpendicular magnetization, as we will show later, it is theoretically possible to use the force pattern, $F(x, y)$, to generate the magnetization pattern, $M(x, y)$, to within a constant. However due to the limited signal–to–noise ratio of a MFM force measurement the better procedure remains to assume a magnetization pattern, calculate its field and then calculate the force on the tip due to this field. This calculated force is then compared with the measured force data. This process is iterated until the agreement between measured and calculated force patterns is optimized.

Magnetic Force Microscope Image Formation

An important task remains the understanding of the image formation process. This can be difficult because of the possibly strong perturbation of the sample magnetization by the magnetization of the tip, or vice versa. Generally the MFM image formation can be divided into three categories: First, the hard magnetic case, where the magnetization of the tip and the sample remains undisturbed by the scanning process of the magnetic tip above the sample. Second, the hysteresis free, soft magnetic case, where the sample or the tip magnetization is uniquely defined by the position of the tip above the sample (and the magnetic structure of the tip and the sample) [41, 55]. And third, the intermediate case, where the magnetization of the tip and the sample influence each other and change in a discontinuous, hysteretic way [56]. The ideal MFM situation, even when measuring soft samples, is a magnetically hard tip with small enough volume that the tip field is so weak that it does not perturb the sample magnetization. The small tip volume also makes higher resolution possible. However, the size of the tip volume is limited by the sensitivity of the force microscope system to detect smaller tip deflections.

In this paper we will restrict the discussion to the hard magnetic case because we believe that the increased sensitivity of future scanning force microscope instrumentation allows the use of hard low magnetic volume tips. To obtain tips with a well defined magnetic moment and direction of the magnetization an electron microscope has been used to grow a deposition tip on a standard pyramid tip of a commercially available cantilever. The needle like tips are then made sensitive to magnetic stray fields by coating one side of the needle with a thin ferromagnetic film. The strong shape anisotropy of the thin film keeps the magnetization along the tip axis [57].

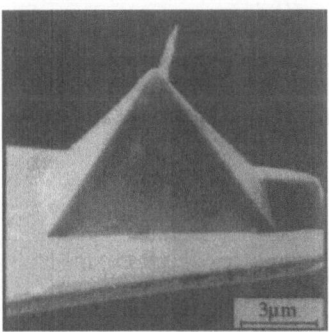

Figure 4: We have used an electron microscope to grow a deposition tip on a standard pyramid tip of a commercially available cantilever. The needle–like tip was intentionally grown at an angle of approximately twelve degrees to the axis of the pyramid tip, so that it is perpendicular to the surface of the sample. The tip is made sensitive to magnetic stray fields by a 25 nm thick Co film which was deposited on the side of the needle–like tip by thermal evaporation.

The calculation of the magnetic field distribution of a sample and the force acting on the tip involves three dimensional spatial integration to get to the field from the magnetization and to the force from the field. We can calculate the magneto–static energy of the magetized tip in the stray field of the sample,

$$E(\vec{t}) = \mu_0 \int \vec{M}_{Tip}(\vec{r}') \cdot \vec{H}_{Sample} \left[\vec{r}' + \vec{t} \right] \, dV' \, , \tag{1}$$

where $\vec{M}_{Tip}$ is the magnetization distribution of the tip and $\vec{H}_{Sample}$ is the stray field of the sample. The integration in Eq. 1 is performed in a coordinate system attached to the tip which is located at the position $\vec{t}$. As common in magnetostatic calculations we define a magnetostatic potential, ϕ, with $\vec{H}_{Sample} = -\text{grad}\phi$. The potential, ϕ, must be a solution of the Laplace equation in the free space outside the sample. The stray field of the sample is then found by solving the Laplace equation using appropriate boundary conditions to describe the stray field at the surface of the sample. The z–component of the force acting on the tip becomes

$$F_z(\vec{t}) = -\mu_0 \int \vec{M}_{Tip}(\vec{r}') \cdot \frac{\partial}{\partial t_z} \vec{H}_{Sample}(\vec{r}' + \vec{t}) \, dV' \ . \tag{2}$$

The other components of the force are easily evaluated by replacing the derivative in Eq. 2 by the corresponding lateral derivatives. At this point it is noteworthy that for an extended tip the force is proportional to the *integral* of the derivative of the stray field and *not* proportional to the derivative of the field. Thus for an extended tip, the field and *not* the derivative of the field is measured.

We can approximate the tip (see Fig. 4) as a long and thin, slab–like object, with (i) a bottom and a top surface, $(b_x b_y)$, (ii) magnetized homogeneously along its long axis, l, and (iii) oriented perpendicular to the surface of the sample . Then Eq. 2 becomes

$$F_z(t_z) = -\mu_0 M_{Tip} \cdot \int_{(b_x b_y)} [H_{z,Sample}(t_z) - H_{z,Sample}(t_z + l)] \, dA' , \tag{3}$$

where the integration is evaluated over the bottom and top surface of the tip only. The derivative along the z–direction of Eq. 2 has become a difference (stray field at the bottom surface of tip minus stray field at the top surface of the tip) when the integration along the z–direction of Eq. 2 is evaluated.

For a tip surface, $(b_x b_y)$, smaller than the grid–resolution [1] during an MFM measurement and a stray field decaying rapidly along the z–direction such that the stray field at the bottom surface is much larger than the one at the top surface, $(H_{z,Sample}(t_z) \gg H_{z,Sample}(t_z + l))$, Eq. 3 finally becomes

$$F_z(\vec{t}) = q_{Tip} \cdot H_{z,Sample}(\vec{t}) , \tag{4}$$

where the tip is modeled by a magnetic point charge, q_{Tip}, given by

$$q_{Tip} = \mu_0 \cdot M_{Tip} \cdot (b_x b_y) . \tag{5}$$

Alternatively, the modeling of MFM data can be accomplished in Fourier space [34, 58, 59, 60]. In the following we assume a perpendicular sample magnetization distribution, $M_z(x, y)$, which is uniform over the film thickness, h. The magnetization, $M_z(x, y)$, gives rise to surface charge distributions at the top surface ($+\sigma$, $z = 0$) and the bottom surface ($-\sigma$, $z = -h$). The magnetic field above the sample is produced by these charges only. If necessary, bulk charges can easily be introduced, for instance by layering the magnetic film into sheets of surface charges.

---

[1] The grid–resolution is defined as the lateral distance between two measurement points.

We then make use of the fact that the Fourier transforms of the magnetization distribution are simply related by multiplicative functions called transfer functions [2]. The Fourier components of the z–component of the stray field, $A_{H_z}^{z,h}(\vec{k})$, of a thin film sample with the top surface at $z = 0$ and a thickness, h, and those of the magnetization, $A_M(\vec{k})$

$$A_{H_z}^{z,h}(\vec{k}) = \frac{e^{-k \cdot z} \cdot (1 - e^{-k \cdot h})}{2} \cdot A_M(\vec{k}) = \text{HTF}_z(\vec{k}) \cdot A_M(\vec{k}) , \qquad (6)$$

where the HTF_z is called the field transfer function. Note that the factor $(1-e^{-k \cdot h})$ originates from the magnetic surface charge density at the lower surface of the film. The relation between the Fourier components of the force, $A_{F_z}^{z,h}(\vec{k})$, and those of the field, $A_{H_z}^{z,h}(\vec{k})$, are given by

$$A_{F_z}^{z,h}(\vec{k}) = A_{H_z}^{z,h}(\vec{k}) \cdot \text{FTF}(\vec{k}) \qquad (7)$$

with

$$\text{FTF}(\vec{k}) = \mu_0 \cdot \left[\int_V \rho_{Tip} e^{-kz'} e^{i\vec{k}\vec{r}'} \, dV' + \oint_A \sigma_{Tip} e^{-kz'} e^{i\vec{k}\vec{r}'} \, dA' \right] , \qquad (8)$$

where Eq. 8 denotes the force transfer function for an arbitrary tip characterized by the magnetic volume charge distribution, ρ, and magnetic surface charge distribution, σ. For the long and thin, slab–like tip, described before, can be simplified to

$$\text{FTF}(\vec{k}) = -\frac{4\mu_0 M_{Tip}}{k_x k_y} \cdot \sin\left(\frac{k_x b_x}{2}\right) \cdot \sin\left(\frac{k_y b_y}{2}\right) \cdot (1 - e^{-k \cdot l}) . \qquad (9)$$

The size of the bottom surface of the tip, $(b_x b_y)$, thus sets a natural limit to the resolution of the stray field. Within this limit the measured force is proportional to the stray field as long as the length of the tip is larger than the decay length of the field, $e^{-k \cdot l} \ll 1$, compare to Eqs. 4 and 5.

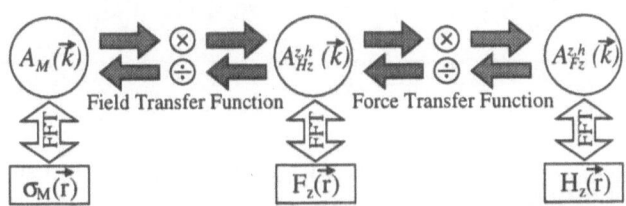

Figure 5: The Fourier components of the force on the MFM tip are calculated from those of the stray field by means of the force–transfer–function, where those of the stray field are calculated from those of the magnetization by the field–transfer–function.

In conclusion (see Fig. 5) the transfer–function approach is an efficient way to calculate the Fourier components of the force (Eq. 9) and the stray field (Eq. 6) from those of the magnetization. A stray field or force calculation becomes similar

---

[2] In this paper we present the force–transfer function methods only in form of a short overview. More details as well as the derivation of all transfer–functions will be described in [61]

to a regular Fourier–filtering procedure. This allows to assume a magnetization pattern, efficiently calculate its field and then calculate the force on the tip due to this field. This calculated force is then compared with the measured force data. Then the parameters such as the tip–to–sample distance and the tip magnetization are varied until the agreement between measured and calculated force patterns is optimized (Fig. 6).

Examples of MFM on Ferromagnetic and Superconducting Samples

We have made an extensive magnetic force microscopy study of the magnetic domain structure in Cu/Ni/Cu/Si(001) over a Ni thickness range of $2\,nm < h < 17.5\,nm$ [39, 62]. The samples studied are a series of MBE-grown, epitaxial Ni/Cu/Si(001) films capped with 2 nm of Cu. The MFM images confirm dramatically that the magnetization is indeed strongly held to a perpendicular direction as indicated by magneto–optic KERR effect measurements [63], vibrating sample magnetometry and quantitative MFM (Fig. 6). In these images the contrast is due to magnetization into and out of the plane; no regions of in–plane magnetization were observed other than the domain walls themselves.

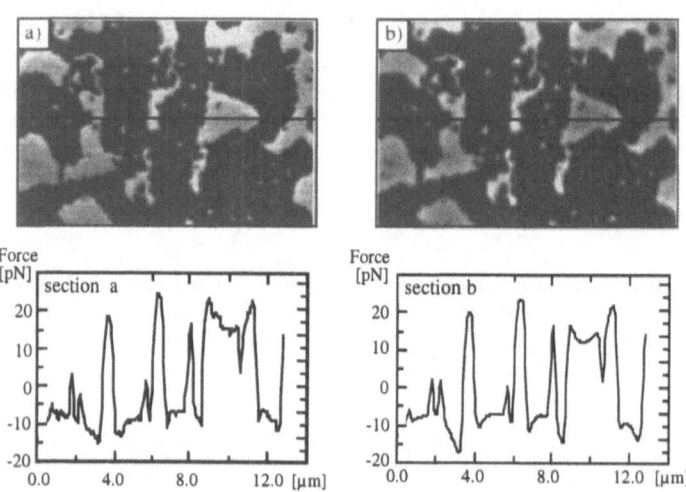

Figure 6: Simulation of MFM data.
(a) MFM measurement of a Cu/Ni(10 nm)/Cu/Si(001) thin film showing perpendicular magnetization.
(b) The simulation of the MFM image is generated from the magnetization pattern using the transfer function theory. The magnetization pattern (not shown) was obtained from the MFM data by a special discrimination process.
(Section a, Section b) Cross sections of the MFM measurement and the simulation.

The domain structure of the as–grown films is characterized by two types of behavior in different thickness ranges. At and below 8.5 nm of Ni the domain patterns are irregularly spotted (Fig. 7b, a). The length scale of this coarse structure is poorly defined but its average decreases with increasing thickness, h. Above 8.5 nm the relatively large serpentine domains begin to break down by internal fragmentation (sub-micron bubble domain formation) as seen in Fig. 7c.

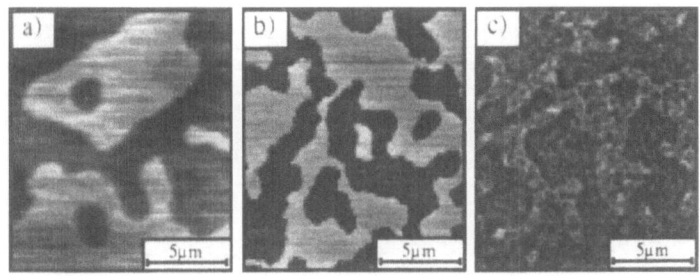

Figure 7: MFM data acquired on Cu/Ni/Cu/Si(100) films containing a) 2 nm, b) 8.5 nm, c) 12.5 nm Ni. (Image size: $13.9\,\mu m \times 15.8\,\mu m$.)

As the Ni thickness increases, magnetostatic energy is reduced by refinement of the interior bubble domain structure rather than by more complex contortions of the existing domain walls. In [39] we have calculated the transition to this finer domain structure and its length scale by extensions of domain theory as outlined by Kittel [64] and developed by others [65, 66]. Furthermore we have discussed the complex transition of the magnetization from a perpendicular to an in-plane direction.

The detection of vortices in supersonductors has been achieved by many of the scanning probe methods. Scanning tunneling microscope (STM) was used to image the vortex lattice of a $NbSe_2$ single crystal by Hess et al. [67]. Recently results have even been obtained on single crystalline $YBa_2Cu_3O_{7-\delta}$ grown in special crucibles [68]. However the imaging of vortices using the extremely surface sensitive STM remains difficult. In contrast, the detection of vortices by measuring their magnetic stray field seems easier, since the penetration depth and thus the magnetic diameter of the vortices is large. Indeed, not only the better technique, but also Lorentz microscopy [69], scanning SQUID microscopy [70, 71], scanning hall probe microscopy (SHPM) successfully imaged single vortices [72, 73] and magnetic force microscopy [42, 43]. The three latter techniques even allow to directly study the reaction of the vortices to a change of the external parameters, such as temperature and magnetic field. However only the MFM allows to image the topography and the magnetic stray field at the same sample area and to modify the micro–magnetic structure of the sample [42].

In laser ablated YBCO thin films we observe a glass–like vortex arrangement (Fig. 8a). Such an non–crystalline vortex structure indicates that the vortex arrangement is dominated by the pinning force rather than by the repulsive inter–vortex forces. The number of the imaged vortices per area coincides well with the one expected from the imaged area and the applied field. Further we have observed that many more pinning centers than vortices are present at such low fields. In [42, 43] we have shown that the MFM images all the vortices in these thin films and that no additional vortices are moved around by the scanning tip. The topography of YBCO thin films is either dominated by screw dislocations or by tower–like islands. The surface roughness usually is around 15 nm where as the thickness of the films used is around 150 nm. From x–ray diffraction data it is known that these growth islands are slightly miss–aligned. In Fig. 8b one single vortex of Fig. 8a and the corresponding topography, Fig. 8c, is imaged. The vortices are always

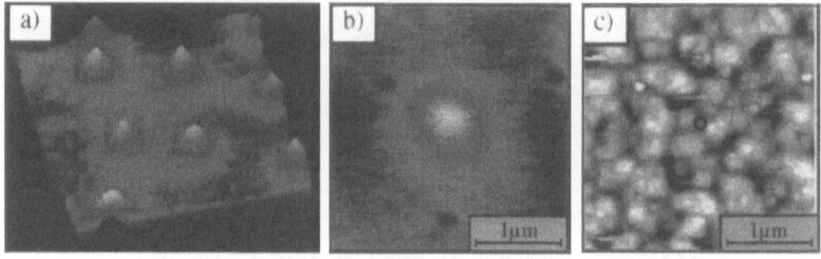

Figure 8: In laser ablated YBa$_2$Cu$_3$O$_{7-\delta}$ (YBCO) thin films a non–crystalline arrangement of vortices is observed (a). The number of the imaged vortices per area coincides well with the one expected the applied field. (b) shows the magnetic stray field of one single vortex. The stray field (b) has to be compared to the corresponding topography image (c). The vortices are pinned between the growth islands but not in the lowest places (position of vortex marked by circle in c). This indicates that the weakened superconducting order parameter between the slightly miss–aligned islands is responsible for the pinning.

pinned between the islands but never in the lowest (black) locations. This indicates that the pinning is not caused by shortening the vortex length but rather due to a weakening of the superconducting order parameter in the small angle grain boundary between two growth islands. The weakening of the order parameter at the grain boundaries has been observed before by transport current measurement in bi–crystalline samples [74]. It was argued that the increasing number of dislocations in the grain boundary with increasing miss–alignment angle limits the maximum transport current in these samples. Furthermore YBCO thin films are subjected to easily loose oxygen via imperfections such as grain boundaries. The measurement shown in Fig. 8 is the first direct proof that the vortices in laser ablated YBCO thin films are pinned by small angle grain boundaries between the growth islands.

4. Sensors based upon scanning probe microscopy

Instrumental developments in STM and especially in AFM promise to become applicable in sensor technology. AFMs are designed to measure small deflections (10^{-4}Å/ $\sqrt{Hz}$) of microfabricated cantilevers. Gimzewski et al. have applied a beam-deflection AFM for calorimetric measurements [75]. A silicon cantilever was coated with an aluminum coating. Due to the bimetallic effect, this coated cantilever became sensitive to small temperature variations ($\approx 10^{-5}K$. This bimetallic cantilever could then be applied to measure the heat production due to a catalytic reaction: Hydrogen and oxygen were introduced into the chamber. A thin layer of platinum, being coated on the bimetallic cantilever, acted as catalysator, where water was formed. The heat evolution was then observed by the deflection of the bimetallic cantilever. Sensitivity studies showed that pico- joules and nanowatts can readily be measured with commercial cantilever [76]. By optimizing the geometry and materials, one expects to measure heat changes as small as 10^{-18}J.

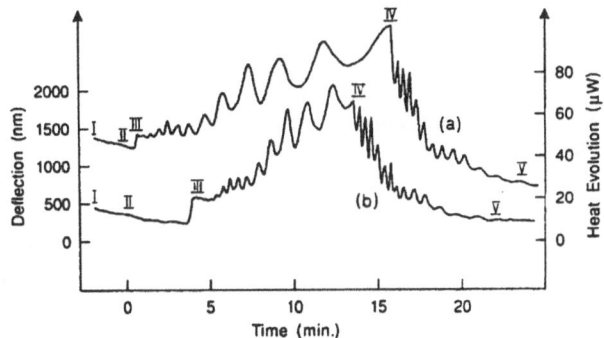

Figure 9: Catalytic reaction of H_2 with O_2 over a Pt sample with a geometric area of $1.4 \cdot 10^{-8} m^2$. The deflection of the bimetallic cantilever (left axis) corresponds to a heat evolution (right axis). From Gimzewski et al. [33].

Besides the high sensitivity, this type of calorimeter has the advantage that quantitative results can be achieved. E.g., enthalpy changes at phase transitions in n-alkanes were determined by Berger et al. [77]. Other modes of operation of AFM-based sensors were introduced recently: 1) Mass changes due to the absorption of molecules on the cantilever surface were observed by frequency changes [78]. 2) Stress changes due to the adsorption of molecules could be measured with a normal cantilever. Berger et al. [79] observed the self-assembly of thio-alkane films as a function of chain length. The time dependence was found to be in agreement with Langmuir-kinetics. A stress sensitivity of the order of 0.001N/m was found, corresponding to variations of molecule concentrations of some attomoles. Ultimately, individual atom reactions might become accessible. Another major advantage of sensors based on micromachined, mechanical sensors is their compactness. Several sensors can be built in close proximity to detect different chemical reactions. This parallel approach appears very attractive for applications in the field of microfabricated electro–mechanical noses.

5. Acknowledgement

We thank A. Baratoff, Ch. Gerber and R.C.O'Handley for stimulating discussions. This work was supported by the Swiss National Science Foundation, the Kommission für Technologie und Innovation, the Swiss National Priority Program NFP30 and the Swiss Priority Program MINAST.

References

[1] *Forces in Scanning Probe Methods*, edited by H.-J. Güntherodt, D. Anselmetti and E. Meyer, NATO ASI Series E: Applied Sciences Vol. 286, Kluwer Academic publishers (1995).

36

[2] *Fundamentals of Friction*, edited by I.L. Singer and H.M. Pollock, Series E: Applied Sciences, Vol. 220, Kluwer Academic Publishers (1992)
Physics of Sliding Friction, edited by B.N.J. Persson and E. Tosatti, Series E: Applied Sciences, Vol. 311, Kluwer Academic Publishers (1996)
Micro/Nanotribology and Its Applications, edited by Bharat Bhushan, Series E: Applied Sciences, Vol. 330, Kluwer Academic Publishers (1997)

[3] G. Binnig,C.F. Quate, and Ch. Gerber, *Phys. Rev. Lett.* **56**, 930 (1986).

[4] G. Binnig,Ch. Gerber, E. Stoll, T.R. Albrecht and C.F. Quate, *Europhys. Lett.* **3**, 1281 (1987).

[5] T.R. Albrecht and C.F. Quate, *J. Appl. Phys.* **62**, 2599 (1987).

[6] J.B. Pethica, *Phys. Rev. Lett.* **57**, 3235 (1986).

[7] E. Meyer, H. Heinzelmann, H. Rudin and H.-J. Güntherodt, *Z. Phys. B.* **79**, 3 (1990).

[8] E. Meyer, H.-J. Güntherodt, H. Haefke and M. Krohn, *Europhys. Lett.* **15**, 319 (1991).

[9] E. Meyer, R. Lüthi, L. Howald, M. Bammerlin, M. Guggisberg and H.-J. Güntherodt in *Micro/Nanotribology and Its Applications*, edited by Bharat Bhushan, Series E: Applied Sciences, Vol. 330, Kluwer Academic Publishers, p. 193, (1997)

[10] S.P. Jarvis, A. Oral, T.P. Weihs and J.B. Pethica, *Rev. Sci. Intstrum.* **64**, 3515 (1993).

[11] R.W. Carpick, D.F. Ogletree and M. Salmeron, *Appl. Phys. Lett.* **70**, 1548 (1997).

[12] M.A. Lantz, S.J. O'Shea, M.E. Welland and K.L. Johnson, *Phys. Rev. B*, **55**, 10776 (1997).

[13] L. Howald, H. Haefke, R. Lüthi, E. Meyer, G. Gerth, H. Rudin and H.-J. Güntherodt, *Phys. Rev. B* **49**, 5651 (1994).

[14] L. Howald, R. Lüthi, E. Meyer, and H.-J. Güntherodt, *Phys. Rev. B* **51**, 5484 (1995).

[15] R. Lüthi, H. Haefke, E. Meyer, L. Howald, H.-P. Lang, G. Gerth and H.-J. Güntherodt, *Z. Phys. B.*, **95**, 1 (1994).

[16] F. Ohnesorge and G. Binnig, *Science* **260**, 1451 (1993).

[17] F.O. Goodman and N. Garci,*Phys. Rev. B* **43**, 4728 (1991).

[18] J.N. Israelachvili, *Intermolecular and Surface Forces*, Academic Press, London (1985).

[19] Y. Martin, C.C. Williams and H.K. Wickramasinghe, *J. Appl. Phys.* **61**, 4723 (1989).

[20] G.M. McClelland, R. Erlandsson and S. Chiang, in *Review of Progress in Quantitative Non-Destructrive Evaluation*, edited by D.O. Thompson and D. E. Chimenti (Plenum, New York, 1987), Vol. 6B, p. 1307.

[20a] M. Nonnenmacher, J. Greschner, O. Wolter, and R. Kassing, *J. Vac. Sci. Technol. B* **9** 1358 (1991).

[21] F.J. Giessibl, *Science* **267**, 68 (1995).

[22] T.R. Albrecht, P. Grütter, D. Horne, and D. Rugar, J. Appl. Phys. **69**, 668 (1991)

[23] P. Güthner, *J. Vac. Sci. Technol. B* **14**, 2428 (1996).

[24] R. Lüthi et al., *Z. Phys. B* **100**, 165 (1996).

[25] M. Guggisberg et al., to appear in *Phys. Rev. B* (1997).

[26] R. Lüthi et al., to appear in Proceedings ACS-Meeting Orlando, August 1996.

[27] R. Perez, M.C. Payne, I. Stich and K. Terakura, *Phys. Rev. Lett.* **78**, 678 (1997).

[28] R. Erlandsson, L. Olsson and P. Martensson, *Phys. Rev. B* **54**, 8309 (1996).

[29] Y. Sugawara, M. Ohta, H. Ueyama and S. Morita, *Science* **270**, 1646 (1995).

[30] M. Bammerlin et al., to appear in *Scanning Probe Microscopy* **1**, 1 (1997).

[31] R.N. Barnett and U. Landman, *Phys. Rev. Lett.* **76**, 2302 (1996).

[32] J.J.Saenz et al., J. Appl. Phys. **62**, 4293 (1987).

[33] Y.Martin and H.K.Wickramasinghe, Appl. Phys. Lett. **50**, 1455 (1987).

[34] C.Schönenberger and S.F.Alvarado, Z. Phys. B **80**, 373 (1990).

[35] A.Wadas, H.J.Hug, A.Moser and H.-J.Güntherodt, J. of Magnetism and Magnetic Materials **120**, 379 (1992).

[36] R.Proksch, S.Foss, E.D.Dahlberg and G.Prinz, J. Appl. Phys. **75**, 5776 (1994).

[37] R.Proksch et al., J. Appl. Phys. **75**, 6892 (1994).

[38] D.Dahlberg et al., J. of Magnetism and Magnetic Materials **140–144**, 1459 (1995).

[39] G.Bochi et al., Phys. Rev. Lett. **75**, 1839 (1995).

[40] Z.Quanmin, Y.Zheng, A.Roytburd and M.Wuttig, Appl. Phys. Lett. **66**, 2424 (1995).

[41] H.J.Hug, T.Jung, H.-J.Güntherodt and H.Thomas, Physica C **175**, 357 (1991).

38

[42] H.J.Hug et al., Physica C **235–240**, 2695 (1994).

[43] A.Moser et al., Phys. Rev. Lett. **74**, 1847 (1995).

[44] A.Wadas and H.-J.Güntherodt, J. Appl. Phys. **68**, 4767 (1990).

[45] C.Schönenberger and S.F.Alvarado, J. Appl. Phys. **67**, 7278 (1990).

[46] H.J.Hug et al., Rev. Sci. Instr. **64**, 2920 (1993).

[47] A.Wadas, H.J.Hug and H.-J.Güntherodt, J. Appl. Phys. **72**, 203 (1992).

[48] G.Matteucci, M.Muccini and U.Hartmann, Phys. Rev. B **50**, 6823 (1994).

[49] G.Matteucci, M.Muccini and U.Hartmann, IEEE Trans. Magn. **133**, 422 (1994).

[50] T.Göddenhenrich, U.Hartmann and C.Heiden, Ultramicroscopy **42-44**, 256 (1992).

[51] T.Ohkubo, J.Kishigami and K. R.Kaneko, IEEE Trans. Magn. **29**, 4086 (1993).

[52] K.Wago, K.Sueoka and F.Sai, IEEE Trans. Magn. **27**, 5178 (1991).

[53] G.A.Gibson, S.Schultz, T.Carr and T.Jagielinski, IEEE Trans. Magn. **28**, 2310 (1992).

[54] G.Persch and H.Strecker, Ultramicroscopy **42–44**, 1269 (1992).

[55] H.J.Hug et al., Physica B **194–196**, 377 (1994).

[56] priv. comm. A.Hubert, Institut für Werkstoffwissenschaften Universität Erlangen, Martenstrasse 7 D-91058 Erlangen, Schluchsee, Germany, 1994.

[57] M.Rührig, S.Porthun and J.C.Lodder, Rev. Sci. Instr. **165**, 3225 (1994).

[58] T.Chang et al., IEEE Trans. Magn. **28**, 3138 (1992).

[59] C.Xiaodong et al., J. Appl. Phys. **73**, 5805 (1993).

[60] I.D.Mayergoyz, A.A.Adly, R.D.Gomez and E.R.Burke, J. Appl. Phys. **73**, 5799 (1993).

[61] H.J.Hug et al., submitted (unpublished).

[62] H.J.Hug et al., , J. Appl. Phys. **79**, 5609 (1996).

[63] G.Bochi et al., Mat. Res. Soc. Proc. **313**, 309 (1993).

[64] C.Kittel, Phys. Rev. **70**, 945 (1946).

[65] B.Kaplan and G.A.Gehring, J. of Magnetism and Magnetic Materials **128**, 111 (1993).

[66] Y.Yafet and E.M.Gyorgy, Phys. Rev. B **38**, 9145 (1988).

[67] H.F.Hess *et al.*, Phys. Rev. Lett. **62**, 214 (1989).

[68] I.Maggio-Aprile *et al.*, Phys. Rev. Lett. **75**, 2754 (1995).

[69] K.Haralda *et al.*, Phys. Rev. Lett. **71**, 3371 (1993).

[70] C.C.Tsuei *et al.*, Phys. Rev. Lett. **73**, 593 (1994).

[71] J.R.Kirtley *et al.*, Nature **373**, 225 (1995).

[72] A.M.Chang *et al.*, Appl. Phys. Lett. **61**, 1974 (1992).

[73] A.M.Chang *et al.*, Europhys. Lett. **20**, 645 (1992).

[74] D.Dimos, P.Chaudhari, J.Mannhart and F.K.LeGoues, Phys. Rev. Lett. **61**, 219 (1988).

[75] J. K. Gimzewski, Ch. Gerber, E. Meyer and R.R. Schlittler, *Chem. Phys. Lett.* **217**, 589 (1994).

[76] E. Meyer, J.K. Gimzewski, Ch. Gerber and R.R. Schlittler, in *Ultimate Limits of Fabrication and Measurement*, edited by M.E. Welland and J.K. Gimzewski, Kluwer Academic Publishers,p. 89 (1995).

[77] R. Berger, Ch. Gerber, J.K. Gimzewski, E. Meyer and H.-J. Güntherodt, *Appl. Phys. Lett.* **69**, 40 (1996).

[78] T. Thundat, E.A. Wachter, S.L. Sharp and R.J. Warmack, *Appl. Phys. Lett.* **66**, 1695 (1995).

[79] R. Berger, E. Delamarche, H.-P. Lang, Ch. Gerber, J.K. Gimzewski, E. Meyer and H.-J. Güntherodt,, *Science* **276**, 2021 (1997).

Single Molecule Force Spectroscopy by AFM Reveals Details of Polymer Structure

Matthias Rief, Patrik Schulz-Vanheyden & Hermann E. Gaub
Lehrstuhl für Angewandte Physik, Universität München

1. Introduction

The AFM [1-4] allows to control small micromanufactured tips with a precision of fractions of Ångströms at a force sensitivity of piconewtons. It is therefore natural to combine nanotechnology and life sciences to investigate the mechanical properties of individual biomolecules [5-12]. Here we report recordings of force extension curves of polysaccharides and polyaminoacids.

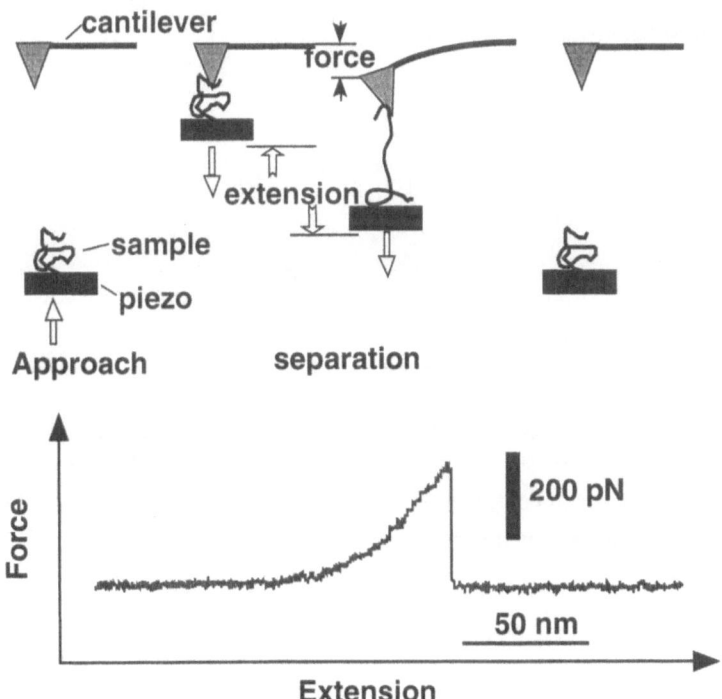

Figure 1. Schematics of single molecule force spectrosopy by AFM

N. García et al. (eds.), Nanoscale Science and Technology, 41–47.

2. Methods

A custom built force spectrometer based on AFM technology was used to stretch individual polymers between the tip of a standard Si_3N_4 AFM-cantilever (Digital Instruments, Santa Barbara) and a glass surface. The polymers cellulose and dextran were attached to a glass surface by drying an aqueous solution of the polymer onto a microscope slide at 70° C. After drying the remaining layer of polymer was thoroughly removed by rinsing with buffer. It turns out that after the rinsing process a thin layer of polymer still tightly adheres to the glass surface. In case of polylysine the polymer in aqueous solution was allowed to incubate onto a freshly evaporated gold surface for 2 h at room temperature before rinsing. A schematics of the experiment is given in Fig. 1. The tip of an AFM cantilever is approached to the polymer coated surface. As soon as the tip gets into contact with the polymer layer the lever is slightly bent up. In order to make the polymer stick to the tip it is pressed onto the layer at forces of several nanonewtons. This kind of attachment holds up to forces in the nanonewton range. One might speculate that in this process a chemical carbon-silicon bond is formed. When the tip is retracted from the surface, adhesion has formed between the tip and a polymer molecule and the molecule is strechted and a force vs. extension curve of the molecule can be recorded. When the maximum adhesion force is exceeded the molecule detaches from the tip and the lever is free again. Because both the pickup of the molecule and the anchorage to the surface can occur at any position along the polymer backbone the length of the extended polymer stretch can vary from zero to the contour length of the polymer.

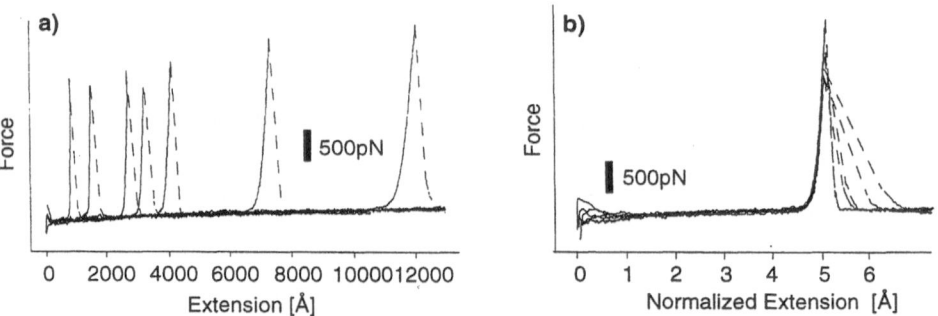

Figure 2. Cellulose: a) Force extension curves of individual molecules.
b) Normalized traces superimposed on top of each other

3. Results and Discussion

Upon separation of tip and sample attractive forces can be registered in many cases up to distances of several micrometers. Fig 2a shows several such force curves of cellulose. Since the majority of these force distance curves can be superimposed with astonishing accuracy when normalized by their length (see Fig. 2b) we interprete theses curves as extension curves of single molecules. In the case of cellulose the polymer deformation can be well described by applying the freely-jointed-chain model with an added elasticity [13] of 5000 pN/Å per glucose monomer. This elasticity corresponds well with results obtained by MD-simulations of such polymer chains [5].

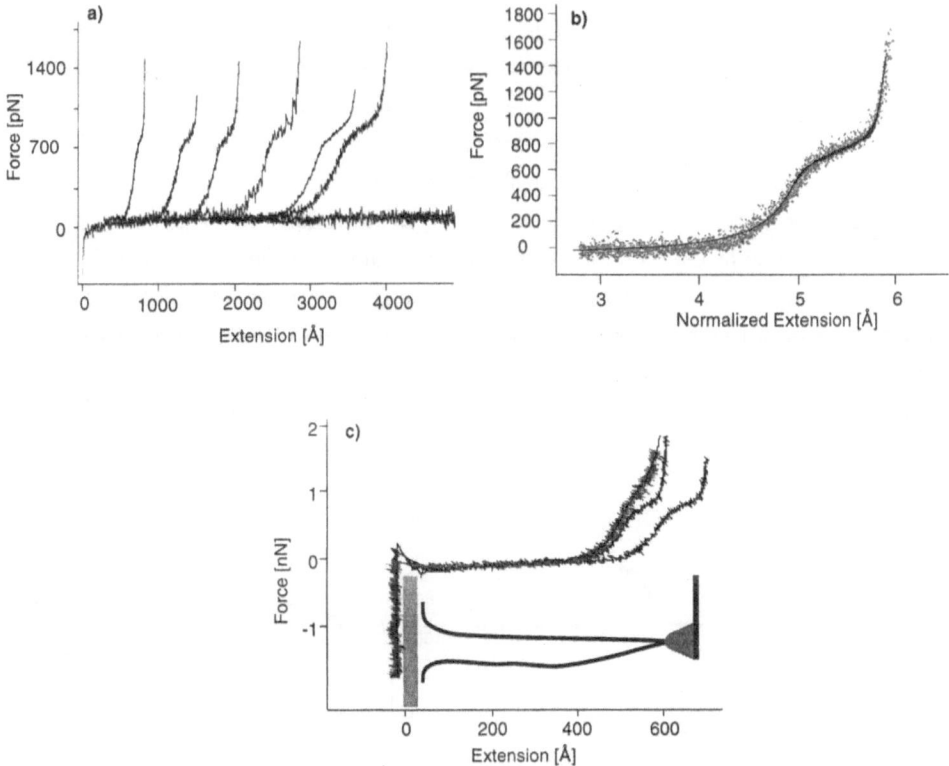

Figure 3. Dextran: a) Force extension curves of individual molecules. b) Normalized traces superimposed with two state model c) In case of stretching two molecules in parallel the force curve can be interpreted as the sum of two single molecule force curves

44

Other polysaccharides show a more complex deformation mechanism including discontinous conformational transitions within the polymer backbone [14]. Force curves of dextran as an example are shown in Fig. 3. Since the connections between the glucose units in dextran are much more flexible than in cellulose the elasticity in the low force regime (<700 pN) is dominated by a monomer spring constant of 750 pN/Å. At forces around 700 pN a pronounced plateau in the single molecule traces appears, which can be interpreted as a length increase of the polymer chain by 0.65 Å/monomer incurred by flipping of the C5-C6 bond. The black curve in Fig. 3b is a fit based on a simple two state model where the two states differ in energy by 13.2 kT and in length by 0.65 Å and the number of monomers in the respective states is governed by the Boltzmann distribution. Since the conformational transition occurs much faster than the timescale of the pulling cycle, the whole experiment is performed in thermodynamic equilibrium and is fully reversible which can be seen in the identical form of the forward and backward traces [5]. In some rare cases, extension traces like the one given in Fig 3c are recorded. As the superimposed curve shows, these can be easily understood as the extension of two molecules of different length attached in parallel.

Since the force at which the transition occurs can be measured with high reliability it can be used as a natural piconewton force standard: Fig. 4 shows two curves taken on dextran at different extension rates using cantilevers of different spring constant. After normalizing the curves to monomer length and multiplying the deflection signal with the spring constants of the two different levers the curves look identical as Fig. 4b shows. The spring constants of all cantilevers used in the experiments presented were determined according to [15].

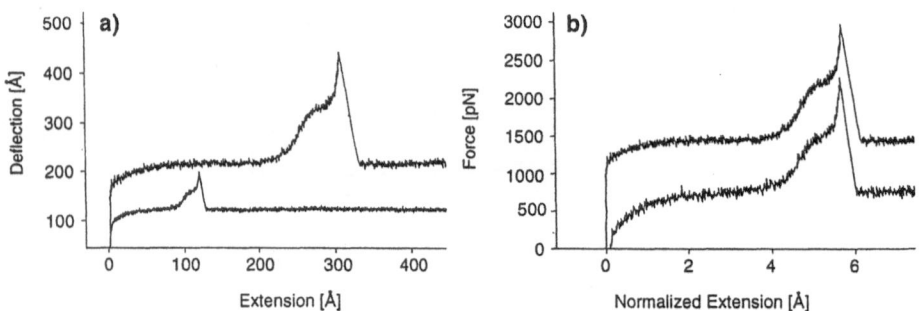

Figure. 4 The conformational transition of Dextran may be used as a pico-Newton force standard a) Curves obtained with different cantilevers at different pulling speeds b) Normalized traces superimposed on top of each other

In cases where the energy difference between the two conformations of a modular polymer aproaches thermal energies the transition blurs out and appears as a more or less well pronounced shoulder as shown in Fig. 5. Here polylysine was streched at pH 11 and room temperature. Under these conditions polylysine is known to form an alpha helix [16].The solid line is the fit based on the two level model (introduced above) with an energy diffence of 1.9 kT and a length increase of 0.47 Å.

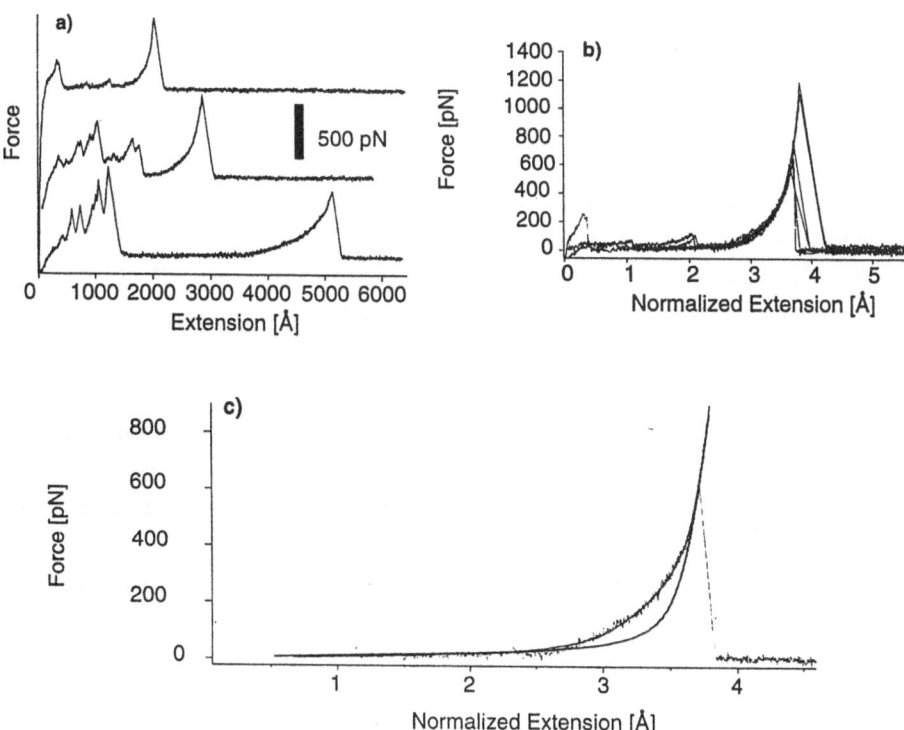

Figure 5. Polylysine: a) Force extension curves of individual molecules. b) Normalized traces superimposed on top of each other. c) Standard freely jointed chain fit with added elasticity (lower solid line) and two state model (upper solid line) superimposed on a typical normalized force curve.

4. Conluding Remarks

Single molecule force specrtoscopy by AFM has evolved into a versatile tool for measuring molecular deformations of modular polymers. Details about the polymer's conformation can be measured. The data from measurements on dextran and polylysine can be explained by the same comparatitively simple two state model despite its very different appearance. The strength of the AFM in comparison with other techniques like optical tweezers lies in the high position accuracy and the simplicity of the experiment. Improvments are to be expected from further miniaturization mainly of the cantilevers leading to a larger bandwidth of the experiments.

5. Acknowledgments

This work was supported by the Deutsche Forschungsgemeinschaft.

6. References

1. Binnig, G. and Rohrer, H., (1987) Scanning Tunneling Microscopy - from birth to adolescence, *Rev. Mod. Phys.* 59 615-625.
2. Alexander, S., *et al.*, (1989) An atomic-resolution atomic-force microsope implemented using an optical lever, *J. Appl. Phys.* 65 (1), 164-167.
3. Binnig, G., Quate, C.F., and Gerber, C., (1986) Atomic force microscope, *Phys. Rev. Lett.* 56 930.
4. Goodman, F.O. and Garcia, N., (1991) Roles of the attractive and repulsive forces in atomic-force microscopy, *Phys. Rev. B (Condensed Matter)* 43 (6), 4728-4731.
5. Rief, M., *et al.*, (1997) Single molecule force spectroscopy on polysaccharides by AFM, *Science* Feb
6. Rief, M., *et al.*, (1997) Reversible Unfolding of Individual Titin Immunoglobulin Domains by AFM, *Science* 276 1109-1112.
7. Florin, E.-L., Moy, V.T., and Gaub, H.E., (1994) Adhesive forces between individual ligand-receptor pairs., *Science* 264 415-417.
8. Moy, V.T., Florin, E.L., and Gaub, H.G., (1994) Intermolecular forces and energies between ligands and receptors, *Science* 266 257-259.
9. Dammer, U., *et al.*, (1995) Specific antigen / antibody interactions observed by atomic force microscopy, *Biophys. J.* 70 (5), 2437-2441.
10. Hinterdorfer, P., *et al.*, (1996) Detection and localization of individual antibody-antigen recognition events by atomic force microscopy, *Proc. Natl. Acad. Sci. USA* 93 3477-3481.
11. Lee, G.U., Chris, L.A., and Colton, R.J., (1994) Direct measurement of the forces between complementary strands of DNA, *Science* 266 771-773.
12. Radmacher, M., *et al.*, (1994) Direct observation of enzyme activity with the atomic force microscope, *Science* 265 1577-1579.
13. Smith, S.B., Cui, Y., and Bustamante, C., (1996) Overstretching B-DNA: the elastic response of individual double-stranded and single-stranded DNA molecules, *Science* 271 795-798.

14. Li, H., *et al.*, (1997) Single molecule force spectroscopy on Xanthan by AFM, *Adv. Ma.* in press

15. Florin, E.L., *et al.*, (1995) Sensing specific molecular interactions with the atomic force microscope, *Biosensors and Bioelectronics* 10 (9-10), 895 - 901.

16. Davidson, B. and Fasman, G.D., (1967) The conformational transitions of uncharged poly-L-lysine., *Biochemistry* 6 (6), 1616-1629.

QUANTUM RESONANCE OF THE MAGNETISATION IN A SINGLE-CRYSTAL OF CLUSTERS IN Mn$_{12}$-ACETATE.

B. BARBARA, L. THOMAS, F. LIONTI

Laboratoire de Magnétisme Louis Néel, CNRS, BP 166, 38042-Grenoble, France.

A. SULPICE

CRTBT, CNRS, BP 166, 38042-Grenoble, France.

A. CANESCHI

Department of Chemistry, University of Firenze, 50144, Italy.

1. Introduction

Observation of the quantum behaviour of a macroscopic variable is a challenging problem [1,2]. Macroscopic systems showing agreement with theoretical predictions are presently related to superconductivity (see eg the case of SQUIDs in [2]). In magnetism, Macroscopic Quantum Tunnelling (MQT) consists in e.g. the rotation of the magnetization of a single domain small particle by tunnelling through its anisotropy energy barrier, or the motion of a small portion of a domain wall through its pinning energy barrier. The first experiments devoted to this question suggested that the total number of spins involved coherently in these MQT processes is of the order of 10^3 to 10^5, depending on the material [3]). Although not at the human scale, this scale is macroscopic in the framework of Quantum Mechanics.

During the last decade Quantum Tunnelling of the Magnetization (QTM) was mostly studied by measuring the temperature dependence of magnetic relaxation after an abrupt change in the applied magnetic field. At high temperatures, when the anisotropy energy barriers (or pinning energy barriers for domain walls) are overcome by Thermal Activation (TA), the magnetic relaxation strongly depends on temperature. At low enough temperatures when the thermal activation becomes extremely weak, the relaxation should in principle vanish. Nevertheless, this is not always the case. Early

N. García et al. (eds.), Nanoscale Science and Technology, 49–64.
© 1998 *Kluwer Academic Publishers.*

experiments devoted to QTM by Barbara and Uehara showed that rare-earth intermetallic alloys (Dy_3Al_2, $SmCo_{3.5}Cu_{1.5}$) relax almost independently of temperature at low temperature, a phenomenon that they attributed to QTM [3]. The crossover temperature T_c between the high and low temperature regimes was a few Kelvin, which is in agreement with to-day 's expectations for highly anisotropic sytems. A QTM model with thermal or quantum domain wall depining was developed by Egami in order to interpret the experiments on Dy_3Al_2 [4].

On the beginning of 90's, several other groups started to work on this field (see some recent references [5-9]). They essentially confirmed the first observations of [3]. However, too naive interpretations of relaxation data in complex systems often led to spurious conclusions. Barbara et al. [10] showed that power-law distributions of energy barriers $n(E) \propto E^{-\alpha}$ can simulate QTM. Several processes giving power-law distributions at low temperatures have been identified : (i) for non-interacting particles with switching field distribution ($\alpha = 1/2$) [11,12] (switching field distributions are always present in assemblies of particles with randomly oriented axes); (ii) for magnetically coupled particles (exchange interactions and/or dipolar interactions), leading to magnetization reversal avalanches ($\alpha > 0$) [13,14] ; (iii) for particles with surface spin disorder (due to frustrated antiferromagnetic interactions) leading to numerous low energy barriers ($\alpha = 1$) [15].

In order to avoid the complications due to energy barrier distributions, two directions of research were taken : the dynamical study of magnetization reversal of (i) individual nanoparticles (or wires) and (ii) arrays of identical magnetic clusters belonging to molecular crystals. These approaches were made possible with recent advances achieved in sample elaboration and measurement techniques, and this allowed this field to develop rapidly these last years.

The first magnetization measurements of individual single-domain nanoparticles at low temperature were done using micro-SQUID detectors [16,6]. These measurements showed unambiguously that the magnetization reversal of single crystalline Co nanoparticles with dimensions between 10 and 30 nm [17] is described by thermal activation (TA) over a single-energy barrier as originally proposed by Néel [18]. In particular, the probability for non-reversal is exponential ($\approx \exp(-t/\tau)$) and the activation volume is very close to the particle volume. No quantum effects were found down to 0.2 K. This is not surprising because the predicted cross-over temperature between TA and QTM for these Co particles is $T_c \approx 20$ mK. However the results of [17] are important because they gave the first demonstration of the Néel-Brown [18] model and now constitute the precondition for the first observation of MQT of magnetization on a single particle with a collective spin as large as $S=10^5$ [19].

Regarding clusters in molecular crystals, QTM was clearly observed in Mn_{12} acetate (Mn_{12}-ac) (Friedman and Hernandez et al. [20], Thomas and Lionti et al. [21]). This system is made of clusters with modest collective spin S=10, so that level quantization plays an important role and QTM occurs when the spin-up and spin-down level schemes are in coincidence (resonant QTM). The same effect was recently found by Sangregorio et al. on the so-called Fe_8 molecule [22].

The First QTM studies on Mn_{12}-ac started with relaxation experiments on oriented polycrystals [6,23] after Sessoli et al. [24] showed that the ac-susceptibility could be fitted above 2 K, to a single relaxation time. Two relaxation times were needed to fit the relaxation experiments between 3.5 and 0.8 K. They both showed a thermal activation behaviour above 2.2±0.2 K and were independent of temperature below [6,23]. This was interpreted as QTM of the collective spin S=10, with a crossover temperature T_c = 2.2±0.2 K. Furthermore, a minimum of the relaxation observed in zero field was attributed to possible resonant QTM [6]. A quantitative link was made between this phenomenon and the observed crossover temperature $3kT_c/2 \approx 20\mu_B H_m$, where H_m is the field at which the relaxation time is maximum ($H_m = \Delta H/2$, where ΔH is the field at which the second resonance should had been observed). Indeed the it is easy to check that $g\mu_B S\Delta H \approx \hbar\omega_0 \approx 4k_B T_c$. A few months later these relaxation experiments were carefully repeated and essentially confirmed these results [25]. The phenomenon of resonant QTM, was also strongly suggested by important dips in the ac susceptibility [26]. Two years later it was quantitatively confirmed by similar but more detailed experiments, also performed on an oriented polycrystal [20]. A striking demonstration was simultaneously given on a single crystal [21].

On the theoretical side, the tunnelling rate of a large spin through its anisotropy barrier was calculated as a function of longitudinal and transverse fields, in the continuous [27-29] or discrete [30-32] level models.

The number of experimental and theoretical studies devoted to the phenomenon of QTM of a large spin grows very rapidly.

In this paper we review some of our recent results on Mn_{12}-ac. Section I gives a short description of the crystal structure. In Section II we recall previous results. Section III is devoted to a study of the stability of the collective spin (S=10). In the third section we discuss on the role of local symmetry in QTM. In sections IV and V we give the first strong arguments in favor of resonant tunnelling between the lowest levels in Mn_{12}-ac.

2. Crystal structure of Mn_{12}-ac

The Mn_{12}-ac crystal, synthetized by Lis [33] in 1980, is built of discrete dodecanuclear $[Mn_{12}(CH_3COO)_{16}(H_2O)_4O_{12}].2CH_3COOH.4H_2O$ molecules with a tetragonal symmetry. Four inner $Mn(1)^{4+}$ (S=3/2) are surrounded by eight Mn^{3+} (S=2). These last are divided in two crystallographic sites with strong tetragonal $Mn(2)^{3+}$ and orthorhombic $Mn(3)^{3+}$ Jahn-Teller distortions with high cystal field anisotropy. Strong antiferromagnetic interactions between the four $Mn(1)^{4+}$ and the four $Mn(3)^{3+}$ lead to a ferrimagnetic ground state, of spin S=10 [24].

3. Resonant tunnelling on a single crystal of Mn_{12}-ac

Here we recall the main facts published in [21]. Isothermal hysteresis loops of a single crystal of Mn_{12}-ac showed staircase behaviour when field was applied along the easy axis of magnetization (fig.1).

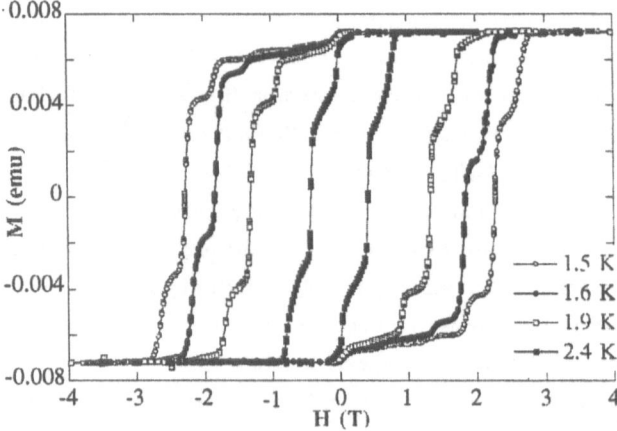

figure 1. Staircase hysteresis loops observed at different temperatures below the blocking temperature.

The quality of our single crystal orientation with respect to the applied field was checked by comparing the remanent magnetization M_R with M_S ; we found $M_R/M_S > 0.99$ (note that in ref. [20], $M_R/M_S \approx 0.5$ is typical for a random distribution of grains orientations which makes difficult observation and to identification of magnetic resonances). In our single crystal experiments, isothermal staircase hysteresis loops were observed below the blocking temperature of 3 K. In the flat portions of the loop, the sample relaxation time is much larger than the measuring time-scale (about 500 s). Conversely the steep portions of the loop correspond to relaxation times of the order of the measuring time-scale. Relaxation-time measurements showed a series of sharp minima occurring precisely at the fields for which the steep magnetization decreases were observed. The striking staircase hysteresis loop and oscillating relaxation time were periodic in field, with a period $\Delta H = 0.44$ T, ($H_n = nD/g\mu_B$, with $D/g\mu_B = 0.44$ T, except below 2 K, where $D/g\mu_B=0.60$ T for n=1). This result was interpreted by assuming that the magnetization of the single crystal relaxes much faster when the the spin-up and spin-down level schemes coincide, as already suggested in [6,25,26] (resonant QTM).

4. Range of stability of the collective spin S=10.

All present interpretations of the magnetic behaviour of Mn_{12}-ac and in particular the phenomenon of resonant tunnelling, assume a well defined collective spin S=10. It is therefore important to know the range of validity of this model.

Magnetization measurements have been performed above the superparamagnetic blocking temperature i.e. between 3 and 300 K, along the c-axis and in the basal plane of a single crystal of Mn_{12}-ac. The M(H) curves were fitted by taking the thermodynamical average of m, with the energy :

$$E_m = -Dm^2 - g\mu_B mH \qquad (1)$$

$-10 \le m \le 10$ and $D = 0.6$ K. The curves measured along the c-axis are compared to the calculated ones fig.2.

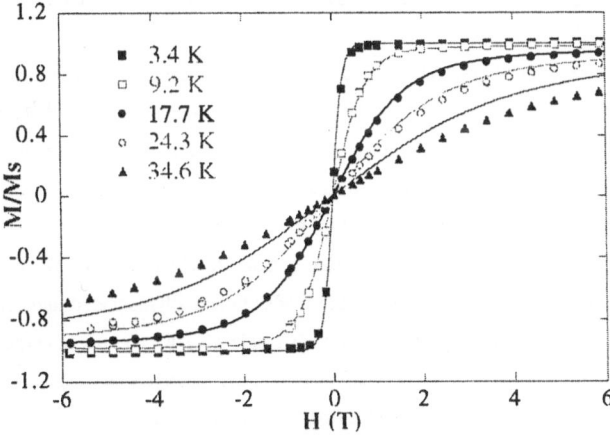

Figure 2. Isothermal magnetisation curves measured along the c-axis of Mn_{12}-ac. Continuous lines are calculated using the energy levels given in formula (1).

Below 10 K, due to their large uniaxial anisotropy, Mn_{12}-ac clusters behave like S=10 Ising spins. Between 10 and 30 K, an Ising to Heisenberg crossover was observed (still with S=10). Above 30 K (see the curves at 24.3 and 34.6 K), the fit with S=10 is no longer valid. Correlatively a fast decrease of the collective spin was observed up to 90 K, which is followed by a flat minimum with $S = 6.25 \pm 0.25$ near 200 K, which is due to persistent antiferromagnetic short range order between Mn^{4+} and Mn^{3+} (fig.3).

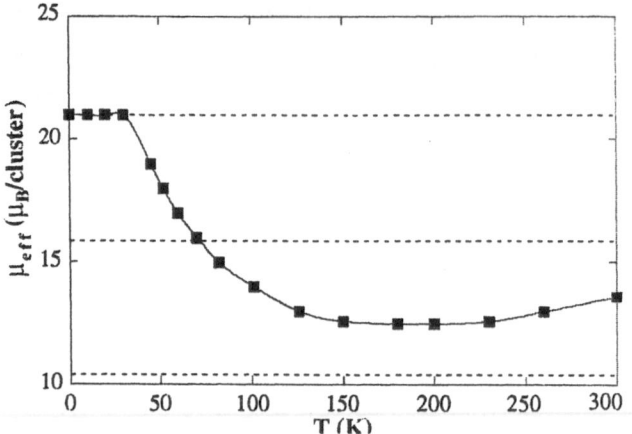

Figure 3. Effective paramagnetic moment per cluster, $g\mu_B\sqrt{S(S+1)}$. This curve is deduced from magnetisation measurements, simply assuming that the collective spin decreases in increasing temperature and that the anisotropy D is constant. The dashed lines indicate the effective moment calculated in the main three regimes : ferrimagnetic cluster ($21\mu_B$), paramagnetic cluster ($16\mu_B$) and paramagnetic cluster with antiferromagnetic short range order ($10.5\mu_B$). Solid line is a guide for the eyes.

The curves measured perpendicularly to the c-axis led to the same conclusions, and also (see Lionti et al in [21]) allowed us to measure directly the anisotropy field $H_a \simeq 9T$.

The crossover at 30 K corresponds to the beginning of the destruction of the ferrimagnetic order at the scale of each cluster, which is completed at 90 K (note that this "Curie temperature", being broadened by a 3-D to 0-D size effect crossover, is not a phase transition). This shows the existence of a S=9 level above 30 K. The presence of levels with S<10 at energies close to those where thermally activated tunnelling takes place on the multiplet S=10, should in principle give a complex superposition of magnetic resonances (one set per multiplet). Neutron scattering experiments showed a continuum of "exchange states" with a first broad level at 60 K [34].

5. Range of validity of the single-spin model

Let us first review the present status of theoretical and experimental studies. The usual single-spin Hamiltonian in tetragonal symmetry contains a part (H_1) which commutes with S_Z and another one (H_2) which does not commute with S_Z :

$$H_1 = -H_Z S_Z - D S_Z^2 - B S_Z^4 + ... \qquad\qquad H_2 = -H_X S_X - C(S_X^4 + S_Y^4)... \qquad (2)$$

This last can be responsible for tunnelling and gives a time-dependent magnetic response.

In Mn_{12}-ac experiments [20,21], resonant tunnelling was observed with $\Delta m=\pm 1$ showing that each molecule experiences a local transverse field H_x (The fourth order transverse anisotropy terms can only mix states with $\Delta m=\pm 4$). Hyperfine and dipolar transverse field components (less than 0.05 T and 0.02 T respectively), as well as fourth order transverse anisotropy have been considered to interpret quantitatively experimental data [32, 35]. It is important to note that hyperfine and dipolar interactions lead to bias and transverse field components, whose dynamical character with different temperature dependences, is essential, in particular because they put levels in resonance even in the presence of static biais field [32]. In the picture of the Landau-Zener model, the probability for magnetisation reversal $p = 1 - \exp(-\omega\Delta^2)$ (where Δ is proportional to the tunnel splitting), the rate ω should be dominated by the faster mechanism of field variation (eg longitudinal and transverse components of hyperfine and dipolar fields). In the absence of an applied bias field and at low temperature, relaxation time minima were calculated at the resonance between the ground state levels ± 10, with the relaxation rates close to those observed experimentally below 2 K [32]. Temperature dependent relaxation rates, calculated in the high temperature regime were also in agreement with experiments [6,25,20,21] above 2K. Resonant tunnelling near the top of the barrier [20,21], certainly plays here an important role [32,36,37]. Off-resonance calculations [32] (with a bias of 0.2 T) gave similar results, but with much larger relaxation rates, as expected.

If tunnelling at lowest energy states cannot be affected by the observed non-conservation of the collective spin S=10 (the level S=9 lies slightly below 40 K), what about tunnelling near the top of the barrier ? The fact that relatively well defined and equally spaced resonances of the S=10 level scheme are observed, suggests that that tunnelling on the multiplet S = 10 is faster than on the other multiplets with S < 10 .

Regarding the effect of large fields, spin-flop evaluated at about 60 T, should involve a not fully developped classical order parameter. Quantum jumps of spin-flop type, occuring in larger fields, have been observed in Mega-Gauss experiments [38].

Local symmetries lower than the cell symmetry and the absence of an inversion center could favour local transverse crystal field directions, anti-symmetrical Dzyaloshinsky-Moriya, anisotropic bilinear and biquadratic and exchange interactions , which could enhance QTM in Mn_{12}-ac. We should also note that contributions to the tunnel splitting of transverse anistropy terms of all orders (higher than four), could be very significant [39].

6. Effect of a transverse field.

In order to investigate the effect of a transverse field on both quantum and classical relaxation, we have tilted a single mono-crystalline grain with respect to the applied field and performed hysteresis loop measurements (Lionti et al. in [21], Sulpice et al [40]). The magnitudes of the magnetization jumps ΔM increase with the transverse field component $H_T = H\sin\theta$, but the positions of the jumps remain at the same longitudinal field $H_{Ln} = H\cos\theta = nD/g\mu_B$. It is not necessary to perform high order expansions to show that, simply because the fields at which crossing occur do not depend on the transverse part of the Hamiltonian, but only on the longitudial one. The effect of the tansverse component is only to open a gap. The increase of ΔM vs H_T must result from (a) easier thermal activation resulting from the lowering of the energy barrier ($\propto (1 - 2H_T/H_A - 2H_L/H_A)$ for H_L, $H_T \ll H_A$), (b) easier QTM resulting from the increase of the tunnel splitting ($\propto (H_T/H_A)^{2m-n}$ + contributions of higher order transvcverse anisotropy terms) of resonant level pairs m and -m+n, where $\dfrac{g\mu_B H_{Ln}}{D} = n$. A comparizon between these two contributions showed that a transverse field favours thermal activation and therefore tunnelling on the top of the barrier (Lionti et al. in [21].

The effect of a transverse field also gives informations on the level structure of Mn_{12}-ac. In our single crystal experiments, level crossings were observed below 2 K for the sequence of fields (in Tesla), $H_0=0$, $H_1=0.62$ for $n\leq1$, and $H_n=0.44n$ for $n\geq2$. Assuming that tunnelling occurs on the ground state at low temperature (this possibility will be discussed in section VI), the resonances with $n\leq1$ and $n\geq2$ give a first excited level (m=9) close to the transitions observed in inelastic neutron scattering at 14.6 K and and 11.6 K respectively (note that this last is observed after a level at about 2 K has been thermally populated) [35]. Their identification give two different g-factors : g=2.06 ($H_n=0.44n$ and $\Delta=11.6$ K) and g=1.85. ($H_1=0.62$ and $\Delta=14.6$ K). These two transitions do not appear to be resolved in EPR experiments [41].

The resonances observed on these magnetisation measurements can be described as (i) ground state resonance when the spin-up level scheme is shifted down (Zeeman energy) by 14.6 K and (ii) Other resonances with the energy scale $D = 0.61$ K (assuming here a second order anisotropy constant only). These last could be thermally activated or result from the ferrimagnetic character of the molecule. The energy barriers corresponding to these two scales are $14.6 + 0.61\times9^2 = 64$ K and $0.61\times10^2 = 61$ K.

Deviations from $D(10^2 - m^2)$ for large m (giving m=9 at 14.6 K) , could result from 4th order uniaxial anisotropy terms [21]. The latest EPR experiments came to this conclusion, with D=0.56 K, B=1.1 mK and C=$\pm$0.03 mK [41]. Using the model Hamiltonian (2) with $H_z=0$, we fitted the magnetization curve measured in a field perpendicular to the c-axis with three sets of parameters (fig. 4).

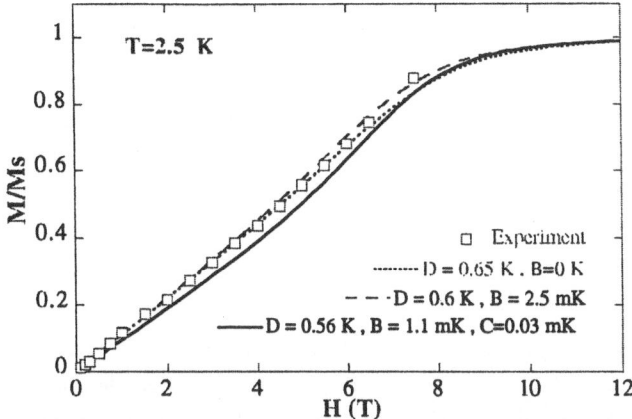

Figure 4. Measured and calculated magnetization curves, for a field perpendicular to the easy axis of magnetization. The best fit is for D = 0.6 K and B= 0.25 mK (see also section 5). Another fit with D = 0.56 K, B = 1.1 mK anc C =0 .03 mK gives a too large initial slope.

Although giving a reasonable description of the observed resonances, the EPR parameters do not lead to a very good fit. The fits with D=0.65 K only or, D=0.6 K and B= 0.25 mK are better but still not completly satisfactory. Note that the effect of C becomes sizable on the M(H) curve only if it is larger than 0.1 mK.

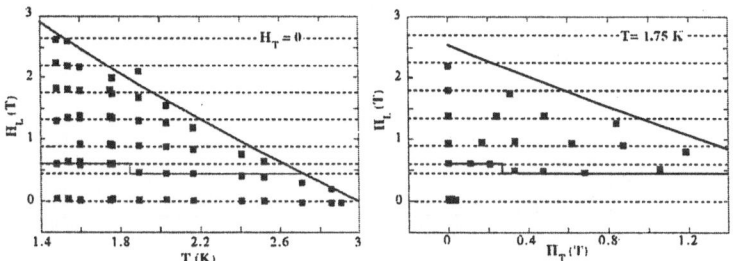

Figure 5. Plots of the bias fields H_L at which magnetization jumps were observed vs temperature (4.a) or transverse field H_T (4.b). The dashed lines show the fields H_n. The solid lines show the evolution of the n=1 resonance (with crossovers at 1.9 K and 2.7 T), and the transition to the regime of thermal fluctuations.

Interestingly the shift of the n=1 resonance is supressed at T>1.9 K [21] or H_T>0.27 T [40] (fig.5). The first, suggests a crossover at Tc ≈2 K, from ground-state to excited-states tunnelling. The transverse field should lead to the same crossover because faster tunnelling paths should move up, due to the increase of resonance splittings near the top of the barrier (for large enough transverse fields, resonance splittings become so important near the top of the barrier, that the energy barrier decreases leading to a quantum deblocking of the magnetization when T→0 [37]). Such a value for the crossover temperature corroborates our first value of 2 K obtained on oriented grains [6, 23, 24] (note that in these experiments relaxation times at temperatures below 1K could have been sometimes underestimated ; however the relaxation times measured at same field and temperature could depend strongly on magnetic history). Using $\hbar\omega_0 \approx 4kT_C$, (where ω_0 is the prefactor of the relaxation rate, which is also of the order of the separation Δ between the ground state and the first excited state), one gets $\omega_0 \approx 10^{12}$ s^{-1} which is not very different from the energy of the first excited level (of the order of 10 K). Note that this value of ω_0 is much larger than the ones deduced in the past, from ac suceptibility (between 10^7 and 10^8 s^{-1}). This discrepency simply results from from the fact that activated tunnelling was not properly accounted for. In more recent ac-susceptibility experiments (Lionti et al in [21]), we have assumed a relaxation time in the thermally activated tunnelling regime given by $\tau^{At} = \omega_0^{At} \exp(-E^{At}(H,T) / kT)$ at resonance and $\tau^{Out} = \omega_0^{Out} \exp(-E^{Out}(H,T) / kT)$ out of resonance (the WKB exponent entering in the prefactor $\omega_0^{At} = \omega_0 \exp(-B^{At})$ or $\omega_0^{Out} = \omega_0 \exp(-B^{Out})$).

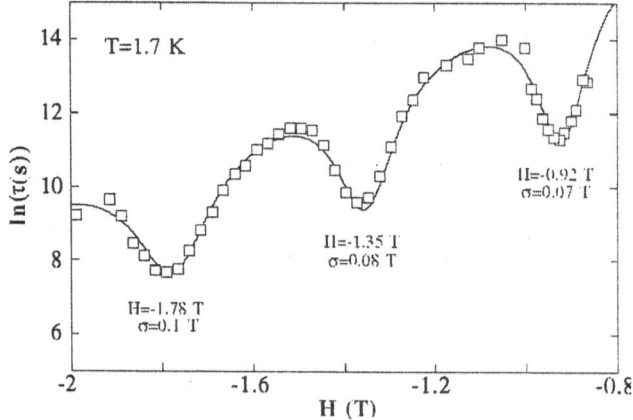

Figure 6. Néperian logarithm of the measured relaxation time vs applied field, obtained from magnetic relaxation experiments at 1.7 K. The fit is the sum of a linear background and lorentzian dips. The fitting parameters for the background (out of resonance) was $\omega_0^{eff} \approx 10^6$ s^{-1}.

Using these forms and the value $\omega_0^{eff} \approx 10^6$ s^{-1} of fig. 6, we find for each peak, the difference in WKB exponents $\Delta B = B^{Out} - B^{At}$ and in the energy difference $\Delta E_0 = E^{Out} - E^{At}$. This difference represents the energy at which tunnelling is most effective (with origin at the top of the barrier). In this simple model, the logarithm of the ratio of the relaxation times, measured on the minimum of the $\tau(H)$ curve and on its background, is equal to $\Delta E = \Delta E_0(1 - \dfrac{2H}{H_A}) - T\Delta B$. A linear plot of $\ln(\tau_{Out}/\tau_{At})$ vs H gives ΔE_0=6.7 K and ΔB=1.1 K for T=1.7 K (fig. 6). Similar curves measured at 1.5 K and 2 K give the values ΔE_0=18 K, ΔB=1.5 and ΔE_0=4.2, ΔB=-1.1 respectively. These values of ΔB are negligibly small, showing that in this range of large bias, the effect of resonant tunnelling does not manifest itself by B values smaller at resonance. On the contrary the transparency of the barrier appears to be nearly the same at resonance and out of resonance, and this is because near the top of the barrier, tunnel splittings are equal or larger than level separations. The corresponding reduction of the energy barrier is of 2 to 18 K, depending on the field and temperature of the experiment. This effect of tunnelling by-pass of thermal activation, near the top of the barrier, is modified in low fields. The same analysis performed at 1.5 K at lower fields gives ΔE_0=35 K and ΔB=10. Here the exponent B is very different at resonance and out of resonance, showing that level separations are much larger than tunnel splittings. Tunnelling does not take place near the top of the barrier (we find m between 7 and 8), suggesting again the possibility of ground state tunnelling in the absence of a biais.

7. Low temperature experiments.

In order to check the existence of ground state tunnelling we have performed a new series of relaxation experiments below and above 2K. The results given fig.7, show relaxation times nearly independent of temperature below 2K and in fields smaller than 0.2 T. The relaxation time τ measured in zero field and plotted vs reciprocal temperature 1/T in a semi-Log scale (fig.7, inset), shows a saturation below 2 K. This saturation depends on the applied field and presents a minimum 25 mT. This value corresponds to the first resonance at n=0, [21]. It is important to remark that the low temperature relaxation time (ie the plateau) decreases by one order of magnitude at the resonance. These experiments strongly suggest the existence of tunnelling on the lowest levels (eg 10 $\rightarrow$ -9). Above 2 K, a thermal activation regime is observed, with still a minimum of relaxation time at the resonance (thermally activated resonance regime[20, 21]). We should note that the time dependence of the magnetisation M(t) as well as the deduced relaxation rate depend critically on the field value near the resonance, and also on the temperature. Moreover we have observed many times that different experimental procedures could modify significantly the M(t) curves [42].

60

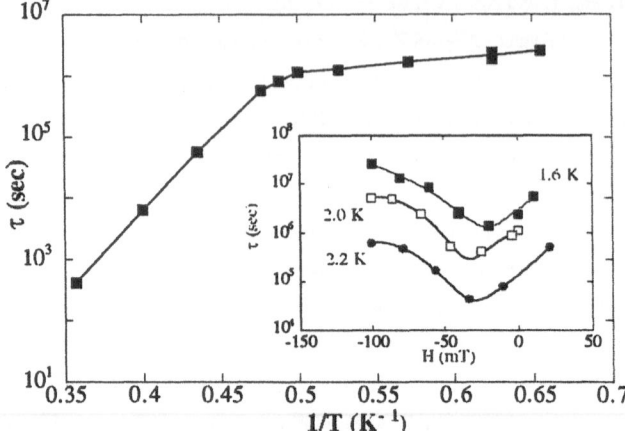

Figure 7. Relaxation time τ at H=0 and plotted vs 1/T, in a semi-log scale. Below 2 K, a saturation of $\log\tau$ is observed. This saturation depends on the field and is minimum at the resonance field of 25 mT (see the insert). Above 2 K, the thermal activation regime is observed, with still a minimum of relaxation time at the resonance (thermally activated resonance).

8.   Conclusion.

Below 2K the resonance n=1 is shifted to larger fields with respect of the other resonances (n$\geq$2). It occurs at a field corresponding exactly to the position of the first excited level m=9, observed in neutron scattering. This shift is supressed above 2K or in the presence of a transverse field larger than 0.2T, showing the existence of a crossover between ground state and excited states tunnelling. Magnetization measurements performed below 2K also show, distinct, on-resonance and off-resonance plateaux in the temperature dependence of the relaxation time. The phenomenon of resonant tunnelling in Mn_{12}-ac does not appear to be limited to resonant thermal activation. This suggests off-diagonal matrix elements not to be very small, in accordance with a relatively high value of the tunnelling prefactor $\omega_0 \approx 10^{11}$ s^{-1}. Finally it is interesting to note that resonant QTM could not have been detected with the poor field resolution of conventional SQUID magnetometers, if resonance lines were not importantly broadened. Phonon broadening being too small by a factor of one hundred [34], broadening of magnetic origin must play the major role [31] (hyperfine and dipolar interactions). Diluted clusters with small isotopic fractions of nuclear spins should give narrow transitions [41], which could be observed with field resolutions of the order of 10^{-4} T.

Ackowledgements : we are very pleased to thank A.L. Barra, B. Canals, D. Fruchart, D. Gatteschi, D. Garanin, L. Gunther, M. Hennion, H. de Raedt, R. Sessoli, I. Tupitsyn, W. Wernsdorfer and A. Zvezdin, for on-going collaborations and useful discussions.

References

[1] A. J. Leggett et al.,. Rev. Mod. Phys., 59 (1987) 1 and Lectures in Physics, Les Houches (1986).

[2] R. F. Voss and R. A. Webb, Phys. Rev. Lett., 47 ₍1981)265 ; J. Clarke, A. N. Cleland, M. H. Devoret, D. Esteve, J. M. Martinis, Science 239 (1988) 992 .

[3] B. Barbara, Proc. 2nd Int. Symp. on Anisotropy and Coercivity 137 (1978) and J. de Phys. 34, 1039 (1972). M. Uehara and B. Barbara, J. Physique 47 (1986)235 .

[4] T. Egami, Phys. Stat. Sol. 20, 157 (1973) and 57 (1973) 211 .

[5] J. Tejada, X. Zhang, J. Magn. Magn. Mater., 140-144 (1995) 1815 .

[6] B. Barbara, W. Wernsdorfer, L. Sampaio, J.G. Park, C. Paulsen, M. Novak, R. Ferré, D. Mailly, R. Sessoli, A. Caneschi, K. Hasselbach, A. Benoit , L. Thomas. J. Magn. Magn. Mater. 1995, 140-144 (1995).1825

[7] R. Sappey et al. "Magnetic hysteresis in novel magnetic materials", Greece 1996, ed. Hadjipanayis, Kluwer Academic Publishers, Dordrecht.

[8] H. Yamazaki, G.Tatara, K. Katsumata , K. Ishibashi , Y. Aoyagi, J. Magn. Magn. Mater. 156 (1996) 135

[9] M.J. O'Shea, P. Perera, J. Magn. Magn. Mater. 156 141 (1996).

[10] B. Barbara et al. Proc. Int. Workschop 'Studies of Magnetic Propoperties of Fine Particles', 235 (1991) Ed. Dormann and Fiorani, (Elsevier, Amsterdam, 1992).

[11] B. Barbara, L. Gunther, J. Magn. Magn. Mater. 128 (1993) 35.

[12] B. Barbara, L. C. Sampaio, A. Marchand, O. Kubo, H. Takeuchi, J. Magn. Magn. Mater. 136 (1994) 183.

[13] R. Ferré, B. Barbara, J. Magn. Magn. Mater., 140-144 (1995) 1861.

[14] R. Ribas, A. Labarta, J. Appl. Phys., 80 (1996) 5192.

[15] R. H. Kodama, A. E. Berkowitz, E. J. McNiff Jr. and S. Foner, Phys. Rev. Lett., 77 (1996) 394.

[16] W. Wernsdorfer, K. Hasselbach, D. Mailly, B. Barbara, A. Benoit, L. Thomas L. J. Magn. Magn. Mater. 145, 33 (1995).

[17] W. Wernsdorfer, E. Bonet Orozco, K. Hasselbach, A. Benoit, B. Barbara , N. Demoncy, A. Loiseau, D. Boivin, H. Pascard, D. Mailly , Phys. Rev. Lett. , 78 (1997) 1791.

[18] L. Néel, Ann. Geophys. 5, 99 (1949); W. F. Brown, Phys. Rev. 130 (1963) 1677.

[19] W. Wernsdorfer, E. Bonet Orozco, K. Hasselbach, A. Benoit, D. Mailly, O. Kubo, H. Nakano and B. Barbara, Phys. Rev. Lett. to appear Nov. 1997.

[20] J. Friedman, M. Sarachik, J. Tejada, J. Maciejewski and R. Ziolo, Phys. Rev. Lett., 76 (1996) 3820 and J. Appl. Phys. 81, 8 (1997) 3978 . J. Hernandez, XX. Zhang, F. Luis, J. Barthomomé, J. Tejada and R. Ziolo, Euro. Phys. Lett., 35 (1996) 301 .

[21] L. Thomas, F. Lionti, R. Ballou, D. Gatteschi, R. Sessoli and B. Barbara, Nature, 383 (1996) 145 . See also the Proceedings of the Colloque Louis Néel, January 1996, Le Mont Saint-Odile. F. Lionti, L. Thomas, R. Ballou, D. Gatteschi, R. Sessoli and B. Barbara, J. Appl. Phys. 81, 8 (1997) 4608 .

[22] C. Sangregorio, T. Ohm, C. Paulsen, R. Sessoli and D. Gatteschi, Phys. Rev. Lett., 78 (1997) 4654 .

[23] C. Paulsen, J.G. Park, B. Barbara, R. Sessoli, A. Caneschi, J. Magn. Magn. Mater. 140-144 (1995) 379 .

[24] R. Sessoli, D. Gatteschi, A. Caneschi and M.A. Novak, Nature, 365 (1993) 141.

[25] C. Paulsen and J.G. Park, page 189 in [29].

[26] M. Novak and R. Sessoli, page 171 in [29].

[27] E. Chudnovsky and L. Gunther , Phys. Rev. Lett. , 60 (1988) 661.

[28] B. Barbara and E.M. Chudnovsky, Phys. Lett. A, 145 (1990) 208 .

[29] "Quantum Tunnelling of the Magnetisation" eds L. Gunther and B. Barbara , NATO ASI Series E: Applied Sciences-Vol. 301 (1995) Kluwer Acad.Publ., Dordrecht

[30] M. Enz and R. Schilling, J. Phys. C19 (1986) L711 .

[31] J.L. van Hemmen and A. Sütö, Europhys. Lett. , 481 (1986) 1.

[32] A. L. Burin et al, Phys. Rev. Lett. 76 (1996) 3040 ; N.V. Prokof'ev and P.C.E. Stamp, J. Low Temp. Phys., 104 (1996) 209.

[33] T. Lis, Acta Cristallo., B36 (1980) 2042.

[34] M. Hennion, L. Pardi, I. Mirebeau, E. Suard, R. Sessoli and A. Caneschi, to appear in Phys. Rev. B 56.

[35] F. Hartmann-Boutron, P. Politi and J. Villain, Int. Jour. Mod. Phys. B, 10 (1996) 2577.

[36] N.V. Prokof'ev and P.C.E. Stamp, preprint.

[37] D.A. Garanin and E.M. Chudnovsky, preprint.

[38] A. Zvezdin et al., ICM at Cairns (Australia), July 1997; JMMM, to appear.

[39] J.L. van Hemmen and A. Sütö, page 19 in [29].

[40] A. Sulpice et al., to be published.

[41] A.L. Barra, D. Gatteschi and R. Sessoli, preprint.

[42] L. Thomas et al., to be published.

[43] D. Garcia-Pablos, P.A. Serena, N. Garcia and H. de Raedt, Phys. Rev. B53 (1996) 741.

QUANTUM TUNNELING OF THE MAGNETIZATION IN NANO-SCALE MAGNETS

D. GARCÍA-PABLOS and N. GARCÍA
Laboratorio de Física de Sistemas Pequeños y Nanotecnología, Consejo Superior de Investigaciones Científicas, Serrano 144, Madrid E-28006, Spain

and

H. DE RAEDT
Institute for Theoretical Physics and Materials Science Centre University of Groningen, Nijenborgh 4, NL-9747 AG Groningen, The Netherlands

1. Introduction

In this paper we discuss some basic aspects of the response of the magnetization of a small, nano-scale, magnetic particle to a time-dependent change in the applied magnetic field. In particular we will focus on the magnetization dynamics at zero temperature, i.e. on the quantum dynamics in the *absence* of thermally-activated relaxation processes. At zero temperature the magnetization of the system can change through the mechanism of Quantum Tunneling of the Magnetization (QTM) [1, 2]. In recent experiments on high-spin ($S = 10$) molecules (Mn$_{12}$-Ac), steps in the magnetization as a function of the time-dependent magnetic field have been observed [3−6]. These steps are characteristic of the resonant QTM [7]. In experiments $T \neq 0$ and the resonant QTM may be thermally assisted [3 − 6].

Theoretical work has shown that at $T = 0$, QTM can only occur for particular values of the applied field, corresponding to the conditions for resonant tunneling [7, 8, 9]. Furthermore it has been shown that the response of the magnetization to a slowly reversing applied field can be understood in terms of Landau-Zener

N. García et al. (eds.), Nanoscale Science and Technology, 65–77.
© 1998 *Kluwer Academic Publishers.*

66

transitions [10, 11, 12].

In this paper we present new results of quantum dynamical simulations of the response of the magnetization to a time-dependent magnetic field. These results support our previous viewpoint [12] that the occurence of steps in the magnetization versus applied field curve, results from successive Landau-Zener transitions between nearly-degenerate eigenstates. We also investigate the effect of replacing the individual magnetic moments of the molecule by one "giant spin" and demonstrate that this approximation can have a large effect on the magnetization dynamics.

2. Models

According to quantum theory, at $T = 0$ the time-evolution of the magnetization can be calculated from the solution of the time-dependent Schrödinger equation (TDSE)

$$i\hbar \frac{\partial}{\partial t}|\Psi(t)\rangle = \mathcal{H}|\Psi(t)\rangle \quad , \tag{1}$$

where $|\Psi(t)\rangle$ denotes the wave function of the spin system at time t. We solve the TDSE (1) by means of exact diagonalization and a Suzuki-product-formula based algorithm [13]. For the present problem product-formula-based algorithms are more suited for solving the TDSE (1) than the standard exact diagonalization technique. This is because $\mathcal{H}$ changes with time through the time-dependent applied field, requiring the exact diagonalization of $\mathcal{H}$ for each value of the time t. Solving (1) by one of the product-formula algorithms is more efficient (in terms of CPU time) than doing the same calculation through exact diagonalization. We have solved (1) for spin-1/2 clusters described by the Hamiltonian

$$\mathcal{H} = - \sum_{i,j \in \mathcal{C}} \left(J_x(i,j)\sigma_i^x\sigma_j^x + J_y(i,j)\sigma_i^y\sigma_j^y + J_z(i,j)\sigma_i^z\sigma_j^z \right) - \Gamma \sum_i \sigma_i^x - H(t) \sum_i \sigma_i^z \quad , \tag{2}$$

where σ_i^x, σ_i^y and σ_i^z denote the x, y and z component of the Pauli-spin matrices, $J_x(i,j)$, $J_y(i,j)$ and $J_z(i,j)$ are the exchange interactions, Γ sets the scale of transverse component of the magnetic field (possibly of dipolar or hyperfine origin), and $H(t)$ represents the time-dependent applied field [8, 9, 12]. Model (2) is sufficiently general, covering most cases of interest. The set $\mathcal{C}$ defines the interactions between pairs of spins in the cluster. As the qualitative features of the results do not depend on the particular choice of $\mathcal{C}$ we will, in this paper, only present results for rings of L spins. Consequently we set $\mathcal{C} = \{(1,2), (2,3), \ldots, (L-1, L), (L, 1)\}$.

A common starting point of most analytical work on QTM is the assumption that the dynamics of the magnetization of the spin cluster can be described in terms of one "giant spin" [14, 15, 16]. The Hamiltonian for this single-spin model reads

$$\mathcal{H} = -K_x \frac{S_x^2}{S^2} - K_y \frac{S_y^2}{S^2} - K_z \frac{S_z^2}{S^2} - 2\vec{H}(t) \cdot \vec{S} \quad , \tag{3}$$

were K_x, K_y, and K_z are the anisotropy constants along the soft, hard and easy axis respectively and $\vec{H}(t) = H(t)(\sin\theta, 0, \cos\theta)$. As in the case of (2) we study the magnetization dynamics of (3) by solving the TDSE (1) [7].

3. Method

In practice we solve the TDSE (1) as follows. First we set the applied field to its minimum value $H(t_0) = -H_0$ and put the system in the corresponding ground state, i.e. $|\Psi(t_0)\rangle = |\phi_0\rangle$ where $\mathcal{H}|\phi_0\rangle = E_0(-H_0)|\phi_0\rangle$. Our convention is such that for large H_0, $|\phi\rangle$ is very close to the state with all spins down. The time-evolution of the wave function is obtained from $|\Psi(t + \tau)\rangle = e^{-i\tau\mathcal{H}}|\Psi(t)\rangle$, where τ denotes the (small) time-step used to integrate the TDSE and $\mathcal{H}$ is taken to be (2) or (3).

During the integration of the TDSE, the applied field changes with time, from $-H_0$ to $+H_0$. It is convenient to introduce two parameters to characterize this change: The field step $\Delta H = 2H_0/m_f$ and the sweep rate $c = \Delta H/\tau m$. The former controls, through m_f, the number of times the field changes as it increases from $-H_0$ to $+H_0$. The latter fixes the amount of time τm during which the system "feels" a constant applied field. The energy and magnetization at a particular value of the field are obtained from

$$E(t) = E(H) = E(H(t)) = \langle\mathcal{H}\rangle = \langle\Psi(t)|\mathcal{H}|\Psi(t)\rangle \quad , \tag{4a}$$

and

$$M(t) = M(H) = M(H(t)) = \langle\Psi(t)|\sum_i \sigma_i^z|\Psi(t)\rangle \quad , \tag{4b}$$

respectively.

4. Results

Earlier work [7, 8, 9, 12] has demonstrated that the salient features of the

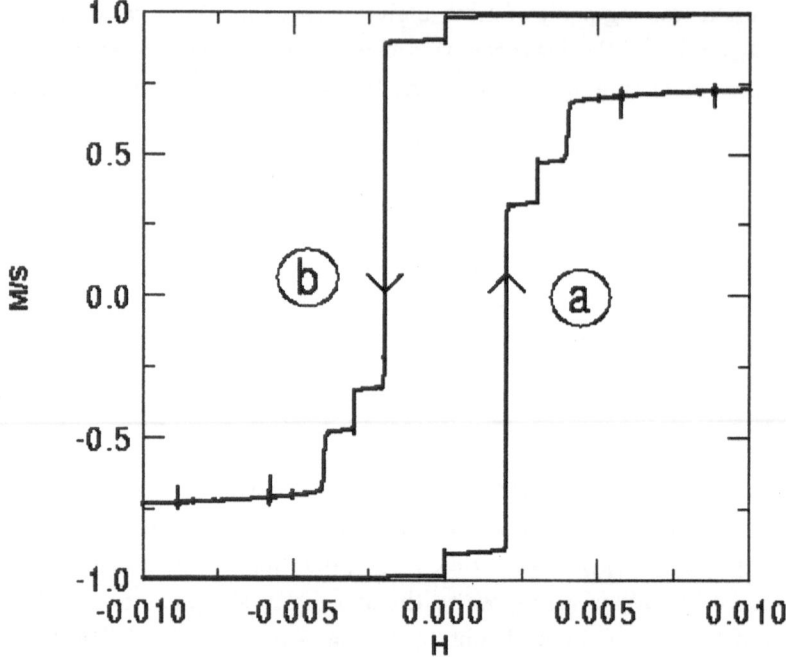

Figure 1. Magnetization M versus applied field H of the single-spin model (3) with spin $S = 10$, $K_x = 0.9$, $K_y = 0.6$, and $K_z = 1$ and $\theta = 1$ degree. The sweep rate $c = 2.86 \times 10^{-11}$ and the field increment $\Delta H = 2.86 \times 10^{-6}$. a: H increases from -0.01 to 0.01. b: H decreases from 0.01 to -0.01.

response of the magnetization to (time-dependent) changes of the applied field are rather universal: For a sudden reversal of the applied field, QTM occurs at particular, so-called resonant, values of the applied field [7,8,9] whereas for a slowly reversing field the magnetization exhibits steps when the field takes the same, resonant, values [12]. At these values at least two eigenstates of the system become nearly degenerate. Depending on the sweep rate of the field, the energy-level splitting magnetization and other, less important factors, the probability for tunneling from one to the other eigenstate may become significant, resulting in a change of the magnetization of the system.

An example of a magnetization-versus-field curve is shown in Figure 1. These results have been obtained by solving the TDSE for the single-spin model (3).

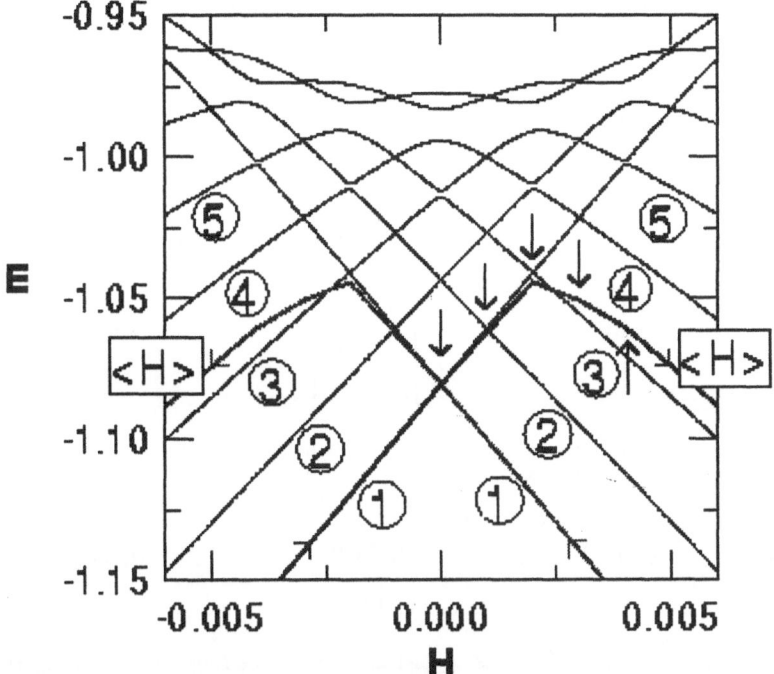

Figure 2. Energy-level scheme for the model that yields the results shown in Figure 1. The arrows pointing down indicate the value of the fields at which eigenstates are nearly degenerate. The lines labeled $\langle \mathcal{H} \rangle$ show how the energy of the system changes with time (or equivalently, applied field).

Starting from the ground state and increasing the field, no steps are observed until $H = 0$. At $H = 0$ the magnetization changes abruptly, suggesting that the system can change its state by tunneling from one eigenstate to another. This is confirmed by Figure 2 where we show the corresponding energy-level diagram. For $H < 0$ the system stays in the ground state (energy level labeled 1) until $H = 0$ where it has the option of staying in the ground state or making the transition to the first excited state. If the probability to stay in the ground state would be close to one the magnetization would reverse almost completely, exhibiting one big step at $H = 0$. Clearly, from Figure 1, this is not the case. For H larger but very close to zero, the state of the system is a linear combination of the ground state and the first excited state, with most weight in the latter.

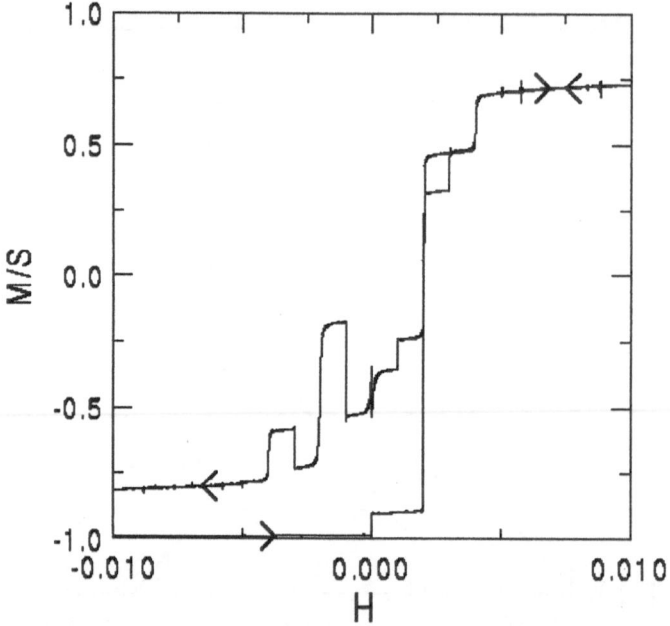

Figure 3. Magnetization M versus applied field H for the same case as in Figure 1. Right arrow: H increases from -0.01 to 0.01. Left arrow: H decreases from 0.01 to -0.01, starting from the state that was obtained by first increasing the field (right arrow). Note that on reversing the sweep direction of the field steps appear at both positive and negative H and that the decrease of the magnetization is not monotonous.

As H increases further a second step appears, exactly at the value at which the first and second excited state are nearly degenerate. A similar process takes place at the third condition for resonance, resulting in a third step in the magnetization curve. In this particular case the energy-level splitting and the sweep rate are such that the system state is no longer scattered into the third and fourth excited state but follows, almost adiabatically, the third excited state. As a direct consequence of the occurrence of several scattering events, the magnetization does not fully reverse. Only if the conditions are such that at $H = 0$ the state of the system evolves adiabatically with H, the magnetization will completely reverse. For low sweep rates, the dynamic behavior of the magnetization can be understood in

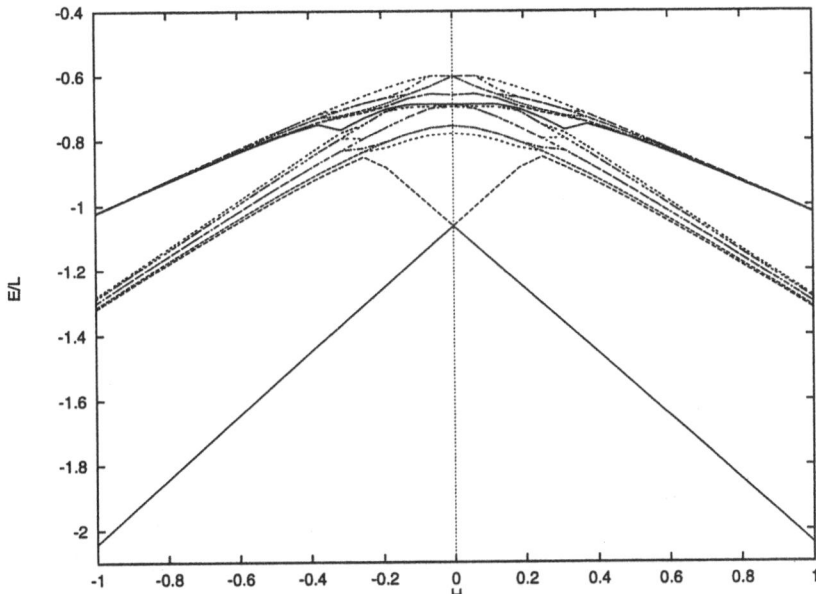

Figure 4. The energy-level scheme for the Ising model in a transverse field as defined by Hamiltonian (2) for $J_x(i,j) = J_y(i,j) = 0$, $J_z(i,j) = 1$ if i and j are neighbors and $J_z(i,j) = 0$ otherwise, and $\Gamma = 0.5$ for a ring of 8 spins.

terms of successive Landau-Zener transitions [12]. From Figures 1 and 2 it is clear that if we start from the ground state for large positive H and let H reverse, the magnetization will only show steps at $H \leq 0$, the magnetization curve (b) being a mirror image of the one (a) obtained by starting from the ground state for negative H. However, if we first follow the procedure that yields curve (a) until we reach (incomplete) saturation and then start to decrease the field, we observe steps at negative and positive H, as shown in Figure 3. Similar behavior is found in experiments on Mn_{12}-acetate molecules [17].

The simplest model for a uniaxial magnet is the Ising model in a transverse field (2). In Figure 4 we show the energy-level scheme for this model ($J_x(i,j) = J_y(i,j) = 0$, $J_z(i,j) = 1$ if i and j are nearest neighbors and $J_z(i,j) = 0$ otherwise, $\Gamma = 0.5$) for a ring of 8 spins ($S = 1/2$). A typical magnetization-versus-field curve is depicted in Figure 5. The general features are the same as those shown in Figure 1: The magnetization exhibits steps at values of the fields at which eigenstates

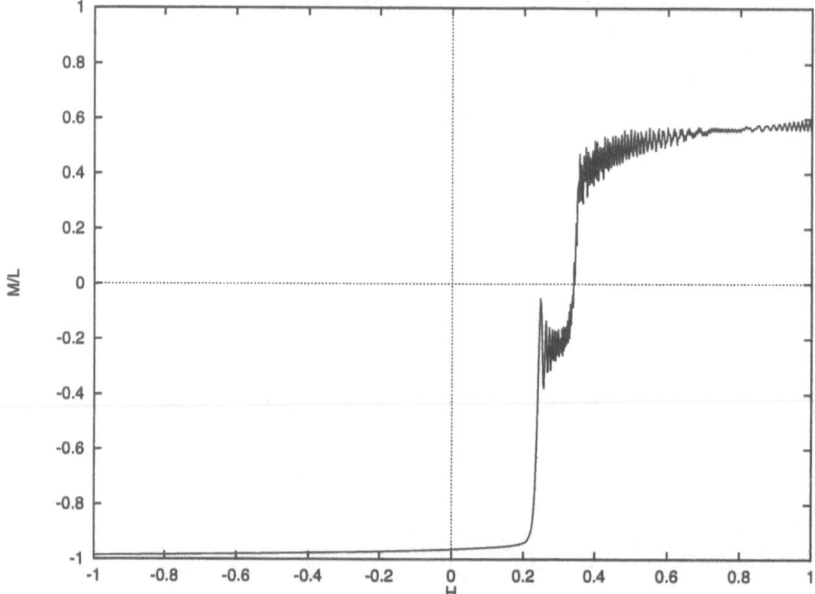

Figure 5. The magnetization M as a function of the time-dependent applied field H for the Ising model in a transverse field with the energy-level scheme shown in Figure 4. The sweep rate $c = 6.1 \times 10^{-4}$ and the step $\Delta H = 1.95 \times 10^{-3}$.

become nearly-degenerate. A systematic study of the $T = 0$ dynamics lead to the conclusion that the response of the magnetization to a sufficiently slow reversal of the applied field can be described, on a quantitative level, in terms of a sequence of Landau-Zener transitions [12]. The energy-level splitting at the resonant field, the sweep rate c, and the magnetization itself determine the size of the step at that particular field [12].

As usual it is worthwhile to attempt to simplify a model without changing the essential physics. The spin-1/2 model

$$\mathcal{H} = - J' \left(\sum_i \sigma_i^z \right)^2 - \Gamma \left(\sum_i \sigma_i^x \right) - H(t) \left(\sum_i \sigma_i^z \right) \quad , \qquad (5a)$$

can formally be written as

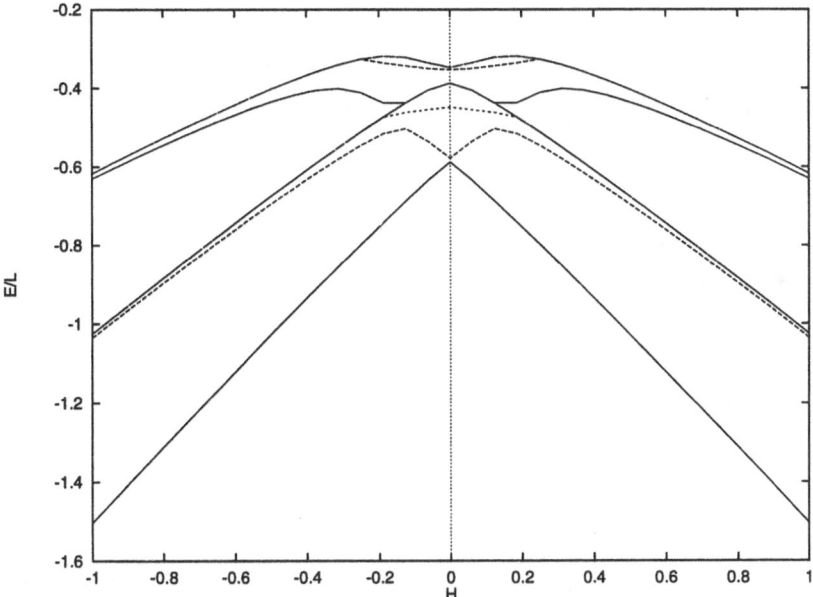

Figure 6. The energy-level scheme of the model defined by Hamiltonian (5) for a cluster of 8 spins, $J' = 0.125$ and $\Gamma = 0.5$ The constant contribution $-J'L$ has been substracted.

$$\mathcal{H} = -J'\sum_{i \neq j}\sigma_i^z\sigma_j^z - \Gamma\left(\sum_i \sigma_i^x\right) - H(t)\left(\sum_i \sigma_i^z\right) - J'L \quad . \qquad (5b)$$

and it is clear from (5b) that up to the irrelevant constant term $J'L$, (5a) is nothing but the Ising model in a transverse field for the special case that $J_z(i,j) = J'$ for all i and j. In the absence of the applied field we expect the energy to be conserved and extensive and to make this explicit we set $J' = J/L$.

The energy-level scheme and the magnetization curve for model (5) are shown in Figure 6 and 7 respectively. For the purpose of comparison we have taken the same model parameters as those used to obtain the data depicted in Figures 3 and 4. From Figure 6 it can be seen that the energy-level splitting at $H = 0$ is (much) larger than in Figure 4. According to the Landau-Zener transition picture [12] this implies a large (compared to the previous case) probability for the system to remain in the ground state (when crossing $H = 0$) and a large step in the magnetization at $H = 0$, in concert with the results shown in Figure 7. Although

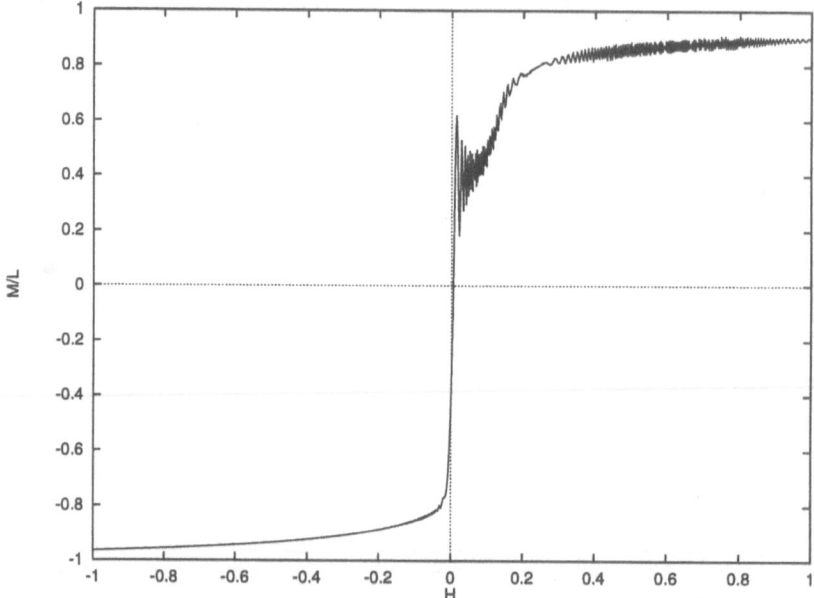

Figure 7. The magnetization M as a function of the time-dependent applied field H for the model (5a) with the energy-level scheme shown in Figure 6. The sweep rate $c = 6.1 \times 10^{-4}$ and the step $\Delta H = 1.95 \times 10^{-3}$.

the energy-level schemes of the nearest-neighbour Ising model and the "mean-field" version (5a) look somewhat similar, there are significant quantitative differences, especially at the resonant values of field. This then leads to large qualitative differences (compare Figures 5 and 7) in the magnetization response.

Model (5a) can be simplified further by replacing the sum of the L spin-1/2 operators $\sum_i \sigma_i^\alpha$ by a single-spin operators S_α for $\alpha = x, y, z$:

$$\mathcal{H} = 2\left(-DS_z^2 - \Gamma S_x - H(t)S_z\right) \quad ; \quad S_z = -\frac{L}{2}, \ldots, \frac{L}{2} \quad , \tag{6}$$

where $D = 2J/L$. An appealing feature of the approximation made in mapping the spin-1/2 model onto a single-spin model is that the size of the Hilbert space is reduced considerably: From 2^L to $L+1$ (note that in our notation the eigenvalues of σ_i^z are ± 1, not $\pm 1/2$).

The energy-level scheme and the magnetization curve for model (6) are shown in Figure 8 and 9 respectively, for values of the model parameters J and Γ identical

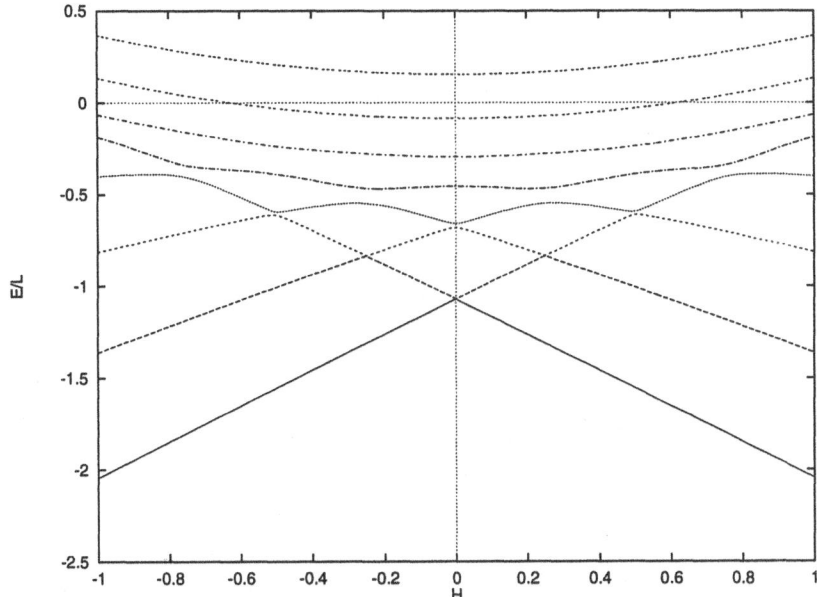

Figure 8. The energy-level scheme of the model defined by Hamiltonian (6) for $L = 8$, $D = 0.25$, and $\Gamma = 0.5$.

to those used to obtain the data depicted in Figures 5,6,7 and 8. Comparing Figures 7 and 9 we conclude that the approximation made to reduce the L-sites spin model to a single-spin model changes completely the qualitative behavior of the magnetization response.

Model (6) is assumed to describe some of the magnetic properties of the Mn_{12}-Ac molecule [3–6]. Using the commonly accepted values for D and Γ ($D \approx 0.60\,K$ and $\Gamma \approx 0.01\,T$) the sweep rate required to observe steps in the magnetization is unrealistically low. This inevitably leads to the conclusion that either QTM is not the mechanism that is responsible for the appearance (in the experiments) of steps in the magnetization at very low temperature or model (6) does not correctly describe the dynamics of the spin cluster. We believe that the energy-level scheme of (6) is too simple compared to that of the Mn_{12}-Ac molecule. A proper description of the latter requires 10^8 eigenstates and it would be remarkable indeed that the quantum dynamics of the magnetization can be modelled correctly by disregarding all but the 21 states of the corresponding single-spin ($S = 10$)

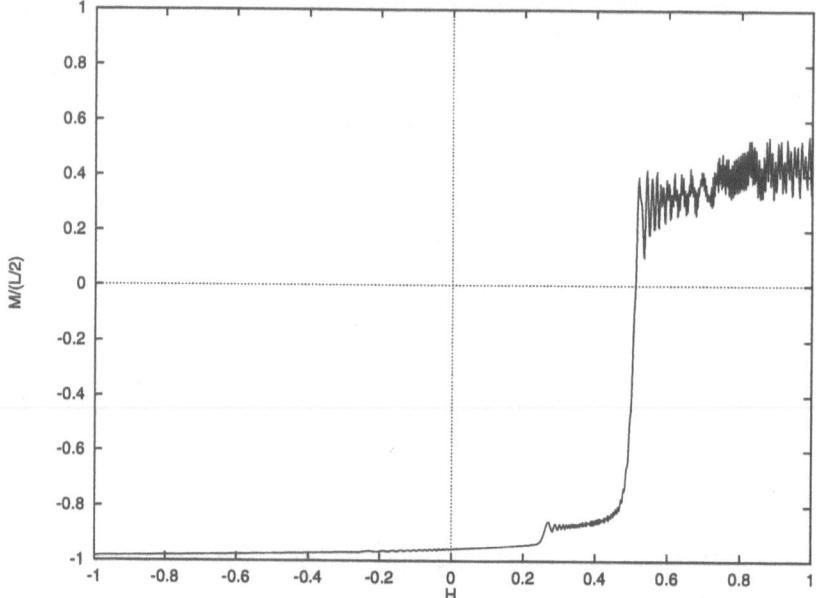

Figure 9. The magnetization M as a function of the time-dependent applied field H for the model (6) with the energy-level scheme shown in Figure 8. The sweep rate $c = 6.1 \times 10^{-4}$ and the step $\Delta H = 1.95 \times 10^{-3}$.

system. Another flaw of the simplified model (6) is that it cannot take into account the local symmetries of the individual spins. In Mn_{12}-Ac molecule these local symmetries may nevertheless affect the details of the spectrum [17], leading to dynamic behavior that is different from that of the simplified model.

ACKNOWLEDGEMENTS

We would like to thank B. Barbara for many useful discussions. This work is partially supported by Spanish and European research contracts, and a grant of the "Stichting Nationale Computer Faciliteiten (NCF)".

References

1. Awschalom, D.D. *et al.*, (1992), Macroscopic Quantum Effects in Nanometer-Scale Magnets, Science **258**, 414 - 421.

2. Barbara, B., and Gunther, L., (eds.), (1995) Quantum Tunneling of the Magnetization-QTM'94, NATO ASI Series E (Kluwer Acad.Publ., Dordrecht), **301**.

3. Friedman, J.R. *et al.*, (1996), Macroscopic Measurement of Resonant Magnetization Tunneling in High-Spin Molecules, Phys. Rev. Lett. **76**, 3830 - 3833.

4. Hernández, J.M. *et al.*, (1996), Field tuning of thermally activated magnetic quantum tunneling in Mn_{12}-Ac molecules, Europhys. Lett. **35**, 301 - 306.

5. Thomas, L. *et al.*, (1996), Macroscopic quantum tunneling of magnetization in a single crystal of nanomagnets, Nature **383**, 145 - 147.

6. Hernández, J.M. *et al.*, (1997), Evidence for resonant tunneling of magnetization in Mn_{12} acetate complex, Phys. Rev. B **55**, 5858 - 5865.

7. García-Pablos, D., García, N., and De Raedt, H., (1997), Dynamical calculations on the reversal of single quantum spins: Quantum coherence, Phys. Rev. B **55**, 937 - 941.

8. García-Pablos, D., García, N., and De Raedt, H., (1996), Quantum dynamical calculations on the magnetization reversal in clusters of spin-1/2 particles: Resonant coherent quantum tunneling, Phys. Rev. B **53**, 741 - 746.

9. García-Pablos, D., García, N., and De Raedt, H., (1997), Resonant coherent quantum tunneling of the magnetization of spin-1/2 systems: Spin-parity effects, Phys. Rev. B **55**, 931 - 936.

10. Miyashita, S., (1995), Dynamics of the magnetization with an Inversion of the magnetic field, J. Phys. Soc. Jpn **64**, 3207 - 3214.

11. Miyashita, S., (1996), Observation of the energy gap due to quantum tunneling making use of the Landau-Zener mechanism, J. Phys. Soc. Jpn **65**, 2734 - 2735.

12. De Raedt, H., *et al.*, Theory of Quantum Tunneling of the Magnetization in Magnetic Particles, preprint subm. to Phys. Rev. B.

13. Suzuki, M., (1991), General theory of fractal path integrals with applications to many-body theories and statistical physics, J. Math. Phys. **32**, 400 - 407 .

14. Chudnovsky, E.M., and Gunther, L., (1988), Quantum Tunneling of Magnetization in Small Ferromagnetic Particles, Phys. Rev. Lett. **60**, 661 - 664.

15. Stamp, P.C.E., Chudnovsky, E.M., and Barbara, B., (1992), Quantum Tunneling of Magnetization in Solids, Int. J. Mod. Phys. B **6**, 1355 - 1473.

16. Politi, P., Rettori, A., Hartmann-Boutron, F., and Villain, J., (1995), Tunneling in Mesoscopic Magnetic Molecules, Phys. Rev. Lett. **75**, 537 - 540.

17. Barbara, B., private communication.

TRANSPORT IN MESOSCOPIC SUPERCONDUCTORS AND SUPERCONDUCTING/NORMAL METAL CONTACTS

T. CLAESON, P. DELSING, Z. IVANOV, S. KUBATKIN, L. KUZMIN,
H. OLIN, V. PETRASHOV[1,2], R.SH. SHAIKHADAROV[1],
and A. TZALENCHUK
*Physics and Engineering Physics, Chalmers University of Technology and
Göteborg University, 41296 Göteborg, Sweden*
[1]*Institute of Microelectronics, Russian Academy of Sciences,
Chernogolovka, Russia*
[2]*Royal Holloway, University of London, Egham, Surrey TW20 0EX, U.K.*

The charging energy is important for a mesoscopic body with small capacitance to the environment. Connecting the small island via low capacitance, high resistance tunnel junctions, the current through the resulting structure will depend upon the charge of the island - charge that can be applied via a gate electrode. For a normal metal island, the current varies periodically with the charge; the periodicity is the electron charge e. If the island is superconducting, the period, e or 2e, depends upon whether the superconducting energy gap is less than or larger than the charging energy. This parity effect can be used to study possible nodes in the gap function of a high-T_c superconductor.

The charge transport through a mesoscopic conductor can be controlled by a phase difference between superconducting contacts to the normal conductor. If the phase difference is applied via a superconducting loop connected to the contacts, the conductance varies periodically with the flux or the current through the loop and the period is the quantized flux unit. The oscillation amplitude is temperature dependent as expected from theoretical predictions.

1. Introduction

Charge and phase are conjugate entities in small superconducting tunnel junctions ($\Delta\phi\Delta Q\approx e/2$). In the high resistance limit, the charge may be well defined (giving rise to, e.g., a Coulomb blockade) while the phase is the well defined entity in low resistance Josephson junctions (for which the number of electron (Cooper) pairs is undetermined). Current and voltage, I and V, are complementary in these effects (no current up to a threshold voltage in a Coulomb blockade device: no voltage drop up to a critical Josephson current in the Josephson junction; voltage and current steps at I=nef (2nef) and V=nΦ_0f (Φ_0=h/2e) occur at microwave radiation of frequency f in the two

N. García et al. (eds.), Nanoscale Science and Technology, 79–89.
© 1998 *Kluwer Academic Publishers.*

types of components; a periodic variation of the conductance occurs with period of e (2e) or Φ_O for the charge on an island intermediate between two high resistance junctions coupled in series and for flux within a loop encompassing two Josephson junctions in parallel, respectively, etc). Transitions from an insulating state to a superconducting state can be encountered as one traverses the phase space of Josephson coupling relatively Coulomb charging energy and junction resistance vs. the quantum resistance.

In this paper, we will discuss aspects of charging effects in such tunnel junction configurations as well as phase controlled conductance in mesoscopic scale structures.

2. Charging Effects

Transport properties in a small tunnel junction will be determined by the value of the small capacitance, C, that is due to the polarization of the barrier between the two tunnel electrodes [1]. If the capacitance is small enough, the voltage fluctuation $\Delta V=Q/C$ caused by a tunneled charge entity (Q=e for a single electron, Q=2e for an electron pair) will be larger than the thermal noise voltage and charging effects will be seen, and in effect become dominant. ($C\approx1$ pF $\Rightarrow \Delta V\approx0.16$ mV ~2 K)

The conditions for the occurrence of charging effects are:
- the charging energy $E_C=Q^2/2C>>k_BT$
- $E_C>>$quantum fluctuation energies $h/2\pi\tau$ determined by the RC time when the junction is effectively isolated from its environment or by $\tau=h/2\pi max(eV,k_BT)$ when the environment determines the impedance of the shunted junction.

Other ways to express these relations are:
- $C<<e^2/2k_BT$ for single electron tunneling
- $R_T>>R_Q=h/4e^2$ (≈6.5 kΩ)

$Z_{environment}>>R_Q$

A certain threshold voltage, V_{thr}, has to be applied before current transport occurs by tunneling:

$QV_{thr}>E_C \Rightarrow V_{thr} = Q/2C.$

If a current is sent through a tunnel junction that has a large resistance and a small capacitance, it will cause charging to the threshold value, induce a tunneling event and discharge the capacitor, charge it again, etc. The oscillation has a period determined by the current and the charge value: f=I/Q.

Charging effects may occur in small tunnel junctions of normal metals or superconductors, in semiconducting structures, in polymer strands between metal electrodes, in small particles contacted by a scanning probe tip, or in a highly resistive conductor itself.

The conductance of two high resistance junctions with an intermediate electrode (island) may be controlled by charge induced on the island via a gate electrode coupled to the island. It is minimum (at small bias) when the additional island charge is zero or a multiple of e; it is maximal when the charge is an odd multiple of e/2. This is the basis of the SET transistor and the electrometer that is able to resolve changes in the charge distribution as small as 10^{-5}e. A typical response of an SET transistor is shown

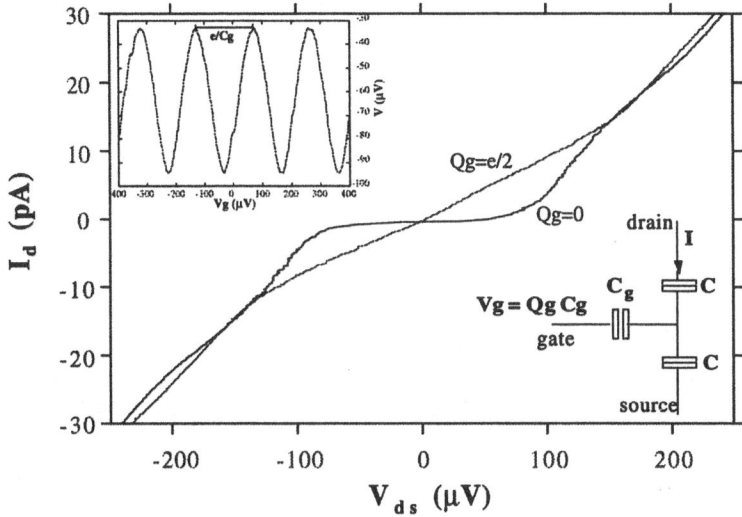

Figure 1. Current-voltage curves of a single electron transistor. The drain-to-source current is shown for two cases: where the induced gate charge is zero (or a multiple of an electron charge) and where it is e/2 (or (n+1/2)e). The Coulomb blockade is lifted by the gate charge. The drain-to-source voltage as a function of gate voltage and at a bias current of about 2 pA is shown in the inset. A periodicity of the electron charge divided by the gate capacitance is evident. The transistor consists of two series coupled Al tunnel junctions that have been deposited by a shadow mask technique on an oxidized Si substrate. The superconductivity of Al is suppressed by a magnetic field at an estimated operating temperature of 100 mK. The resistance of each junction was of the order of 3 MΩ.

in figure 1.

If the intermediate island is a superconductor, the transport remains periodic in gate charge. It is still e-periodic for small values of the superconducting energy gap (Δ->0). However, when Δ is large compared to the charging energy, it becomes energetically unfavorable to tunnel quasiparticles into the superconductor (each costing at least an energy of Δ). Thus the conductance becomes 2e-periodic for the superconducting circuit [2]. The I-V curve varies characteristically not only with gate voltage, but also with bias, temperature and magnetic field. An example [3], showing a 2e periodic dependence on V_g at low bias, going over to an e-periodic behavior at large bias, is given in figure 2.

There are strong indications that the superconducting state in most high-T_C cuprate superconductors has a d-wave symmetry rather than the s-wave symmetry that is common for conventional, low-T_C superconductors. Δ is strongly anisotropic and there are nodes in the energy gap with phase shifts in the gap function (change in sign) both in d-wave and extended s-wave symmetries.

The parity effect (e or 2e periodicity) may give a possibility to probe the symmetry of the order in a high T_C superconductor. It should be a thermodynamic probe of the energy spectrum, being characteristic of the island as a whole rather than depending on interface properties [4]. If the superconductor were of isotropic s-wave symmetry, we

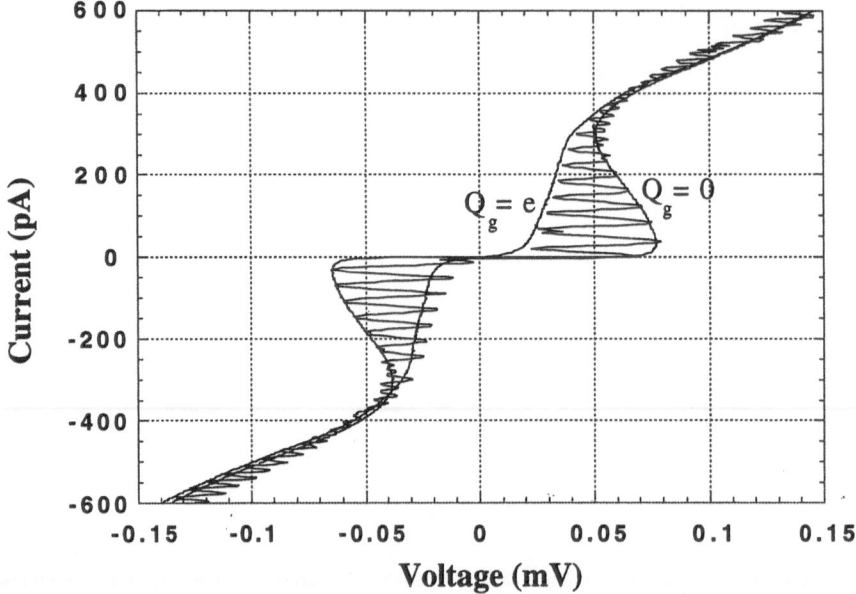

Figure 2. The current-voltage characteristic for a superconducting transistor modulated by a gate voltage. This demonstrates the parity effect of a small superconducting island: whether it is favorable to add a Cooper pair with a larger Coulomb charging energy or a quasiparticle with a smaller charging energy but with an additional gap energy. The current is swept with a period much larger than the one of the oscillating gate voltage. Hence, one sees how the I-V curve oscillates between the two extremal I-V curves corresponding to the Coulomb blockade and the lifted blockade. At low bias, the periodicity in the gate response corresponds to a charge value of 2e. The superconducting gap energy is comparatively large compared to the charging energy and it costs too large an energy to tunnel a single electron into the superconductor. At large bias, the period is e. In the transition region, there is a mixed behavior and the modulation amplitude goes down to a small value. Similar switch-overs from 2e to e periodicity have also been noted for increasing temperature or magnetic field. The junctions of the transistor were made of Al and the resistance of each was about 110 kΩ. The curve was traced at a temperature of 40 mK. From ref. 3.

may expect a 2e periodicity in the transport through the high-T_c island. However an e-periodic behavior may be less conclusive as a $\Delta=0$ state at the boundary, like a non-superconducting inclusion, may be a competing explanation.

The configuration of a structure to test the parity effect in a high temperature superconductor is shown in figure 3. A 100 nm wide microbridge of YBa$_2$Cu$_3$O$_{7-\delta}$ (YBCO) remained of good quality after patterning: an unaffected transition temperature (T_c) and a high critical current density ($j_c \approx 10^7$ A/cm^2 at 4 K). Gold counterelectrodes contacted the small island that had been chemically treated to form a barrier [5] giving a resistance of the order of 12 MΩ (for the device of figure 2). The current through the two junctions and the island was controlled by a gate electrode nearby. Envelopes of I-V curves for different gate voltages are also shown in figure 2. The set of curves goes from a well pronounced Coulomb blockade to a nearly linear dependence. Contrary to a

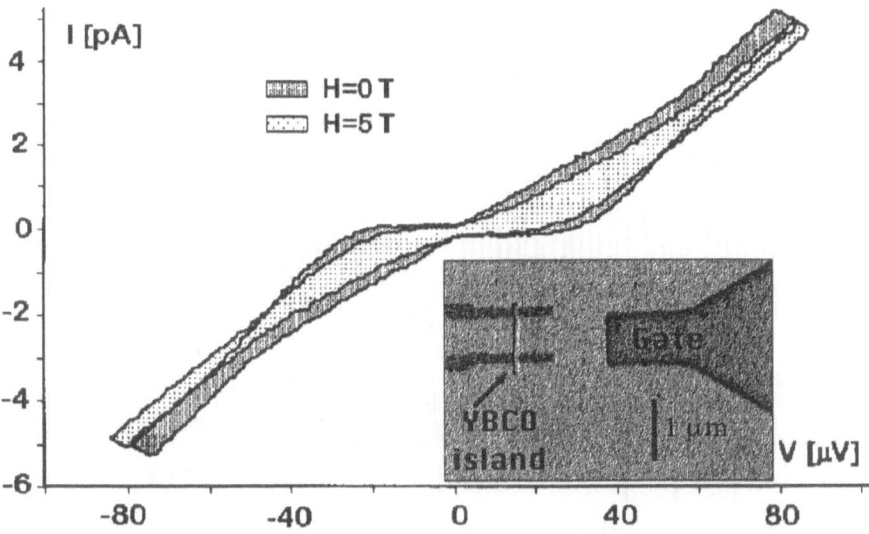

Figure 3. A parity type, single charge tunneling experiment was designed to test the symmetry, or rather the occurrence of zero or small energy gaps, in a high temperature superconductor by a method sensitive to the thermodynamic state of the particle. A small island of YBCO was patterned using electron beam lithography (width 200 nm, thickness 150 nm, length 1000 nm). A procedure described in Ref. 5 was used to form a native tunnel barrier at the subsequent deposition of two 200 nm wide Au counter electrodes. The resistance of the device was about 12 MΩ. A gate electrode was placed approximately 2 μm from the superconducting island in order to control the current going through the device. The envelopes of two sets of I-V curves are shown in the diagram. The gate voltage oscillated such that I-V varied between the blockade state and the maximally lifted blockade state. The sets were taken for two applied magnetic flux densities of 0 and 5 T and at a base temperature of 30 mK. The inset shows an SEM image of the electrometer.

low-T_c superconductor, like Al, the I-V sets were not affected to any larger extent by such a high magnetic field as 5 T.

The voltage oscillations across the double junction structure as a function of gate voltage and at a constant bias current is shown in figure 4. The amplitud $\langle$t the oscillation was somewhat affected by a 5 T magnetic flux density, but not the period. Even if such a field is well below the critical value for YBCO, it is expected to create a large amount of normal core regions. The periodicity was neither seen to change with bias nor temperature within investigated ranges.

To really assure that an e-periodic response is present, one should compare with a standard. As noted above, the superconducting state of a high-T_c island could not be suppressed by an available magnetic field to obtain a normal island standard. Therefore, similar N-S-N and S-N-S structures will be fabricated within the same deposition cycle and be compared to see if there is any difference in periodicity.

84

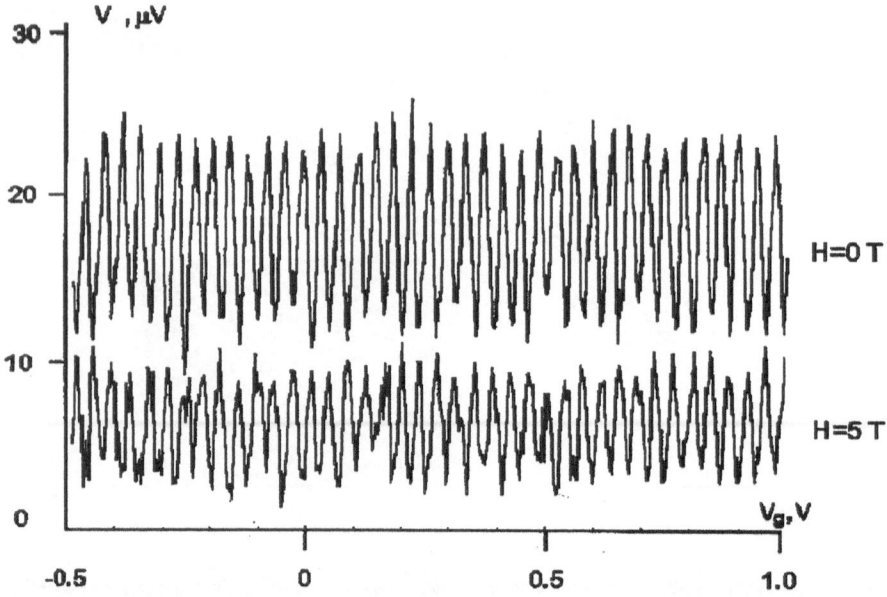

Figure 4. The dependence of the drain-to-source voltage on gate voltage for the current biased superconducting transistor / electrometer described in figure 3. The curves have been displaced vertically in voltage so that they do not overlap. The period in gate voltage does not change for magnetic fields up to at least 5 T but the amplitude of the oscillation goes down. The curves were registered at a base temperature of 30 mK.

3. Phase Dependent Conductance

Electron interference may give rise to magnetoconductance oscillations in solid state circuits, like the Aharonov-Bohm effect (AB, period $\Delta\Phi=h/e$) in clean metal rings, or the AAS (Altshuler, Aronov, Spivak) interference (period $\Delta\Phi=h/2e$) in the dirty limit. The oscillation amplitude is generally small, of order e^2/h in conductance or about 10^{-5} of the normal state conductance in a typical component.

Electrons in a normal metal (or semiconductor) may be reflected against an NS interface (N stands for normal metal, or semiconductor, S for superconductor). It can be reflected either as an electron, with a high probability if there is a barrier at the interface, or as a phase reversed hole (particularly for energy less than Δ) retracing the electron orbit before loosing its phase coherence by inelastic scattering. As charge has to be conserved, an electron pair continues in the superconductor. This is named an Andreev reflection [6]. It usually gives rise to an increase in the current transport as an incoming electron gives rise to an outgoing electron pair.

If a conductor is connected to two superconductors, that can be placed along the conductor (like being contacts) or well outside the straight conducting path, an electron can be Andreev reflected into a hole by the first superconductor and the hole can then be reflected back into an electron quasiparticle by the second superconductor. The quasi-particles pick up phases ϕ_1 (electron-to-hole) and $-\phi_2$ (hole-to-electron), respectively, as

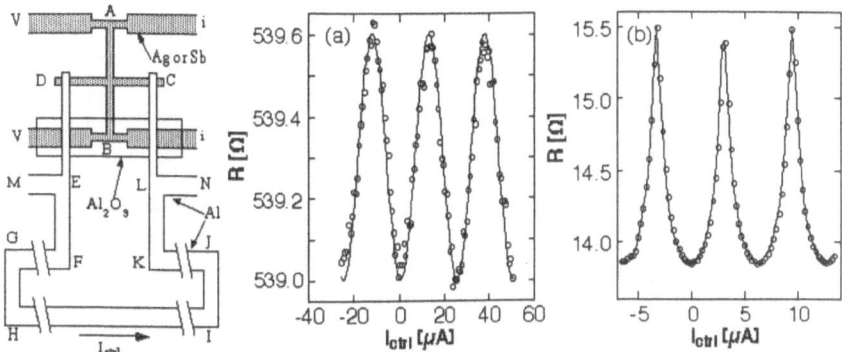

Figure 5. Phase dependent resistance (conductance) is measured in a cross like mesoscopic structure. The resistance over one arm of the cross, R_{AB}, is determined in a four point measurement. The cross is made of Ag or Sb, 50 nm thick film, with arms 2 μm long and 100 nm wide. Superconducting Al strips cross the other arm of the cross, at C and D. They are part of a loop which has a relatively large area (56 ± 2 μm^2) and which is isolated from the cross by Al$_2$O$_3$, except at points C and D. The phase difference between the superconducting points C and D can be varied by applying a perpendicular magnetic field or sending a control current through the loop. Diagram (a) shows the variation of R_{AB} with I_{ctrl} for the Sb/Al structure and (b) the corresponding variation for the Ag/Al structure. Measuring currents were 0.2 μA and 0.5 μA, resp., at 30-300 Hz. The base temperature was 20 mK. Similar dependencies were registered with the magnetic field as the variable. R_{AB} for the Sb cross increased when Al became superconducting (a so called anomalous proximity effect) while it decreased for the Ag cross. There are pronounced resistance oscillations for both cases with the period determined by a magnetic flux in the loop equal to the flux quantum. The oscillation has a sinusoidal shape for Sb and its relative amplitude is much smaller than for Ag. Both amplitudes are much larger, though, than expected from normal conductance fluctuations.

they are scattered against the superconductors characterized by these phases, ϕ_1 and ϕ_2, at the points of reflection. Thus, the electrical conductance may vary periodically with the phase shift $\Delta\phi=\phi_1-\phi_2$.

Experiments on a cross type configuration, where the resistance (or conductance) was measured across one arm and superconducting strips were connected to the other arm, did indeed show resistance oscillations depending upon the phase shift between the superconductors [7-9]. However, the oscillation amplitude was much (up to several orders of magnitude) larger than the typical conductance fluctuations, e^2/h. Figure 5 shows examples of the resistance variations for Ag/Al and Sb/Al configurations. The phase difference could be varied by changing the magnetic flux in a loop connecting the two superconducting strips across the side arms of the cross or by sending a control current through the loop. A non-sinusoidal dependence and an oscillation amplitude as large as about 10% of the normal state resistance could be seen for the Ag/Al case, probably due to a screening effect in the presence of a finite critical current in silver [9]. The magnitude was dependent on the contact resistance between Ag and Al. The Sb/Al structure gave a sinusoidal dependence and an oscillation amplitude of the order of 1 per mille.

The oscillation amplitude depends on phase difference, bias, and temperature. The bias dependence of the differential resistance of an Ag/Al structure is seen in figure 6. The phase difference is the parameter of the different curves in the figure, running from $\Delta\phi=0$ to $\Delta\phi=\pi$. The normalized resistance varies considerably with bias for no phase

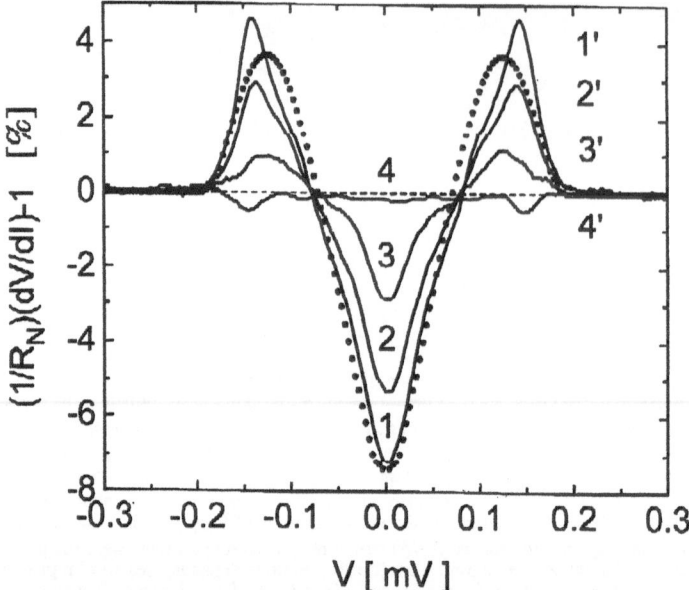

Figure 6. Normalized differential resistance versus voltage for a sample similar to the one shown in figure 5. Curves are given for different superconducting phase differences. 1-1': $\Delta\phi = 0$; 2-2': $\Delta\phi = 0.6\pi$; 3-3': $\Delta\phi = 0.8\pi$; 4-4': $\Delta\phi = \pi$, $T = 0.58$ K.. The arm lengths $L_{AB}=2$ μm and $L_{CD}=0.5$μm. $R_N=10.5$ Ω. $D=84$ cm^2/s. $L_\phi=1.5$ μm. l≈25 nm. ξ_N≈110 nm at the lowest temperature (0.5 K) of the experiment. The largest resistance variation, $\Delta R/R$, is about 7.3%. Dotted and dashed lines are the theoretical curves for $\Delta\phi = 0$ and $\Delta\phi = \pi$, respectively. From reference 10.

difference (or a multiple of 2π) between the superconductors while there is no variation for $\Delta\phi=\pi$. Note that all the curves go through zero at the same bias (V≈0.08 mV). Calculated variations by Volkov [10] give quantitative agreement at $\Delta\phi=0$ and π.

Figure 7 shows the dependence of the differential resistance, of the same device, on phase difference between the two superconductors on the crossing arm and with the bias voltage as a parameter. There is a 2π period for all the curves with a large resistance variation at low and large bias and a diminishing amplitude (with phase reversal) in between (V≈0.08 mV). The same behavior is also evident in figure 6.

The temperature dependence of the normalized resistance at zero bias is given for two values of $\Delta\phi$, namely 0 and π, in figure 8. There is practically no variation with temperature for the resistance at $\Delta\phi=\pi$ while it decreases steadily with decreasing temperature, within the investigated temperature interval, for $\Delta\phi=0$. This is according to theoretical expectations [11-13]. A proximity effect induced conductance peak is predicted to occur at an energy (temperature) of the order of the Thouless energy (temperature) for $\Delta\phi=0$ (which in our case would correspond to 15 to 250 mK, depending upon the distance between the superconducting mirrors). On the other hand, there should be no additional conductance (nor temperature variation) for $\Delta\phi=\pi$ [11]. Indeed, we do see a resistance minimum at a temperature agreeing with the expected value (in a series of experiments with an extended temperature range) and a return to the

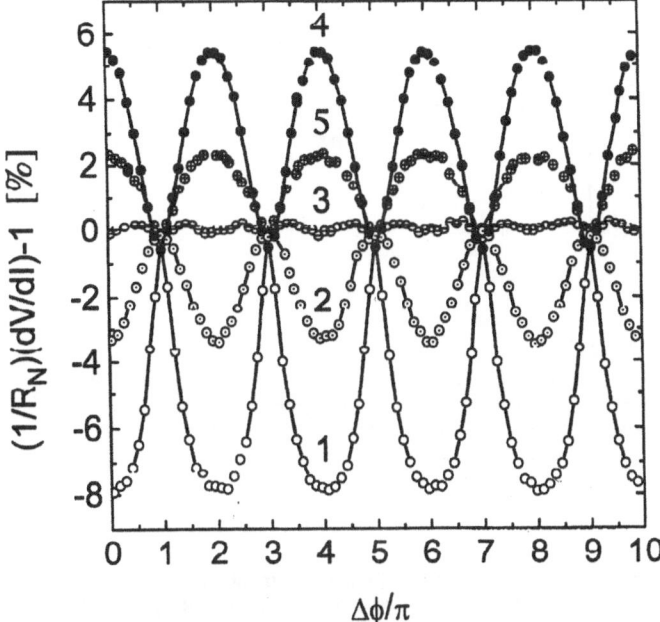

Figure 7. Normalized differential resistance as a function of superconducting phase difference at different applied dc voltages for the same Ag/Al structure as in figure 6. 1: $V = 0$; 2: $V = 0.045$ mV; 3: $V = 0.08$ mV; 4: $V = 0.14$ mV; 5: $V = 0.17$ mV; $T = 0.58$ K.

non-affected normal state value, R_n, at the lowest temperatures for $\Delta\phi=0$ [14]. (There may also be a sharp minimum in dV/dI for $\Delta\phi=\pi$ at temperatures below the minimum for $\Delta\phi=0$, but it has to be clarified if this variation is an experimental artifact.)

4. Applications as High Frequency Detectors

The related effects, single charge tunneling and Andreev interference in mesoscopic structures, may be applied to detect photons with high sensitivity in the sub-millimeter or far infrared ranges.

Experiments at low temperature show that high resistance tunnel junctions have to be well shielded also from thermal radiation in order to develop a full Coulomb blockade. Even 4 K radiation gives rise to a considerable smearing of the I-V curve at the threshold voltage.

Photons that are absorbed in doped silicon can produce free charge carriers. These can be collected on an electrode coupled to a single electron transistor. Measurements at Saclay [15] estimate a noise equivalent power, NEP, of 2×10^{-21} W/√Hz at $\lambda=30$ μm for the photodetector operated at 20 mK.

An additional quasiparticle, that can only be added to an intermediate superconducting island by photon assisted tunneling across one of the junctions, may

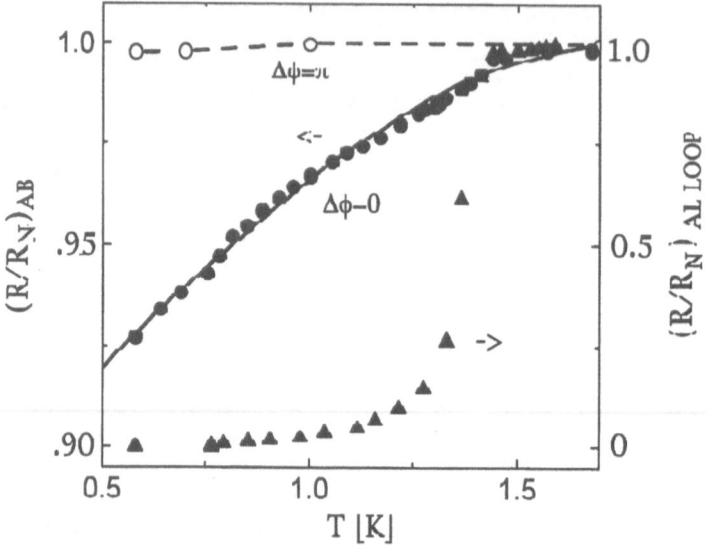

Figure 8. Normalized zero bias resistance for the AB arm as a function of temperature at $\Delta\phi=0$ (filled circles) and $\Delta\phi=\pi$ (open circles). The sample is the same as in figure 6. The variation is nil for $\Delta\phi=\pi$ while the resistance decreases steadily with lower temperature within the range given for $\Delta\phi=0$. The resistance of the superconducting loop is also given. Solid and dashed curves are calculated resistance variations for the two extremes of phase difference. From reference 10.

open the superconducting transistor (parity effect) such that a large number of charge carriers (at least a hundred) can traverse the two junctions before the additional quasiparticle decays (within about one μs) and the transistor returns to its state of blockade. A NEP of the order of 3×10^{-20} W/√Hz for 80 GHz photons was estimated by Hergenrother et al. [16] when the current was measured with a commercial current amplifier.

Non-equilibrium quasiparticles, that are heated by absorbed photons in a normal metal nanobridge, will relax back towards the equilibrium state via phonon cooling into the substrate, via diffusive cooling into the contacts, or via the thermometer that is used to measure the temperature rise. By contacting the normal metal strip by superconductors, Andreev reflections against the two NS contacts can increase the trapping time, the temperature rise, and, hence, the sensitivity. Nahum and Martinis [17] used an SIN tunnel junction, where the normal metal absorber was one of the electrodes, to read the increase of the temperature of the bolometer and they estimate (by feeding in electrical power) a temperature rise of about 10 mK/fW, a responsivity of about 10^9V/W and an amplifier limited NEP $\approx 3 \times 10^{-18}$ W√Hz at an operating temperature of 100 mK.

5. Conclusion

Number (charge) and phase (flux) are conjugate quantum entities. We have studied some of their quantum mechanical consequences in two cases: charging effects in single charge tunneling in small junctions and phase dependent transport in mesoscopic conductors coupled to superconducting mirrors.

The experimental results are, to a large extent, in good accord with existing theories where those for the mesoscopic transport are actively under development. In particular, the variation in the amplitude of the phase dependent oscillations of the conductance seems to agree with the theoretically predicted diminishing amplitude at high and low energies (temperatures).

A couple of applications of the phenomena were touched upon in the Forum presentation: the hitherto less conclusive investigation of gap nodes in a high temperature superconductor by a thermodynamic parity effect in a small superconducting island and sensitive bolometers designed for the difficult far infrared range. The competitiveness of these new detectors, either based upon single charge effects or Andreev confinement, has still to be proven.

References,

1. See, e.g., Averin, D.V.,and Likharev, K.K.. (1991) in B.L. Altshuler, P.A. Lee, and R.A. Webb (eds.) *Mesoscopic Phenomena in Solids,* Elsevier, Amsterdam, Chap.6; and several chapters in H. Grabert and M. H. Deboret (eds.) (1992) *Single Charge Tunneling: Coulomb Blockade Phenomena in Nanostructures,* Plenum Press, New York, NATO ASI Series B: Physics Vol. 294.
2. Tuominen, M.T., Hergenrother, J.M., Tighe, T.S., and Tinkham, M. (1992) *Phys. Rev. Lett.* **69**, 1997.
3. Pashkin, Yu., Haviland, D., Kuzmin, L., and Zorin, A. (1997) in H. Koch and S. Knappe (eds.) *ISEC'97, 6th Int. Supercond. Electronics. Conf., Extended Abstracts,* PTB, Braunschweig, Vol. 2, 397.
4. Mineev, V.P. (1994) *JETP Lett.* **60**, 876.
5. Valles, J.M., Dynes, R.C., Cucolo, A.M., Gurvitch, M, Schneemeyer, L.F. Garno, J.P., and Waszczak, J.V. (1991) *Phys. Rev. B* **44**, 11986.
6. Andreev, A.F. (1965) *Zh. Eksp. Teor. Fiz.* **49**, 655.
7. Petrashov, V.T., Antonov, V.N., Delsing, P., and Claeson, T. (1994) *Pis'ma Zh. Eksp. Teor. Fiz.* **60**, 589; (1995) *Phys. Rev. Lett.* **74**, 5268.
8. Pothier, P.G.N., Gueron, S., Esteve, D., and Devoret, M.H. (1994) *Phys. Rev. Lett.* **73**, 2488.
9. Van Wees, B.J., den Hartog, S.G., and Morpugo, A.F. (1996) *Phys. Rev. Lett.* **76**, 1402.
10. Petrashov, V.T., Shaikhaidarov, R.Sh., Sosnin, I.A., Delsing, P., Claeson, T., and Volkov, A. (1997) Phase-Periodic Proximity Effect Compensation in Symmetric Normal/Superconducting Mesoscopic Structures, to be publ.
11. Nazarov, Yu. V. and Stoof, T.H. (1996) *Phys. Rev. Lett.* **76**, 823.
12. Volkov, A., Allsopp, N., and Lambert, C.J. (1996) *J. Phys.:Condens. Matter* **8**, L45.
13. Zaitsev, A.V. (1995) *Pis'ma Zh. Teor. Fiz.* **61**, 755.
14. Shaikhaidarov, R.Sh. *et al.* (1997) to be publ., see also refs. 11 and 12.
15. Cleland, A.N., Esteve, D., Urbina, C., and Devoret, M.H. (1992) *Appl. Phys. Lett.* **61**, 2820.
16. Hergenrother, J.M., Lu, J.G., and Tinkham, M. (1995) *IEEE Trans. Appl. Supercond.* **5**, 2604.
17. Nahum, M. and Martinis, J.M. (1993) *Appl. Phys. Lett.* **63**, 3075.

GIANT IN-PLANE OPTICAL ANISOTROPY OF SEMICONDUCTOR HETEROSTRUCTURES WITH NO-COMMON-ATOM

Olivier KREBS, Paul VOISIN and Michel VOOS

Laboratoire de Physique de la Matière Condensée
de l'Ecole Normale Supérieure
24 rue Lhomond, F75005 Paris, France

1. Introduction

The envelope function theory (EFT) originally proposed by G. Bastard [1] for the calculation of quantum well electronic properties has become very popular because, in spite of its remarkable simplicity, no quantitative or qualitative breakdown of its predictions has been reported until very recently. The EFT has been widely used not only to evaluate bare « first order » electronic properties, which was its original motivation, but also to predict many refined effects, such as QW polaritons [2], second order optical non-linearities [3] or spin-relaxation phenomena [4,5]. The classical EFT is based on a somewhat simplified version of the 8x8 k.p hamiltonian describing the electronic properties of bulk III-V or II-VI semiconductors having the zinc-blend structure [6], to which scalar potentials describing the shifts of the band extrema at interfaces, and an eventual external potential are added [1]. Here, we consider exclusively the situation of heterostructures grown along the (001) axis. A characteristic feature of the EFT is that the projection J_z of the angular momentum on the quantification axis z is a good quantum number at the zone center, i.e. when the in-plane wavevector $k_t = 0$. In other words, the heavy hole states $| 3/2 \pm 3/2 \rangle$ do not couple to the $J_z = \pm 1/2$ light particle states at $k_t = 0$. This remains true even if the quantum well potential is asymmetric, as for instance when an axial electric field is present. Biaxial strain due to a possible lattice mismatch does not change this result [7], which contrasts with the classical group theoretical result stating that (neglecting effects associated with the integer or half-integer character of the layer thicknesses [8]) the point groups of symmetric and asymmetric quantum wells are respectively D2d and C2v [9-11]. Indeed, in the first case, heavy and light holes with opposite symmetries (eg H2 and L1) should be coupled at the zone center, and in the second case, all heavy and light holes levels should be coupled at $k_t=0$. In this sense, the classical EFT is « oversymmetric ». This excessive symmetry is partly removed at $k_t \neq 0$ when considering the full 8x8 **k.p** Pikus and Bir hamiltonian including the interaction with remote bands up to the second order. Heavy holes and light particles are indeed coupled by various k_t-dependent terms, including the classical Luttinger terms which give rise to the now well understood complex in-plane dispersion of the valence subbands [1], and the inversion asymmetry matrix elements recently taken into account by Zhu and Chang [12] in an attempt to explain the quantum well Pockels effect. The main consequence of these symmetry considerations is that the optical properties of symmetric quantum wells for light propagating along the growth axis should be isotrope with respect to the in-plane polarization. The lifting of this polarization isotropy by an electric field F applied along the growth axis, or longitudinal Pockels effect, is expected to be a small effect [12].

N. García et al. (eds.), Nanoscale Science and Technology, 91–105.
© 1998 *Kluwer Academic Publishers*.

In contrast with these predictions, small polarization anisotropy was first reported by Gourdon et al. [13] in unbiased GaAs-AlAs short period superlattices, and a rather large emission anisotropy was reported by Kwok et al. [14] for biased GaAs-AlGaAs MQWs and by Vashori [15] for asymmetric GaAs-AlGaAs MQWs inserted in an optical microcavity. More recently, a very large polarization anisotropy of the optical absorption was observed in heterostructures of the C1A1-C2A2 type in which the host materials share no common atom (NCA), such as InP-(AlIn)As [16] and (GaIn)As-InP [17] MQWs. These observations imply a major breakdown of the EFT. The leading cause is the subtle loss of the rotoinversion symmetry which occurs in the vicinity of interfaces: as illustrated in Fig. 1, interface anions A1 are bound with cations C1 with chemical bonds all lying in the (110) plane, while they are bound to the other type of cations C2 by chemical bonds lying in the perpendicular (-110) plane. This asymmetry, which breaks the equivalence of the (110) and (-110) directions, is or is not compensated at the other interface, depending on the degree of symmetry of the interface themselves [10], and on the symmetry of the external potential. The resulting effects are particularly large in the NCA systems, because the interfaces involve specific chemical bonds (resp. A1-C2 and A2-C1) which do not exist in the bulk of the layers. Three theoretical approaches presently cope with these problems: the tight binding method (TBM) [17], which naturally takes into account the local symmetry properties, the generalized boundary condition theory (GBC) discussed by Ivchenko et al. [18], and the H_{BF} model introduced by Krebs and Voisin [19]. Here, we present a set of experimental observations concerning the optical anisotropy and its electric field dependence in InGaAs-based heterostructures, namely (GaIn)As-InP NCA MQWs and (GaIn)As-(AlIn)As Common Anion (CA) MQWs, and we compare the results with theoretical predictions.

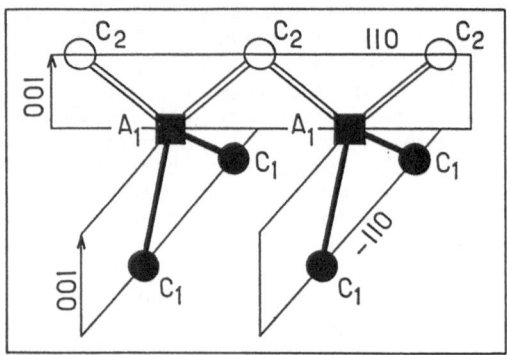

Fig. 1: scheme of the geometrical arrangement of chemical bond for the interface M1-M2 between NCA materials M1=C1A1 and M2=C2A2.

2. Samples and experiments

The three samples reported here are state-of-the-art quality 50-period GaInAs-based multi-quantum wells forming the intrinsic region of n-i-p diodes. They were grown respectively by MOCVD (S1, $Ga_{0.47}In_{0.53}As$ (120 Å)-InP(100 Å)), by Gas-Source MBE (S2, $Ga_{0.46}In_{0.54}As$ (104 Å)-InP(70 Å)) and by standard solid source MBE (S3, $Ga_{0.47}In_{0.53}As$ (100Å)-$Al_{0.48}In_{0.52}As$(70 Å). Among many others, a 80-period $Ga_{0.46}In_{0.54}As$ (45 Å)-InP(70Å) MOCVD sample S0 with narrow wells was also investigated, and its parameters used for theoretical discussions [17]. X-ray diffraction, luminescence, PLE and photocurrent spectroscopies were used to assess the sample quality, but here we focus on the results of transmission measurements at 77 K under linearly polarized light, using the experimental arrangement shown schematically in Fig. 2. We use the parallel-polarizers transmission (PPT) or the crossed-polarizers transmission (CPT) configurations.

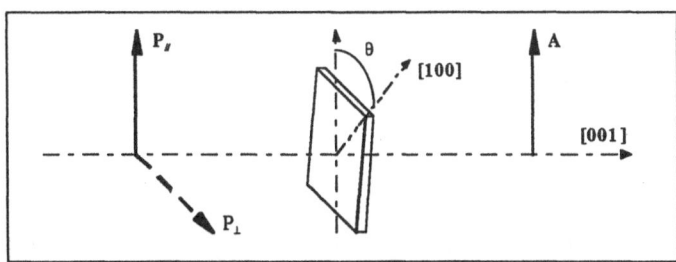

Fig. 2 : Scheme of the polarization resolved optical transmission set-up

Unexpectedly, as can be observed in Fig. 3, we obtain in both cases an oscillatory behavior indicating that the optical properties are anisotropic with respect to the in-plane polarization. The optical axis are along the [110] and [-110] directions, i.e. along the sample cleaved edges. In the PPT configuration (upper trace), the dependence of the absorption spectra with respect to the angle θ between the analyzer and the sample [001] axis is directly measured, but spectra recorded for different θ's correspond to different runs, and the measurement of small differences requires a severe control of slow drifts of the apparatus. Relative differences in the 1% range can be reliably determined over the investigated spectral range of 100 meV. Conversely, in the CPT configuration (lower trace), one measures directly and with great sensitivity a difference between two orthogonal directions, but it is no longer possible to separate the contributions of the real and imaginary parts of the optical index: instead, one gets $I_\perp/I_{//} = |\Delta n|^2 (\pi L/\lambda \cos 2\theta)^2$. Hence, PPT spectra reveal dichroïsm, i.e. variations of the imaginary part of the optical index with θ, while CPT spectra show with greater sensitivity the optical anisotropy through $|\Delta n(\theta)|^2$, at the expense of loosing information on the phase.

We display in Fig. 4a,b the absorption spectra deduced from PPT (upper trace), their relative difference or polarization spectra $P = (\alpha_{110} - \alpha_{-110})/(\alpha_{110} + \alpha_{-110})$ (medium trace) and the optical anisotropy spectra directly measured from CPT (lower trace), for samples S1 and S2 respectively, in the vicinity of the MQW bandgap. A significant optical anisotropy is observed in the spectral range between the H1-E1 and L1-E1 transitions. The same phenomenon is observed on extended spectra between the H2-E2 and L2-E2 transitions. From the comparison of polarization and anisotropy spectra, it is clear that the contribution of the birefringent index dispersion plays a secondary role, while the dichroïsm is the driving effect.

To summarize the observations, a very large dichroïsm corresponding to a polarization rate up to 12% (larger than the optical anisotropy of quartz !) is observed in the spectral range between the H1-E1 and L1-E1 transitions. In terms of index difference,

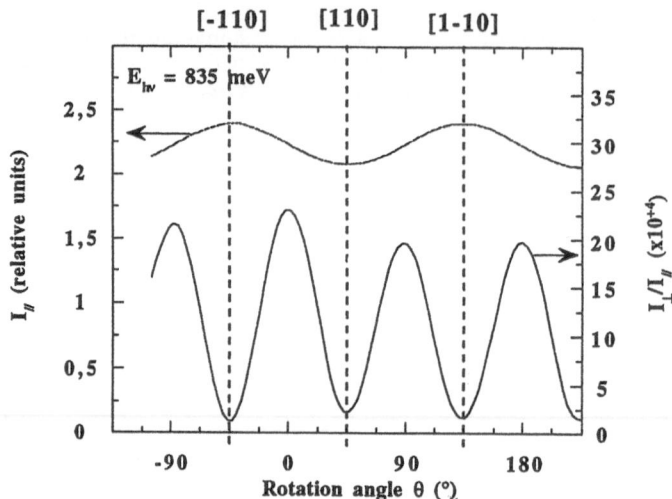

Fig. 3: Signals versus the angle θ between the analyzer and sample [100] axis measured at hν = 835 meV in sample S1 in the PPT (upper trace) and CPT (lower trace) configurations.

this corresponds to $|\Delta n| = 5 \ 10^{-2}$, which is rather surprising for materials built on a square lattice. There are qualitative differences between S1 and S2, namely the anisotropy peak is near H1-E1 in S1, and is closer to L1-E1 in S2, but the order of magnitude of the peak value is the same. Noteworthyly, the InGaAs-AlInAs sample S3 which has similar parameters and equivalent quality shows no effect at all at the scales of Figs. 3 and 4. These observations clearly point to the asymmetry of interfaces as the origin of the anisotropy effect: the common anion S3 has necessarily symmetric interfaces and shows no polarization (because the effects of adjacent interfaces compensate), while the NCA samples S1 and S2 show strong polarization effects due to the specificity of the chemical bonds for interface anions. However, due to the existence of anion exchange during the growth interruption which is required when commuting the four chemical species (C1, A1 $\Rightarrow$ C2, A2), the actual stoechiometry of such interfaces depends on the details of the growth procedure. MOCVD does not require long growth interruption times, because of the fast reactor sweep-out, while ultra-high vacuum epitaxy usually needs longer growth interruption to let the partial pressure of the undesired anion specie decrease to small enough values. Whatever the details of interface chemistry are, there is, to some extent, a possibility of acting on the interface composition through the growth interruption sequence. In other words, CA systems necessarily have symmetric interfaces, while NCA systems may have more or less asymmetric interfaces. In addition to that, it is of course possible to generate asymmetric quantum wells through the growth of more complex

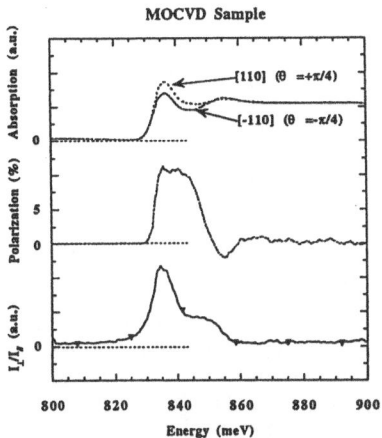
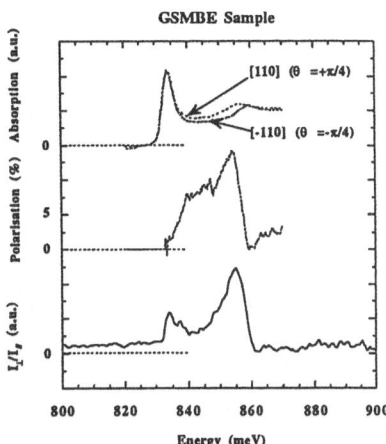

Fig. 4: Polarization resolved absorption spectra (upper trace), polarization spectra (medium trace) and optical anisotropy spectra (lower trace) measured at 77 K in sample S1 (a) and sample S2 (b). GaInAs-AlInAs MQW sample S3 shows no polarization at all at the scale of this figure.

sequences of layers. At this point, it is worth mentioning another consequence of interface asymmetry which is the non-commutativity of band offsets. Indeed, the interface dipole contribution to the band offset depends on the nature of the chemical bonds. Recent measurements in the InGaAs-InP and InP-AlInAs systems [20] have demonstrated this effect, in agreement with self-consistent band offset calculations [21].

3. Group theory explanation and tight binding calculation [17].

In Common Anion systems like C1A-C2A, only one sort of interface exists, formed by a plane of anions A between C1 and C2 cation planes. The point group is D_{2d} and its symmetry operations can be divided into those changing the coordinate along the growth axis z into -z and those leaving z unchanged. In NCA systems, the existence of two different anion types leads to two inequivalent interfaces corresponding to the C1-A1-C2 and C2-A2-C1 sequences. For these systems only the operations leaving z invariant are conserved and form the point group C_{2v}. The remaining symmetry operations centered at the midpoint of the layers like for instance the roto-inversion operations leave the system invariant except for the interface planes. The crystal potential of these superlattices is the sum of the contribution V_0 originating from all the atoms not belonging to the interface planes and the two contributions V^{A1}_I of the A1 anions of interface plane I and V^{A2}_J of the A2 anions of the other interface plane J. This sum can be decomposed in two parts, symmetric and antisymmetric with respect to the roto-inversion operations. The symmetric part: $V_s = V_0 + 1/2 \, (V^{A1}_I + V^{A2}_J + V^{A2}_I + V^{A1}_J)$ describes a system with D_{2d} symmetry, corresponding to a CA superlattice with the virtual anion A having the average potential of A1 and A2. The antisymmetrical part: $V_a = 1/2 \, (V^{A1}_I - V^{A2}_I + V^{A2}_J - V^{A1}_J)$ is invariant under the symmetry operations of the C_{2v} group and is multiplied by -1 under the others. It transforms according to the representation Γ_4 of the D_{2d} group. The effect of this small potential V_a can be evaluated by means of the perturbation theory and symmetry considerations. Indeed, the Γ_6 and Γ_7 states are coupled together via an interaction with Γ_4 symmetry. For instance, the H1 state (Γ_6 symmetry) is mixed at the mini-zone center with the L1 state (Γ_7 symmetry) to give the ground hole state |V1>. This point is essential because the magnitude of the in-plane anisotropy is proportional to the mixing coefficient and consequently to the anion dissymmetry. This approach thus gives an easy qualitative insight into the microscopic origin of the anisotropy. Unfortunately, it is inadequate to obtain quantitative results because the perturbation matrix elements cannot be written in a simple closed form. Precise theoretical results to be compared with experimental measurements have been obtained from tight-binding calculations of InGaAs-InP superlattice electronic structure, using a sp3s* nearest neighbor tight-binding model including the spin-orbit coupling [17]. The results corresponding to narrow GaInAs (45 Å)-InP(70 Å) quantum wells are illustrated in Figs. 5,6. It can be observed in Fig. 5 that this ab-initio calculation containing no ajustable parameter does predict optical anisotropy spectra at least qualitatively similar to the experimental ones. The quantitative agreement is not very good in this particular case, as the observed polarization rate is significantly larger (about a factor 3) than the predicted one.

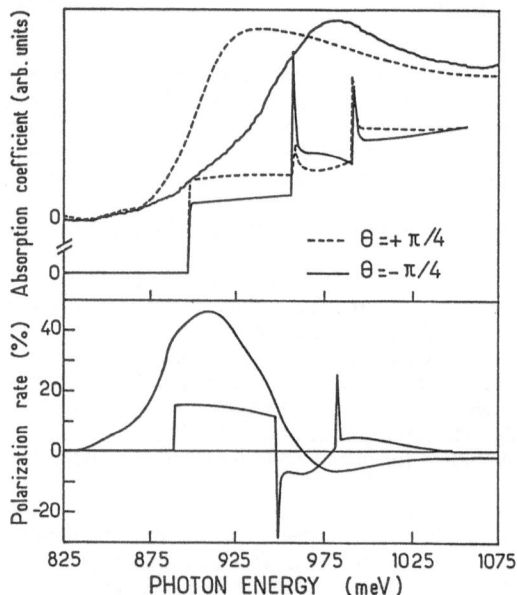

Fig. 5: Absorption spectra for a photon polarization along {1,1,0} ($\theta=\pi/4$) and along {-1,1,0} ($\theta=-\pi/4$) (upper panel) and relative difference of these spectra, or "polarization spectrum" of the sample S0 (lower panel). Corresponding theoretical curves (tight binding) are also shown.

In the case of sample S1, the calculated polarization rate is 4%, which is again less than the value observed in Fig. 4a. This discrepancy is presently unexplained: it could reflect the effect of the nearest neighbor approximation, or the role of Coulomb corrections (« excitons ») to the optical spectra, or more likely, the mere fact that actual interfaces are by far more complicated physical objects than considered in these simplified models. The projections of the $k_t=0$ valence band ground state V1 onto the $J_z= 3/2$ (« H ») and $J_z= 1/2$ (« L ») anion orbitals are displayed in Fig. 6 and illustrate that optical anisotropy is associated with zone center mixing of the heavy and light hole characters, an effect which cannot exist in the classical envelope function theory.

While bringing convincing explanation of the observations, these tight binding calculations suffer from some lack of versatility. This is why it is highly desirable to implement the local symmetry considerations discussed above into the envelope function theoretical framework. This was done simultaneously following two different approaches [18,19] which we summarize in the next sections.

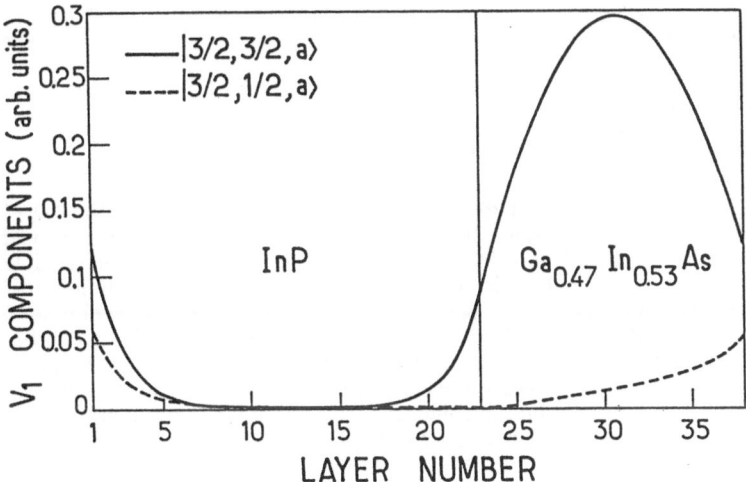

Fig. 6: Projections of the valence band ground state on the $J_z= 3/2$ (« H ») and $J_z= 1/2$ (« L ») anion orbitals

4. Generalized boundary condition approach.

In the approach proposed by Ivchenko et al. [18, 22], the reduction of crystal symmetry at interfaces is included in the envelope function theory through generalized boundary conditions. For instance, a heavy hole incident on the interface should be partly reflected/transmitted as a light hole. More precisely, these authors start with the hamiltonian in the absence of spin-orbit interaction and show that, in order to fulfill the group-theoretical requirements of the C_{2v} symmetry, the X and Y orbitals must be coupled by adding in the hamiltonian a term having the form:

$$H_{X-Y} = \pm \frac{\hbar^2 t_{X-Y}}{m_0 a_0} \begin{pmatrix} \{I_X I_Y\} & 0 \\ 0 & \{I_X I_Y\} \end{pmatrix} \delta(z - z_{int})$$

where I is the orbital momentum operator, and the free electron mass and lattice constant are introduced in order to characterize the mixing hamiltonian with the dimensionless parameter t_{X-Y}. The sign $\pm$ refers to the M1-M2 and M2-M1 interface respectively. Taking into account this coupling term, the boundary conditions for the six component vector F formed by the envelopes F_i (where i runs through the $X_{\uparrow\downarrow}$, $Y_{\uparrow\downarrow}$, $Z_{\uparrow\downarrow}$ basis) are:

$$F(z_{int}^+) = F(z_{int}^-) \quad \text{and} \quad (v_z F)(z_{int}^+) = (v_z F)(z_{int}^-) + 2i\, H_{X-Y} \colon F$$

where v_z is the velocity operator along the growth axis, $v_z = 1/m_z\, \partial/\partial z$. The spin-orbit interaction $-2/3\, \Delta\, (s\,.I)$ is then reintroduced and the unitary transformation from the X, Y, Z basis to the classical Γ_8, Γ_7 basis is performed. In the limit of large spin-orbit interaction, one finally gets the generalized boundary conditions for the envelope functions f_H, f_L associated with a given valence state V:

$$f_{H,L}(z_{int}^+) = f_{H,L}(z_{int}^-)$$

$$[1/m_H\, \partial\, f_H/\partial z]\,(z_{int}^+) = [1/m_H\, \partial\, f_H/\partial z]\,(z_{int}^-) + t_{HL}/m_0 a_0\, f_L(z_{int}^-)$$

$$[1/m_L\, \partial\, f_L/\partial z]\,(z_{int}^+) = [1/m_L\, \partial\, f_L/\partial z]\,(z_{int}^-) + t_{HL}/m_0 a_0\, f_H(z_{int}^-)$$

where $t_{HL} = t_{X-Y}/\sqrt{3}$, and $m_{H,L}$ are the heavy and light hole masses along the growth axis. In first approximation, the difference between common anion and NCA systems is that in the second case, t_{HL}^{left} differs from t_{HL}^{right}. In this case, the contributions of the two interfaces do not cancel each other, and each valence state has simultaneously non-zero f_H and f_L envelopes. This approach gives a perfect fit of the tight-binding « envelopes » shown in Fig. 6, using $t_{left}=0.8$ and $t_{right}=1.5$ [22]. Furthermore, the t-coefficients can be precisely related to the parameters of the tight-binding model. However, a completely rigorous treatment of the NCA situation requires an evenmore general formulation of the boundary conditions in order to take into account the specific interface chemical bonds [22]. The complete calculation of the optical spectra implies the generalization of the formalism to the case of non-zero in-plane wavevector k_t, which has not been reported yet, but it is rather easy to calculate the polarization rate P of the ground optical transition V1-E1: $P = 2/\sqrt{3}\, \langle f_L|\, f_E \rangle / \langle f_H|\, f_E \rangle$. This formula of course gives a polarization identical to the tight-binding result, once the t-coefficients have been fitted to reproduce the wavefunctions themselves.

5. Perturbation approach: the « H_{BF} » model (si non e vero, e bene trovato)

The leading idea of the « H_{BF} » model is to treat the symmetry breaking as a perturbation of the classical EFT. Schematically, the starting point is the same as in the Generalized Boundary Condition theory, that is the consideration of a potential localized at interfaces and mixing the heavy and light holes according to the prescription of the C2v symmetry. Then the idea is that this localized potential can be treated as a perturbation in the basis spanned by the classical envelope function solutions. In this sense, the two approaches are simply two different technical means of treating the same problem. In practice, the model was developed from a naive geometrical point of view which allows a very intuitive presentation. The scheme of a semiconductor interface (Fig. 1) can be described as the succession of a « half-monolayer » of the M1= C1A1 material with all the chemical bonds pointing « Backward » and lying in the (1,1,0) plane and another half-monolayer of the M'2 = C2A1 material with all the bonds pointing « Forward » and lying in the (-1,1,0) plane. The geometrical properties of the « backward » (B) or « forward » (F) bonds can be characterized by projection operators: we define operators P_j (j=1,4) [23] projecting the coodinates (or the corresponding atomic orbitals) X, Y and Z onto the actual bond directions [24], and we pair them to get the B and F operators whose matrix elements in the X, Y, Z basis are: $\langle i|$ B or F$|i\rangle$ = 1/2, $\langle X$ or $Y|$ B$|Z\rangle$ = $\langle X$ or $Y|$ F$|Z\rangle$ = 0, $\langle X|$ F$|Y\rangle$ = 1/2 and $\langle X|$ B$|Y\rangle$ = -1/2.

As B+F= Identity, the usual envelope-function valence band potential in an arbitrary system may be written as $V_{QW}(z) + V_{ext}(z) = V(z) = \Sigma_n$ (B+F) $V(z_n)$ h(z-z_n), where h(z) is equal to 1 in the [-a/4, +a/4] segment and zero outside, and the summation runs over the anion plane positions z_n. When we consider an external potential having no inversion symmetry (for instance electrostatic) or a valence band discontinuity corresponding to an interface, the Backward and Forward chemical bonds differ or experience different polarizations, which breaks the fourfold rotoinversion invariance around the z-axis. In simpler words, the [110] and [-110] directions are no longer equivalent. The corresponding anisotropy is introduced in the theory by writing:

$$V(z) = \Sigma_n \{V(z_n) - a/4 \, dV/dz\}h(z-z_n) \, B + \{V(z_n) + a/4 \, dV/dz\}h(z-z_n) \, F$$

This simply amounts to associate to each half-monolayer the corresponding average potential. Similarly, in the case of a compositional discontinuity involving a C1-A1-C2 monolayer, the valence band energy of the C1A1 material (V_1) is affected to the B operator and that of the C2A1 material (V'_2) to the F operator [24]. The sequence of C-A-C' monolayers in presence of the external potential $V_{ext}(z)$ is fully described by the succession of B and F operators with their associated potentials. This is illustrated schematically in Fig. 7 which shows the classical EFT quantum well and the corresponding H_{BF} representation. Note that the situation of a bulk material in presence of an electric field can be approached following the same ideas (traces d and e in Fig. 7). The classical EFT is recovered if one replaces the operators B and F by (B+F)/2 = 1/2, and ignores the peculiarities of the interface bonds.

The differences between the present hamiltonian H_{BF} and the envelope function hamiltonian H_0 are actually small and rapidly varying at the scale of the envelope functions. They can be treated in a somewhat unusual perturbation scheme by diagonalizing $\delta H = H_{BF}-H_0$ in a truncated set of solutions of H_0. Reminding that the total wavefunction Ψ is the product of a slowly varying envelope function f(z) by a rapidly varying atomic-like Bloch function u(r) (more precisely, spinors built with such products), one notes that the rapidly varying perturbation will affect only the atomic part of the wavefunction, which contrasts with the traditional situation of slowly varying perturbation potentials affecting only the envelopes. Here, we restrict the algebra to the Γ_8 basis, i.e. to the heavy- and light-hole states H$\pm$ = $|3/2,\pm3/2\rangle$ and L$\pm$ = $|3/2,\pm1/2\rangle$. B and F have diagonal matrix elements all equal to 1/2 and in addition off-diagonal matrix elements which couple the heavy- and light-hole states:

$$\langle H+| \text{ F} | L-\rangle = \langle H-| \text{ B} | L+\rangle = i/2\sqrt{3} \text{ and } \langle H-| \text{ F} | L+\rangle = \langle H+| \text{ B} | L-\rangle = - i/2\sqrt{3}.$$

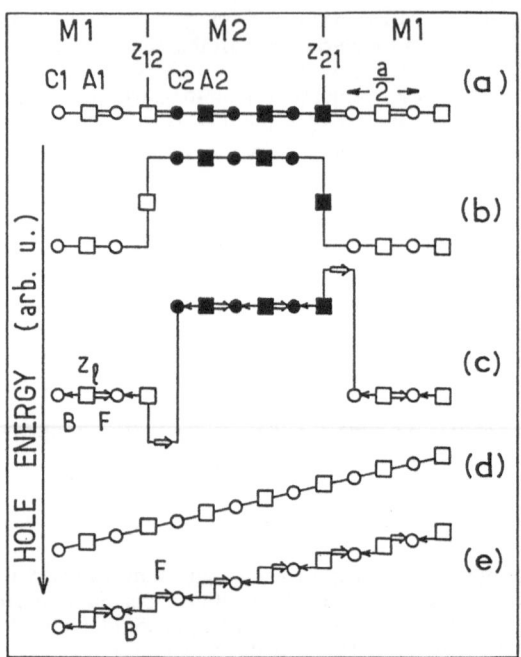

Fig. 7: The succession of cation and anion planes for an heterostructure grown along (0,0,1) (a), the associated envelope function potential (b) and the corresponding representation in terms of B and F operators (c). Bulk layer submitted to an electrostratical potential: envelope function potential (d) and its B and F operators representation (e).

The perturbation matrix element is $M^{i,j}_{\mu,\nu} = \langle f^i_\mu \; \mu | \; \delta H \; | \; f^j_\nu \; \nu \rangle$, where i (j) is a subband index and $\mu \; (\nu) = H\pm, L\pm$. Let us stress that the envelopes f^i_μ used here simply are the normalized envelope functions forming the solutions of the classical EFT problem. In the spirit of the envelope function formalism, the envelopes are slowly varying at the scale of a/2, which allows a factorization of the integrals. In the cases of a slowly varying potential $V_{ext}(z)$ or a C1-A1-C2 interface respectively we get :

$$M^{ij}_{\mu\nu} = \{ \; a/4 \int f^i_\mu(z) f^j_\nu(z) \; \partial V_{ext}/\partial z \; dz \; \} \; \langle \mu | \; F\text{-}B | \; \nu \rangle$$

$$M^{ij}_{\mu\nu} = a/2 \; f^i_\mu(z_{12}) f^j_\nu(z_{12}) \; x\{(V'_2\text{-}(V_1\text{+}V_2)/2) \; \langle \mu | \; F | \; \nu \rangle + (V_1\text{-}(V_1\text{+}V_2)/2) \; \langle \mu | \; B | \; \nu \rangle\}$$

The perturbation matrix has diagonal elements producing small shifts of the valence band energy levels, and off-diagonal terms which couple the heavy and light hole states at $k_t=0$. The diagonal correction is proportional to the squared amplitude of the corresponding wavefunction at the interface. This is usually negligible for the heavy hole states, but may be significant (a few meV) for the light hole states. The observed L1-E1 transitions in S1 and S2 tend to indicate that the dominant interface potential is attractive, as the measured H1-L1 spittings are about 5 meV smaller than calculated in the EFT (X-ray diffraction shows that these sample are basically strain-free).

To illustrate the « mechanics » of the model, we restrict the basis to the $H1\pm$ and $L1\pm$ levels, and we describe the in-plane motion within the "diagonal" approximation where the in-plane dispersions are parabolic with masses m^H_t and m^L_t. Although it is a crude approximation of the valence band structure, it has been proved that the diagonal model is quite sufficient to calculate accurately the band to band absorption spectrum [25]. In addition, valence subband mixing at finite k_t by the Luttinger matrix does not produce any optical anisotropy. The 4x4 hamiltonian matrix separates in 2 nearly identical 2x2 matrices corresponding to the H1+, L1- and H1-, L1+ subsets respectively. The first one is:

$$\begin{pmatrix} H1 + \hbar^2 k_t^2/2m_t^H & M_{H+L-} \\ M^*_{H+L-} & L1 + \hbar^2 k_t^2/2m_t^L \end{pmatrix}$$

and the second one differs by the sign of the off-diagonal term, $M_{H-L+} = - M_{H+L-}$. We have chosen the "hole energy" notation with positive confinement energies and in-plane masses. The corresponding eigenfunctions are $\Psi_{1,2} = a_{1,2} f_{H1} | H+ \rangle + b_{1,2} f_{L1} | L- \rangle$. Optical transitions to the first conduction subband induced by a photon propagating parallel to the z-axis and polarized at an angle θ with respect to the [100] axis can easily be calculated. As expected, the absorption spectrum $A(h\nu, \theta)$ is polarization dependent. As M_{HL} (hence b_1 and a_2) are pure imaginary, we have, using the notations $\underline{a}_i = a_i \langle f_{E1} | f_{H1} \rangle$, $\underline{b}_i = b_i \langle f_{E1} | f_{L1} \rangle$:

$$A/A_0 = \{ \underline{a}_1^2 + 1/3 | \underline{b}_1 |^2 - 2/\sqrt{3} \, \underline{a}_1 | \underline{b}_1 | \sin 2\theta \} \, Y(h\nu - (Eg+E1+H'1)) +$$
$$\{ | \underline{a}_2 |^2 + 1/3 \, \underline{b}_2^2 + 2/\sqrt{3} \, \underline{b}_2 | \underline{a}_2 | \sin 2\theta \} \, (\mu^L_t / \mu^H_t) Y(h\nu - (Eg+E1+L'1))$$

where $Y(x)$ is the step function, $A_0 \approx 6 \cdot 10^{-3} N_w$ is the absorption by the H1-E1 transition calculated in the diagonal approximation, and N_w the number of quantum wells. $\mu^{H(L)}_t$ is the in-plane reduced mass for the corresponding electron-hole pair, and H'1 and L'1 are the modified eigenenergies.

Assuming that $| M_{HL} |$ remains small compared to the L'1-H'1 splitting in the energy range of interest and taking into account that $\langle f_{E1} | f_{H1} \rangle$ and $\langle f_{E1} | f_{L1} \rangle$ are close to unity, the polarization rate at the absorption edge is immediately obtained by the perturbation formula as:

$$P = (A_{max} - A_{min})/(A_{max} + A_{min}) = 2/\sqrt{3} | M_{HL} | /(L1-H1)$$

Clearly, M_{HL} matrix element in the meV range and a H1-L1 splitting of 20 meV correspond to a polarization rate of the order of 5%, and give the polarized absorption spectrum shown in Fig.8. These simplified calculations hence explain, at least qualitatively, the observation of strong optical anisotropy in unbiased NCA quantum wells. However, such reduction of the valence band to a mere two level basis is inadequate to describe the effects induced by an external electric field F, because the field breaks down the parity selection rules on the envelope function overlaps which is explicitly used in the above formula, and couples at the first order in F the H1 and H2, L1 and L2 or E1 and E2

102

levels for instance. More generally, use of an extended basis is required for quantitative analysis, because the perturbation effect converges relatively slowly with the size of the basis (the M_{HL} matrix elements tend to increase with the level indexes). Calculations using a basis extended to all the QW bound states are technically straightforward: the EFT eigenstates are first calculated, then the perturbation matrix $M^{ij}_{\mu\nu}$ is readily evaluated, and diagonalized in the truncated EFT basis. There is no particular difficulty in expanding the in-plane motion terms using the same technique [1], which allows one to treat « exactly » the complex in-plane dispersion of the valence subbands. For the situation of ideal NCA quantum wells, the ultimate results depend only on two parameters which are the potentials dV_1 and dV_2 at the M1-M2 and M2-M1 interfaces. Finally, it is noteworthy that the matrix representation of the F-B operator in the X,Y, Z basis is identical to that of the $\{I_X I_Y\}$ operator, so the H_{BF} hamiltonian is strictly equivalent to the one introduced by Ivchenko and co-workers.

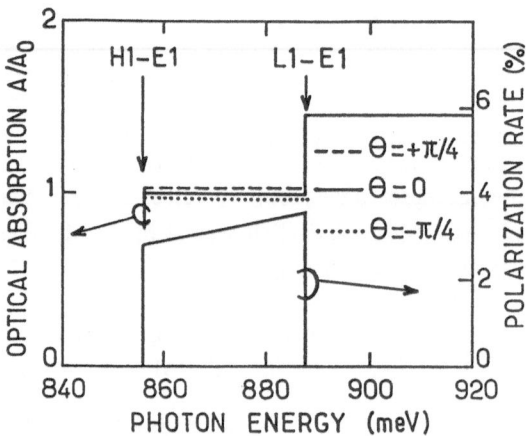

Fig. 8: Absorption spectra for various polarization angles θ and associated polarization spectrum. This calculation uses the two-band H1, L1 basis and the parameters of a 100 Å GaInAs-InP quantum well (having supposedly symmetrical band offsets) with $(V'_2+V'_1) - (V_1+V_2)= 1$ eV, which corresponds to $M_{HL} = 0.74$ meV.

6. The quantum-confined Pockels effect

In the following, we examine the main features of the field-induced modification of the optical anisotropy, or Quantum Confined Pockels Effect. We shall discuss the evolution of the optical anisotropy spectra (obtained from CPT measurements) with applied electric field, because they are more easily obtained than the polarization spectra. The data recorded in S1 and S2 are displayed in Fig.5. Again, the data measured in S3 are not shown for the simple reason that the observed effects are over one order of magnitude smaller that those observed in S1 and S2. The characteristic features in S1 are the strong decrease of optical anisotropy near the H1-E1 transition, and the slow increase of the second anisotropy peak observed close to the nearly degenerate H2-E1 and L1-E1 transitions. Conversely, in S2, a strong increase of the second anisotropy peak is observed, while the first one is nearly constant. These behaviors are characteristic of strong interface

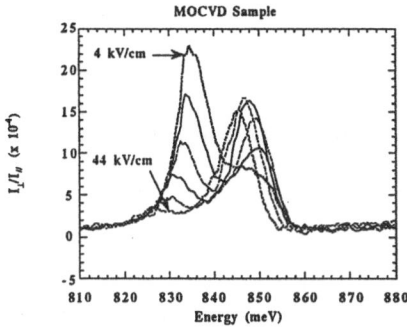 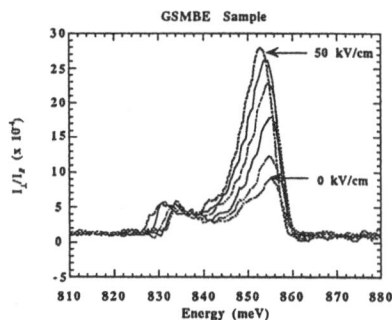

Fig. 9 : Optical anisotropy spectra recorded at 77 K in sample S1 (a) and S2 (b) for different values of the electric field.

potentials in both samples, with very asymmetric interfaces in S1 and more symmetric interfaces in S2. Let us insist on the fact that only one direction of the applied electric field can be investigated, due to the n-i-p diodestructure of the samples. Equivalent samples grown in an inverted (p-i-n) diode would allow the exploration of the other field direction. As we expect a linear dependence, at least at low field, this imply that the native anisotropy could be strongly enhanced in a sample like S1. These data can be compared to the prediction of the H_{BF} model, calculated using a H1, H2 and L1 three level basis. This turns out to be enough for a first approach, because the parity breaking effects are dominated by the Quantum Confined Stark shift of the heavy hole, i.e. by the field-induced mixing of H1 and H2. Theoretical polarization spectra obtained for samples S1 (left panel) and S2 (right panel) are displayed in Fig. 10. They correspond respectively to $dV_1 = -2eV$, $dV_2 = 0$ (S1), and $dV_1 = -1$ eV, $dV_2 = -1$ eV (S2). These values do correspond to attractive potentials whose diagonal contribution lowers the energy of L1 by a few meVs. As for S3, we find that no Pockels effect should be expected because, for this range of sample parameters, the interface contribution is almost perfectly balanced by the bulk contribution [19]. It can be observed in Fig. 10 that the quantitative agreement is quite good for the H1-E1 transition, and somewhat less satisfactory for the H2-E1 / L1-E1 transitions: this should be expected at least because the restricted basis does not account for the light hole Stark shift. Also, it should be reminded that the theoretical spectra of Fig. 10 should not be too directly compared to the anisotropy spectra of Fig. 9 which contain a contribution from the real part of the optical index: forthcoming improvements of the data reduction and of the calculations themselves (extended basis, complete treatment of the in-plane motion) will allow a more pernickety comparison of theory and experiment.

The values of the « interface potentials » obtained from the fit of the flatband anisotropy and of the Quantum Confined Pockels effect are obviously too large to be interpreted as band offsets between the host materials C1A1 and C2A2 on the one hand and the virtual « half monolayer » of interface materials C2A1 and C1A2 on the other hand. The main reason appears to be the restriction of the basis, which eliminates the contribution to upper-lying valence subbands. Preliminary calculations using an extended basis show equivalent fits using significantly smaller interface potentials. Yet, several reasons may be invoked to explain the difference between reality and the naive presentation of the H_{BF} model. The first one is that the self-consistent dipole screening correction part of band offsets between bulk materials certainly does not apply at the scale of a half monolayer. On the same line, the specific interface bonds in NCA systems are highly strained, which raises the question of the absolute valence band deformation potentials, which can be much larger than the more well known bandgap deformation potentials. In other words, even if the H_{BF} rule of affecting to each half monolayer the corresponding average potential turned out to be an exact result, the average energy of interface half monolayers would not follow straightforwardly from standard calculations of bulk materials. The other argument is that the consideration of ideal growth sequences is certainly an oversimplification of the underlying reality: real interfaces, especially in the case of NCA systems, probably extend over several monolayers, due to the mechanisms of ion exchange, to the non-local nature of the dipole contribution to band offsets, etc... A more complete analysis would require the consideration of much more than one « perturbed » half-monolayer at each interface, and would involve a proportionally larger number of potential parameters. Given all these reasons for underlying complications, it can be considered that our simple-minded theory gives a fairly consistent description of the observations.

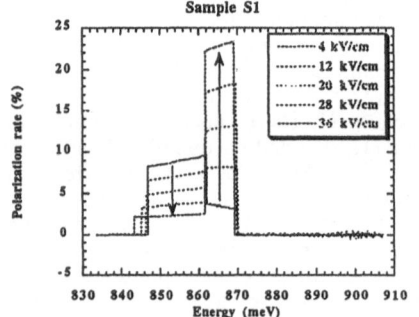 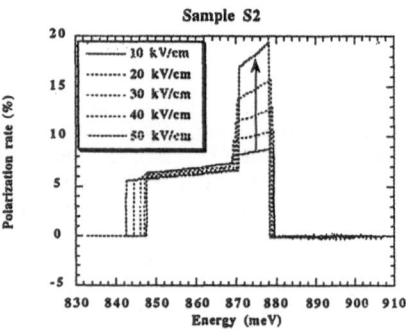

Fig. 10 : Calculated polarization rate spectra in the diagonal model, using the truncated basis H1, H2 and L1. The computation uses two parameters, the « interface potentials » dV1 and dV2. They are respectively equal to -2 eV and 0 eV for Sample S1 whereas they are both equal to -1 eV for Sample S2

7. Conclusion.

We have observed new optical polarization anisotropy phenomena, which imply a breakdown of the classical EFT. Parameter-free tight binding calculations show that these effects are intrinsic and associated with the reduced symmetry of No-Common-Atom heterostructures. The envelope function theory can be completed by considering additional interface perturbation potentials, which mix the heavy and light hole states at the mini-zone center and restore the correct symmetry properties of the hamiltonian. These corrections can be treated, technically speaking, in the generalized boundary condition approach of Ivchenko et al. [18] or in our «H_{BF}» perturbative approach [19] which has the definite advantage of very high flexibility.

Acknowledgment: We are indebted to Drs J.P. André (LEP), D. Rondi (LCR-Thomson-CSF), J.L. Gentner and L. Goldstein (Alcatel-Alsthom Research) and J.C. Harmand (CNET-Bagneux) who have grown the high quality samples used in this study. We also express our gratitude to Prs. D. Bertho and C. Jouanin for their essential contribution on the tight-binding approach, to Pr. E.L. Ivchenko and Drs A. Toropov, G. Bastard and R. Ferreira for a number of stimulating discussions. This work was supported in part by the OFCORSE IT-program of the European Community.

references

1   G. Bastard, Phys. Rev. B24, 5693 (1981) and Phys. Rev. B25, 7584 (1982).
    See also the textbook « wave mechanics applied to semiconductor heterostructures » (les Editions de Physique, les Ulis, 1992)
2   See for instance L. Andreani, A. d'Andrea and R. del Sole, Phys. Lett. A168, 451 (1992)
3   J. Khurghin, Phys. Rev. B38, 4056 (1988)
4   T. Uenoyama and L.J. Sham, Phys. Rev. Lett. 64, 3070 (1990)
5   R. Ferreira and G. Bastard, Phys. Rev. B43, 9687 (1991)
6   See G.E. Pikus and A.N. Titkov, « Spin relaxation under optical orientation in semiconductors », in « Optical Orientation », edited by F. Meier and B.P. Zakharchenya (Elsevier, 1984)·
7   See for instance J.Y. Marzin, J.M. Gérard, P. Voisin and J.A. Brum, «Optical studies of strained III-V heterolayers » in Semiconductors and Semimetals vol. 32 (Academic Press, 1990)
8   Yu. E. Kitaev, A.G. Panfilov, P. Tronc and R.A. Evarestov, J. of Physics Condensed Matter 9, 257 (1997)
9   D.L. Smith and C. Mailhiot, Rev. Mod. Phys. 62, 173 (1990)
10 P.V. Santos, P. Etchegoin, M. Cardonna, B. Brar and H. Kroemer, Phys. Rev. B50, 8746 (1994)
11 D. Vakhshoori, Appl. Phys. Lett. 65, 259 (1994)
12 B-F. Zhu and Y.C. Chang, Phys. Rev. B 50, 11932 (1994)
13 C. Gourdon and Ph. Lavallard, Phys. Rev. B46, 4644 (1992)
14 S.H. Kvok, H.T. Grahn, K. Ploog and R. Merlin, Phys. Rev. Lett. 69, 973 (1992)
15 D. Vakhshoori and R.E. Leibenguth, Appl. Phys. Lett. 67, 1045 (1995)
16 W. Seidel, P. Voisin, J.P. André and F. Bogani, Solid State Electronics 40, 729 (1996)
17 O. Krebs, W. Seidel, J.P. André, D. Bertho, C. Jouanin and P. Voisin, to appear in Semicond. Sci. Technol. Lett. (1997)
18 E. L. Ivchenko, A. Yu. Kaminski and U. Rössler, Phys. Rev. B54, 5852 (1996)
19 O. Krebs and P. Voisin, Phys. Rev. Lett.77, 1829 (1996)
20 W. Seidel, O. Krebs, P. Voisin, J.C. Harmand, F. Aristone and J.F. Palmier, Phys. Rev. B55, 2274 (1997)
21 Y. Foulon and C. Priester, Phys. Rev. B45, 6259 (1992)
22 E.L. Ivchenko and A. Toropov, preprint
23 More precisely, we define $|j\rangle = 1/2|\pm(X+Y)+Z\rangle$, $1/2|\pm(X-Y)-Z\rangle$ and $P_j = |j\rangle\langle j|$
24 This is obviously an over-simplification, in fact only an image, since dipole corrections and local strain effects are not taken into account. The central idea is to affect to each half-monolayer the average valence band potential.
25 S. Chelles, R. Ferreira and P. Voisin, Semicond. Sci. Technol. 10, 105 (1995)

STRONG ELECTRON TUNNELING IN MESOSCOPIC TUNNEL JUNCTIONS

JÜRGEN KÖNIG[1], HERBERT SCHOELLER[1], GERD SCHÖN[1] AND
ANDREI D. ZAIKIN[1,2]
[1] *Institut für Theoretische Festkörperphysik,*
Universtität Karlsruhe, 76128 Karlsruhe, Germany
[2] *I.E. Tamm Department of Theoretical Physics, P.N. Lebedev*
Physics Institute, Leninskii pr. 53, 117924 Moscow, Russia

Abstract. We describe electron transport through small metallic grains
with Coulomb blockade effects beyond the perturbative regime. For this
purpose we study the real-time evolution of the reduced density matrix of
the system. In the first part of the paper we present a diagrammatic expansion for not too high junction conductance, $h/4\pi^2 e^2 R_t \lesssim 1$, in a basis
of charge states. Quantum fluctuations renormalize system parameters and
lead to finite lifetime broadening in the gate-voltage dependent differential conductance. We derive analytic results for the spectral density and
the conductance in the limit where only two charge states play a role. In
the second part of the paper we consider junctions with large conductance,
$h/4e^2 R_t \gtrsim 1$. In this case contributions from all charge states become important, and a description in the phase representation is more useful. In
this case the effective capacitance of the junction is strongly renormalized.

1. Introduction

Electron transport through mesoscopic grains is strongly influenced by the
large charging energy, $E_C = e^2/2C$, associated with the low capacitance
C of the system [1, 2, 3, 4]. An interesting example is the "single-electron
transistor" where a small metallic island is coupled via tunnel junctions to
leads and via a capacitor to a gate voltage. At low temperatures, $T \ll E_C$,
a variety of single-electron phenomena have been observed in this system,
including the Coulomb blockade and oscillations of the conductance as a
function of a gate voltage. If the dimensionless conductance of the tunnel

N. García et al. (eds.), Nanoscale Science and Technology, 107–126.

junctions between the island and the lead electrodes,

$$\alpha_t \equiv \frac{R_K}{4\pi^2 R_t} = \frac{h}{4\pi^2 e^2 R_t}, \tag{1}$$

is low, on a scale defined by the quantum resistance $R_K \simeq 25.8$ kΩ, the charge of each island is a well-defined variable. In the limit $\alpha_t \ll 1$, the sequential single-electron tunneling can be studied in perturbation theory [1, 3]; and descriptions based on a master equation or equivalent simulations of the stochastic dynamics are sufficient to account for the dominant features observed in single-electron devices.

Recent experiments beyond the perturbative regime show deviations from the classical description, e.g. a broadening of the conductance peaks much larger than temperature [5, 6]. This indicates that, in general, quantum fluctuations and higher-order coherent processes should be considered. Even in the limit of weak tunneling, $\alpha_t < 1$, nontrivial features appear in the vicinity of the Coulomb blockade threshold, when two charge states become nearly degenerate and perturbation theory fails. Several theoretical papers [8, 9, 10, 11, 12, 13, 14] dealt with the problem of higher-order processes, exploiting the physical picture of electron tunneling via discrete charge states. This includes "inelastic cotunneling" [7, 14], where in a second-order process in the parameter α_t electrons tunnel via a virtual state of the island. An extension of this process, which gains importance near resonances, is "inelastic resonant tunneling" [10, 13], a process where electrons tunnel an arbitrary number of times between the reservoirs and the islands. The term "inelastic" indicates that with overwhelming probability different electron states are involved in the different steps of the higher order processes. The description can been extended to describe strong tunneling through single level quantum dots [15].

If the conductance of tunnel junctions is not small, $\alpha_t \gtrsim 1$, the physical picture changes. In this case the inverse lifetime $\Gamma = 1/R_t C$ and, hence, the broadening of the excited charge states due to quantum fluctuations exceed the typical level spacing of excited island states, $\hbar\Gamma \gtrsim E_C$. Thus charge levels overlap and the concept of tunneling via discrete charge states becomes ill-defined, raising the question whether charging effects survive under such conditions or whether they are washed out completely by strong quantum fluctuations. In Refs. [2, 16, 17, 18, 19, 12] it was demonstrated that at sufficiently low temperatures even for large values of α_t quantum fluctuations of the charge *do not* destroy Coulomb blockade of tunneling, but they lead to a strong renormalization of the effective junction capacitance, $C_{\text{eff}} \propto C \exp(2\pi^2 \alpha_t)$. The exponential dependence on α_t had been derived independently by renormalization group arguments [16, 12], instanton techniques [18], and Monte Carlo studies [12, 20]. One important consequence of the strong capacitance renormalization with increasing $\alpha_t \gtrsim 1$ is the exponential reduction of the temperature limit below which charging effects can be observed.

This article is devoted to the calculation of the conductance of a SET transistor beyond perturbation theory in α_t, in a range of parameters which is accessible to experiments. The island contains a large number of electrons which are coupled strongly by Coulomb interactions. We, therefore, cannot proceed with ordinary perturbation theory. Rather, we reformulate the quantum mechanical many-body problem of these electrons in a real-time path-integral representation. In order to handle the Coulomb interaction we perform a Hubbard-Stratonovich transformation which introduces a phase as a collective variable. We trace out all microscopic degrees of freedom and arrive at an effective action of the system [22, 2], similar in structure to that known from the studies of Ohmic dissipation in quantum mechanics [23]. This procedure is addressed in Section 2.

After a change from the phase to a charge representation we are able to perform for $\alpha_t \lesssim 1$ a diagrammatic expansion of the time evolution of the reduced density matrix. In a charge representation we can identify sequential, co- and resonant tunneling processes with certain classes of diagrams. A restriction to two charge states allows us to evaluate the spectral function and the conductance of the system analytically. The results will be presented in Section 3. At higher temperatures more charge states play a role, which in general requires a numerical study of the diagrammatic expansion.

In the opposite limit of strong tunneling, $\alpha_t \gtrsim 1$, many charge states play a role, and a formulation in terms of the phase, which is canonically conjugated to the charge, is more convenient. This limit is discussed in Section 4. We analyze the quantum dynamics of the phase variable in a semiclassical (saddle-point) approximation and obtain an expression for the system conductance valid at not too low temperatures $T \gtrsim e^2/2C_{\text{eff}}$. We further review results obtained earlier within different imaginary time techniques, e.g. renormalization group and instanton methods, and compare these results with those of our real time analysis.

2. Formulation of the Problem

We consider a metallic island coupled by two tunnel junctions (L,R) to two leads and capacitively to an external gate voltage V_g. An applied transport voltage $V = V_L - V_R$ drives a current. A microscopic description of this single-electron transistor is based on the Hamiltonian, $H = H_L + H_R + H_I + H_{ch} + H_{t,L} + H_{t,R}$. Here $H_r = \sum_{k\sigma} \epsilon_{k\sigma r} a^{\dagger}_{k\sigma r} a_{k\sigma r}$ describes noninteracting electrons in the left and right lead, r= L,R, and $H_I = \sum_{q\sigma} \epsilon_{q\sigma} c_{q\sigma}^\dagger c_{q\sigma}$ models the island states. The Coulomb interaction is accounted for in a capacitance model

$$H_{ch} = E_C \left(\sum_{q\sigma} c_{q\sigma}^\dagger c_{q\sigma} - n_g \right)^2 . \tag{2}$$

The energy scale $E_C \equiv e^2/(2C)$ of the transistor depends on the total island capacitance, $C = C_L + C_R + C_g$, determined by the left and right tunnel junction and the gate capacitance. The charging energy can be tuned continuously by the "gate charge"

$$Q_g \equiv -en_g = -(C_L V_L + C_R V_R + C_g V_g) . \tag{3}$$

The tunneling Hamiltonian $H_{t,r} = \sum_{kq\sigma} \left(T^{\sigma r} a_{k\sigma r}^\dagger c_{q\sigma} + \text{h.c.} \right)$ describes tunneling between the island and the left and right leads. The matrix elements are related to the tunnel conductances by $R_r^{-1} = (e^2/h) \sum_\sigma N_r^\sigma(0) N_I^\sigma(0) |T^{\sigma r}|^2$, where $N(0)$ denotes the densities of states of the island and the leads, respectively. In the following we will consider "wide" metallic junctions with $N \gg 1$ transverse channels. Extending the spin summation they can be labeled by the index $\sigma = 1, ... N$. In the following we will put $\hbar = 1$ (except when it enters the quantum of resistance).

Our aim is to study the time-evolution of the density matrix. We shortly sketch the main steps of the derivation of this description:
– The time evolution of the density matrix introduces two propagators, a forward and backward propagator, which get coupled when we trace out electron degrees of freedom of the reservoirs. The procedure is known from the work of Caldeira and Leggett [23] who, generalizing earlier work of Feynman and Vernon, studied the influence of Ohmic dissipation on a quantum system. Similarly the influence on electron tunneling was described in Refs. [22, 2]. Here, we generalize the later work from a single tunnel junction to the transistor.
– In order to describe the Coulomb interaction between electrons we introduce via a Hubbard-Stratonovich transformation the electric potential of the island $V(t)$ as a macroscopic field. The interaction between electrons is replaced in this way by an interaction with the collective variable.
– We treat the leads as well as the electrons in the island as large equilibrium reservoirs. The electrochemical potentials of the reservoirs are fixed, $\mu_r = -eV_r$ for r = L,R. The only fluctuating field is voltage of the island $V(t)$. The definition $eV(t) \equiv -\dot\varphi(t)$ relates $V(t)$ to a phase $\varphi(t)$. Its quantum mechanical conjugate is the number of excess electrons $n(t)$ on the island. As a consequence of the procedure outlined so far, the macroscopic

field $n(t)$ is independent of the microscopic degrees of freedom described by $c_{q\sigma}$ and $c_{q\sigma}^\dagger$. At this stage, the electronic degrees of freedom can be traced out.

– The time evolution of the reduced density matrix $\rho(t; \varphi_1, \varphi_2)$, which depends only on the phase variable φ, can thus be expressed by a double path integral over the phases corresponding to the forward and backward propagators φ_j $(j = 1, 2)$

$$\rho_c(t_f; \varphi_{1f}, \varphi_{2f}) = \int_{-\infty}^{\infty} d\varphi_{1i} \int_{-\infty}^{\infty} d\varphi_{2i} \int_{\varphi_{1i}}^{\varphi_{1f}} \mathcal{D}[\varphi_1(t)] \int_{\varphi_{2i}}^{\varphi_{2f}} \mathcal{D}[\varphi_2(t)]$$
$$\exp\left(iS[\varphi_1(t), \varphi_2(t)]\right) \rho_c(t_i; \varphi_{1i}, \varphi_{2i}) . \tag{4}$$

– The form (4) describes the situation where charges can take any continuous value and the phase is an extended variable. However, in our physical system the charge on the island is quantized in units of the electron charge e. In this case the phase variable is compact (i.e., the states φ and $\varphi + 2\pi$ are equivalent), and we rewrite (4), introducing integer winding numbers $m_1, m_2 = 0, \pm 1, \pm 2, \ldots$,

$$\rho_d(t_f; \varphi_{1f}, \varphi_{2f}) = \sum_{m_1, m_2} \int_{-\infty}^{\infty} d\varphi_{1i} \int_{-\infty}^{\infty} d\varphi_{2i} \int_{\varphi_{1i}}^{\varphi_{1f}+2\pi m_1} \mathcal{D}[\varphi_1(t)] \int_{\varphi_{2i}}^{\varphi_{2f}+2\pi m_2} \mathcal{D}[\varphi_2(t)]$$
$$\exp\left(iS[\varphi_1(t), \varphi_2(t)]\right) \rho_d(t_i; \varphi_{1i}, \varphi_{2i}) . \tag{5}$$

The two integrations can be combined to a single integral along the Keldysh contour, which runs forward and backward between t_i and t_f along the real-time axis. As a result the reduced propagator Π is written as a single path integral along this contour

$$\Pi = \text{tr}\left[\rho_0 \, T_K \exp\left(-i \int_K dt \, H(t)\right)\right] = \int \mathcal{D}[\varphi(t)] \exp\left(iS[\varphi(t)]\right) . \tag{6}$$

Here the collective variable $\varphi(t)$ and the time integral are defined on the Keldysh contour K, and the time-ordering operator T_K orders the following operators accordingly.

The effective action entering the propagator is $S[\varphi(t)] = S_{ch}[\varphi(t)] + S_t[\varphi(t)]$. The first term represents the charging energy

$$S_{ch}[\varphi(t)] = \int_K dt \left[\frac{C}{2}\left(\frac{\dot\varphi(t)}{e}\right)^2 + n_g \dot\varphi(t)\right] . \tag{7}$$

Electron tunneling is described by $S_t[\varphi(t)]$, which, in the case of wide metallic junctions, is expressed by the simplest electron loop connecting two times,

$$S_t[\varphi(t)] = 2\pi i \sum_{r=L,R} \int_K dt \int_K dt' \, \alpha_r^K(t,t') e^{i\varphi(t)} e^{-i\varphi(t')} . \qquad (8)$$

The kernels $\alpha_r^K(t,t') = \alpha_r^{\pm}(t-t')$ for $t < t'$ ($t > t'$) depend on the order of the times along the Keldysh contour. Their Fourier transforms are [2, 10, 13]

$$\alpha_r^{\pm}(\omega) = \pm \alpha_{t,r} \frac{\omega - \mu_r}{\exp[\pm(\omega - \mu_r)/T] - 1} . \qquad (9)$$

They are proportional to the dimensionless tunneling conductance $\alpha_{t,r} = h/(4\pi^2 e^2 R_r)$ between the island and the leads r = L,R.

For large systems, the phase behaves almost like a classical variable while its conjugate variable, the charge, fluctuates strongly. A natural basis is then the phase representation. In the presence of strong Coulomb interaction, however, the situation is different: the phase underlies strong fluctuations while the time evolution of the charge is almost governed by classical rates. For this reason, it may be useful to change from the phase to the charge representation. The time evolution of the density matrix in a charge representation depends on the propagator from n_1 forward to n_1' and on the backward branch from n_2' backward to n_2. It is given by the matrix element of the reduced propagator [13]

$$\Pi_{n_2,n_2'}^{n_1,n_1'} = \int d\varphi_1 \int d\varphi_1' \int d\varphi_2' \int d\varphi_2 \, e^{in_1\varphi_1} e^{-in_1'\varphi_1'} e^{in_2'\varphi_2'} e^{-in_2\varphi_2} \qquad (10)$$

$$\int_{\varphi_2,\varphi_2'}^{\varphi_1,\varphi_1'} \mathcal{D}[\varphi(t)] \int \mathcal{D}[n(t)] \exp\left(-iS_{ch}[n(t)] + iS_t[\varphi(t)] + i\int_K dt \, n(t)\dot{\varphi}(t) \right) .$$

In the charge representation the charging energy is simply described by $S_{ch}[n(t)] = \int_K dt \, E_C [n(t) - n_g]^2$.

3. Expansion in the tunneling conductance

A diagrammatic description is obtained by expanding the tunneling term $\exp(iS_t[\varphi(t)])$ in the reduced propagator and integrating over φ. Each of the exponentials $\exp[\pm i\varphi(t)]$ describes tunneling of an electron at time t. These changes occur in pairs in each junction, r=L,R, and are connected by tunneling lines $\alpha_r^K(t,t')$. Each term of the expansion can be visualized by a diagram. Several examples are displayed in Fig. 1. The value of a diagram is calculated according the rules which follow from the expansion of Eq. (10) and are presented in detail in Ref. [13].

The propagator from a diagonal state n to another diagonal state n' is denoted by $\Pi_{n,n'}^{n,n'} = \Pi_{n,n'}$. It is the sum of all diagrams with the given

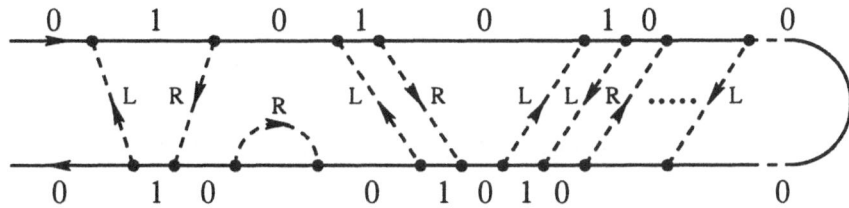

Figure 1. A diagram showing from left to right: sequential tunneling in the left and right junction, a term preserving the norm, a cotunneling process, and resonant tunneling.

states at the ends and can be expressed by an irreducible self-energy part $\Sigma_{n,n'}$, defined as the sum of all diagrams in which any vertical line cutting through them crosses at least one tunneling line. The propagator can be expressed as an iteration in the style of a Dyson equation, $\Pi_{n,n'} = \Pi_n^{(0)} \delta_{n,n'} + \sum_{n''} \Pi_{n,n''} \Sigma_{n'',n'} \Pi_{n'}^{(0)}$. The term $\Pi^{(0)}$ describes a propagation in a diagonal state which does not contain a tunneling line. The stationary probability for state n follows from $P_n = \sum_{n'} P_{n'}^{(0)} \Pi_{n',n}$ (in which $P_n^{(0)}$ is the initial distribution) and is *not* the equilibrium one if a bias voltage is applied. Our diagram rules then yield

$$0 = \sum_{n'} [-P_n \Sigma_{n,n'} + P_{n'} \Sigma_{n',n}]. \tag{11}$$

We recover the structure of a stationary master equation with transition rates given by $\Sigma_{n',n}$. In general, the irreducible self-energy Σ yields the rate of all possible correlated tunneling processes. We reproduce the well-known single-electron tunneling rates by evaluating all diagrams which contain no overlapping tunneling lines. Similarly cotunneling is described by the diagrams where two tunneling lines overlapping in time, as shown in Fig. 1.

We calculate the current I_r flowing into reservoir $r = L, R$ by adding a source term to the Hamiltonian and then taking the functional derivative of the reduced propagator with respect to the source. The result $I_r = -ie \int d\omega \, \{\alpha_r^+(\omega) C^>(\omega) + \alpha_r^-(\omega) C^<(\omega)\}$ is expressed by the correlation functions $C^>(t, t') = -i\langle e^{-i\varphi(t)} e^{i\varphi(t')} \rangle$ and $C^<(t, t') = i\langle e^{i\varphi(t')} e^{-i\varphi(t)} \rangle$ describing charge transfer at different times. These are related to the spectral density for charge excitations on the island by $2\pi i A(\omega) = C^<(\omega) - C^>(\omega)$.

For sequential tunneling, the current reduces to

$$I_r = \frac{e}{h} 4\pi^2 \int d\omega \sum_{r'} \frac{\alpha_{r'}(\omega) \alpha_r(\omega)}{\alpha(\omega)} A(\omega)[f(\omega - \mu_{r'}) - f(\omega - \mu_r)] \tag{12}$$

with

$$A^{(0)}(\omega) = \sum_{n=-\infty}^{\infty} [P_n + P_{n+1}]\delta(\omega - \Delta_n) \tag{13}$$

and $\Delta_n = E_{ch}(n+1) - E_{ch}(n) = E_C[1 + 2(n - n_g)]$, where the probabilities follow from $P_n\alpha^+(\Delta_n) - P_{n+1}\alpha^-(\Delta_n) = 0$.

At the minima of the Coulomb oscillations the system is in the Coulomb blockade regime, and cotunneling processes determine the conductance [7]. But also at the resonance higher order terms are important, and the complete theory of cotunneling [14] has to cover both regimes. The second order processes are described by diagrams as shown in Fig. 1. For definiteness, we concentrate on situations where only two charge states, $n = 0, 1$, need to be considered. This is the case when the energy difference of the two states $\Delta_0 \equiv E_{ch}(1) - E_{ch}(0)$, the bias voltage $eV = eV_L - eV_R$, and the temperature T are low compared to E_C.

Using the notations $\alpha_r(\omega) = \alpha_r^+(\omega) + \alpha_r^-(\omega)$ and $\alpha(\omega) = \sum_r \alpha_r(\omega)$, and defining

$$R_{\pm}(\omega) = \frac{1}{\omega - \Delta_0 + i0^+} - \frac{1}{\omega - \Delta_{\pm 1} + i0^+} \tag{14}$$

we obtain for the "cotunneling" contribution $I^{(2)}(\Delta_0) = \sum_{i=1}^{3} I_i^{(2)}(\Delta_0)$ with

$$I_1^{(2)}(\Delta_0) = \int d\omega\, I^{(1)}(\omega)\alpha(\omega)\mathrm{Re}\left[P_0 R_-(\omega)^2 + P_1 R_+(\omega)^2\right], \tag{15}$$

$$I_2^{(2)}(\Delta_0) = -I^{(1)}(\Delta_0)\int d\omega\, \mathrm{Re} \sum_{\sigma=\pm} \alpha^\sigma(\omega)R_\sigma(\omega)^2, \tag{16}$$

$$I_3^{(2)}(\Delta_0) = -\frac{\partial I^{(1)}(\Delta_0)}{\partial \Delta_0}\int d\omega\, \mathrm{Re} \sum_{\sigma=\pm} \alpha^\sigma(\omega)R_\sigma(\omega). \tag{17}$$

Here, $I^{(1)}(\Delta_0)$ is the sequential tunneling result (i.e. Eq. (12) with $A^{(0)}(\omega) = \delta(\omega - \Delta_0)$). The poles at $\omega = \Delta$ are regularized in a natural way (it comes out of our theory and is *not* added by hand) as Cauchy's principal values and their derivative.

In the Coulomb blockade regime, only the first term of $I_1^{(2)}$ contributes. At $T = 0$, the integrand is zero at the poles, and we can omit the term $+i0^+$. This gives the well-known result of inelastic cotunneling[7]. At finite temperature, however, the regularization scheme is needed which is not provided by previous theories. Our result is also well-defined for $T \neq 0$.

Furthermore, we are able to describe the system at resonance. In this regime, $I_2^{(2)}$ and $I_3^{(2)}$ become important. The origin of the second term may intuitively be interpreted as the reduction of the first order contribution $I^{(1)}(\Delta_0)$ since quantum fluctuations lead to an occupation of the adjacent

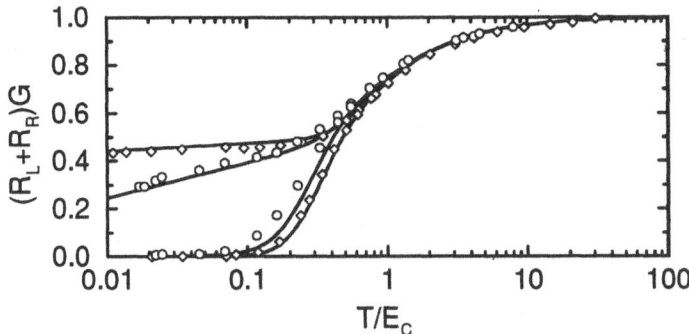

Figure 2. The sequential plus cotunneling contribution to the maximal and minimal linear conductance for $E_C = 1.47K$ and $\alpha_0 = 0.015$, and $E_C = 1K$ and $\alpha_0 = 0.063$. The data points are experimental data from Ref. [5].

charge states $n = -1$ and 2. Therefore, the probability of the system to be in state $n = 0$ or 1 is decreased. The third term may indicate the appearance of a renormalization of the excitation energy Δ_0 [10, 13, 8, 12]. Due to this renormalization the system is effectively "closer" to the resonance as the original parameters would suggest. The current would then, in second order, be roughly given by the derivative of the first order term times the renormalization.

In Fig. 2 we compare our results with recent experiments [5]. The temperature dependence of the Coulomb oscillations were measured for two samples with different conductances. For one with $\alpha_0 = 0.015$, our results in second-order perturbation theory agree perfectly in the whole temperature and gate voltage range. Also for the other sample with $\alpha_0 = 0.063$ the agreement is very good.

For still stronger tunneling higher-order effects are relevant. To describe this regime we include processes of arbitrary high order, but we restrict ourselves to matrix elements of the density matrix which are at most two-fold off-diagonal [13]. In this case we can evaluate – in a conserving approximation – the irreducible self-energy analytically. The following results are derived in this limit.

We find $P_0 = \lambda_-$ and $P_1 = \lambda_+$ with $\lambda_\pm = \int d\omega \, \alpha^\pm(\omega)|\pi(\omega)|^2$ and

$$\pi(\omega) = [\omega - \Delta_0 - \sigma(\omega)]^{-1} \quad , \quad \sigma(\omega) = \int d\omega' \frac{\alpha(\omega')}{\omega - \omega' + i0^+} . \quad (18)$$

Again, the current is given by Eq. (12), but the spectral density becomes

$$A(\omega) = \frac{\alpha(\omega)}{[\omega - \Delta_0 - \mathrm{Re}\,\sigma(\omega)]^2 + [\mathrm{Im}\,\sigma(\omega)]^2} . \quad (19)$$

The following results depend on the parameter

$$\alpha_t = \sum_r \alpha_{t,r} = \frac{h}{4\pi^2 R_t}, \qquad (20)$$

which also defines the parallel tunneling conductance $1/R_t = \sum_r 1/R_r$. In lowest order in α_t we have $A^{(0)}(\omega) = \delta(\omega - \Delta_0)$, and the classical result is recovered. In general, quantum fluctuations yield energy renormalization and broadening effects, which enter in the spectral density via the complex self-energy $\sigma(\omega)$ given in Eq. (18). In order to evaluate $\sigma(\omega)$ we introduce a Lorentzian cut-off which we choose equal to E_C (since the energy difference to charge states which are not taken into account here is of the order of the charging energy). In this case we find

$$\text{Re}\,\sigma(\omega) = -\sum_r \alpha_{t,r}(\omega - \mu_r)\left[2\ln\left(\frac{E_C}{2\pi T}\right) - 2\text{Re}\,\Psi\left(i\frac{\omega - \mu_r}{2\pi T}\right)\right] \qquad (21)$$

and $\text{Im}\,\sigma(\omega) = -\pi\alpha(\omega)$. The effect of the quantum fluctuations can be estimated from the spectral density in the limits $T \gg eV, |\omega|$ or $eV \gg T, |\omega|$. Then, the spectral density is

$$A(\omega) = \frac{Z^2\alpha(\omega)}{[\omega - Z\Delta_0]^2 + [\pi Z\alpha(\omega)]^2}, \qquad (22)$$

with

$$Z^{-1} = 1 + 2\alpha_t \ln(E_C/\max\{eV/2, 2\pi T\}). \qquad (23)$$

We observe a renormalization of Δ_0 and α_t by Z and a broadening given by $\pi Z\alpha(\omega)$. From this result we conclude that lowest order perturbation theory is sufficient for $\alpha_t \ln(E_C/\max\{eV/2, 2\pi T\}) \ll 1$. At larger values, our results for resonant tunneling show clear deviations from sequential tunneling.

A pronounced signature of quantum fluctuations is contained in the differential conductance $G = \partial I/\partial V$. In Figs. 3 and 4 we present our results for the differential conductance in the linear response regime ($V = 0$). They clearly display the effect of resonant tunneling:

– For comparison, we show on the left hand side of Fig. 3 plots which are obtained from the master equation description of sequential tunneling,

$$\frac{G(T, n_g)}{G_{as}} = \frac{\sum_n \exp\left[-\frac{E_C}{T}(n - n_g)^2\right] \dfrac{\frac{E_C}{T}(1+2(n-n_g))}{\exp\left[\frac{E_C}{T}(1+2(n-n_g))\right]-1}}{\sum_n \exp\left[-\frac{E_C}{T}(n - n_g)^2\right]}. \qquad (24)$$

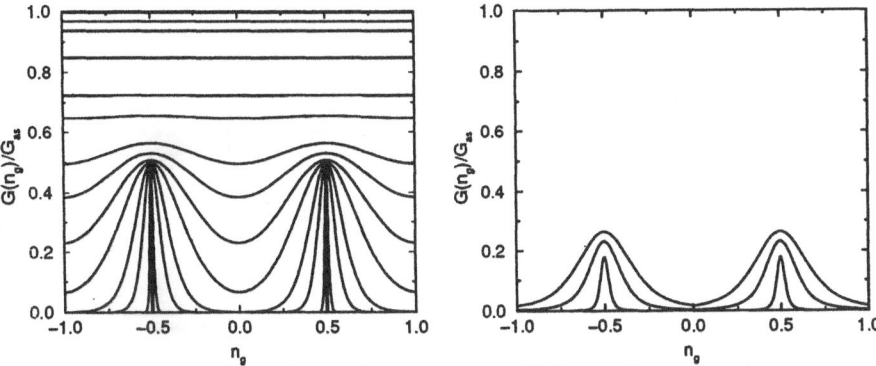

Figure 3. The linear differential conductance normalized to the high temperature limit. Left hand side: result from a master equation in lowest order perturbation theory with $T/E_C = 0.01, 0.05, 0.1, 0.2, 0.3, 0.4, 0.5, 0.75, 1, 2, 5,$ and 10. In this limit the scaled conductance is independent of α_t. Right hand side: result of resonant tunneling with $\alpha_t = 0.2$ and $T/E_C = 0.01, 0.05,$ and 0.1.

The asymptotic high-temperature conductance is $G_{as} = 1/(R_L + R_R)$. At low temperatures, when processes involving only two charge states dominate, the maximal classical conductance saturates at one half of the asymptotic conductances at high temperatures. The width of the peaks scale linearly with temperature.

– The situation changes when resonant tunneling processes are taken into account (see the plots on the right hand side of Fig. 3). The maximal conductance and the peak width are renormalized by Z and Z^{-1} which depend logarithmically on temperature. For this reason, the conductance peak does not reach one half of the high temperature limit and decreases with lower temperatures, while the peak width is increased compared to the lowest order perturbation theory result. For an estimate of the maximal conductance, we use can the spectral density in the form of Eq. (22) and perform the integral Eq. (12) analytically,

$$\frac{G_{\max}(T)}{G_{as}} \approx Z \left[\frac{1}{2} - \frac{1}{\pi} \arctan \left(\frac{(\pi Z \alpha_t)^2 - 1}{2 \pi Z \alpha_t} \right) \right]. \tag{25}$$

(The results shown in Fig. (4), however, were obtained by numerical analysis based on Eq. (19).)

Recent experiments [5, 6] in systems with junctions with small barriers show, indeed, a broadening and decreasing height of the linear conductance peaks, which cannot be explained by thermal smearing and qualitatively agrees with our theory.

118

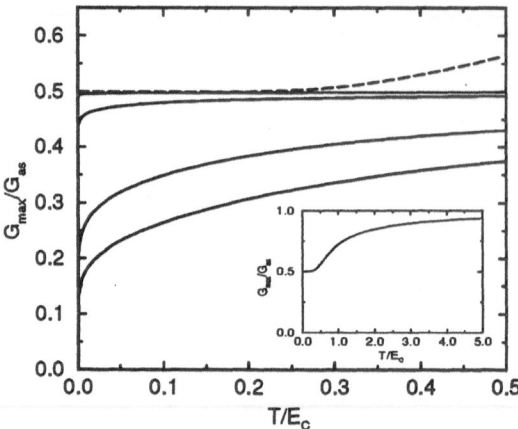

Figure 4. The maximum linear differential conductance normalized to the high temperature limit for $\alpha_t = 0.001, 0.01, 0.1, 0.2$ (from top to bottom). For comparison we also show the result obtained from lowest order perturbation theory (dashed line and inset).

The effects of quantum fluctuations are even more pronounced in the nonlinear differential conductance when the transport voltage dominates over temperature. In Fig. 5 we compare the results of perturbation theory and resonant tunneling at $T = 0$ assuming that for $eV < 2E_C$ only two charge states $n = 0, 1$ are involved.
– The sequential tunneling result for a symmetric transistor with $\alpha_{t,L} = \alpha_{t,R}$ and $C_L = C_R$ is

$$\frac{G(V, n_g)}{G_{as}} = 2\frac{E_C^2(1 - 2n_g)^2 + (eV)^2/4}{(eV)^2} \; \Theta\left(\frac{eV}{4E_C} - \left|n_g - \frac{1}{2}\right|\right) . \qquad (26)$$

As a function of n_g it shows a series of structures of width CV/e with vertical steps at its edges. The width scales linearly with bias voltage.
– Resonant tunneling leads to a renormalization of the height and width by Z and Z^{-1} respectively, which depends now logarithmically on the voltage (see Fig. 5). For this reason, the height of the structure is below the sequential tunneling result and further decreases at lower voltages, while the width is enhanced. Furthermore, the sharp edges are smeared out even in the absence of thermal fluctuations (since $T = 0$).

4. Strong tunneling

If the junction conductance is high and hence the fluctuations in the charge are strong the phase representation outlined above is a more suitable start-

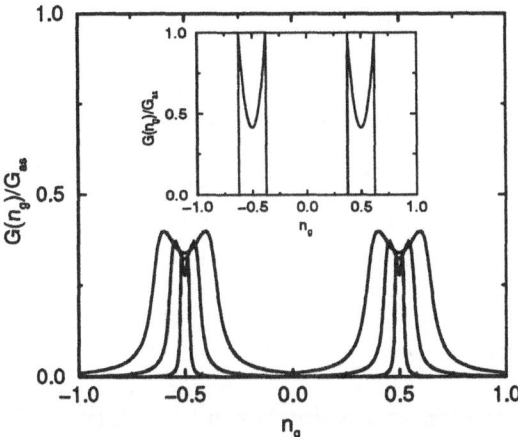

Figure 5. The normalized nonlinear differential conductance for $\alpha_t = 0.1$ and $eV/E_C = 0.05, 0.2, 0.5$ at zero temperature. The inset shows the result from a master equation in lowest order perturbation theory for $eV/E_C = 0.5$.

ing point for the analysis of the problem. It turns out that the dimensionless conductance appears in the form

$$\tilde{\alpha}_t = \frac{h}{4e^2 R_t} = \pi^2 \alpha_t , \qquad (27)$$

which differs from the expansion parameter α_t of the weak tunneling expansion by a factor π^2. The real-time path-integral technique discussed above provides an expression for the reduced density matrix $\rho(\varphi_1, \varphi_2)$. It has been analyzed by Golubev et al. [26, 27] in a quasiclassical approximation, which is sufficient in the limit

$$\max\{eV, T\} \gg w_0 = \frac{2\tilde{\alpha}_t E_C}{\pi^2} \exp(-2\tilde{\alpha}_t + \gamma), \qquad (28)$$

(here $\gamma = 0.5772...$ is Euler's constant). In this case the linear conductance of the SET transistor becomes

$$\frac{G(T)}{G_{\text{as}}} = 1 - f(T) - g e^{-F(T,0)} \cos\left(2\pi n_g\right) , \qquad (29)$$

where $g = 1.22/\tilde{\alpha}_t + 11.29$, and $\tilde{\alpha}_t = h/(4e^2 R_\Sigma)$ with $1/R_\Sigma = 1/R_L + 1/R_R$. Furthermore we introduced

$$f(T) = \frac{1}{2\tilde{\alpha}_t} \left[\gamma + \frac{2\tilde{\alpha}_t E_C}{\pi^2 T} \Psi'\left(1 + \frac{2\tilde{\alpha}_t E_C}{\pi^2 T}\right) + \Psi\left(1 + \frac{2\tilde{\alpha}_t E_C}{\pi^2 T}\right)\right] , \qquad (30)$$

$$F(T,0) = 2\tilde{\alpha}_t \left[1 + 2\ln\left(\frac{2\tilde{\alpha}_t E_C}{\pi^2 T}\right) - 2\Psi\left(1 + \frac{2\tilde{\alpha}_t E_C}{\pi^2 T}\right)\right] + \frac{2\pi^2 CT}{e^2} \ . \quad (31)$$

Here $\Psi(x) = \Gamma'(x)/\Gamma(x)$ is the digamma function.

These results are displayed in Fig. 6 in the temperature range $T \gtrsim 10w_0$, where we estimate the approximations used above to be justified. In the high-temperature limit the conductance becomes independent of the gate charge, but due to charging effects it is still reduced below the asymptotic value by

$$\frac{G(T)}{G_{as}} = 1 - \frac{E_C}{3T} + \frac{6\zeta(3)}{\pi^4}\tilde{\alpha}_t\left(\frac{E_C}{T}\right)^2 - \dots \quad (32)$$

For high temperatures this expression is valid for all (including small) values of $\tilde{\alpha}_t$. The first nontrivial term in this expansion does not depend on $\tilde{\alpha}_t$. At lower temperatures the conductance is further suppressed by charging effects and modulated by the gate charge Q_g. In the figure the minimum and maximum conductance values are presented corresponding to $Q_g = 0$ and $Q_g = e/2$, as well as the Q_g-averaged conductance. The modulation with Q_g becomes more pronounced as the temperature is lowered, however, it is exponentially suppressed with increasing $\tilde{\alpha}_t$ (cf. Fig. 6). For $\tilde{\alpha}_t \gtrsim 4$ the modulation effect can hardly be resolved while the overall suppression of the system conductance G is very pronounced.

In a number of earlier papers [16, 18, 19, 12] the combination of charging and strong tunneling effects in metallic junctions has been analyzed within imaginary time approaches. In the limit of strong tunneling, $\tilde{\alpha}_t \gg 1$, a renormalization group equation for $\tilde{\alpha}_t$ can be derived [16, 2]

$$d\tilde{\alpha}_t / d\ln\omega_c = \beta(\tilde{\alpha}_t) \ , \quad (33)$$

where in the lowest order in $\tilde{\alpha}_t$ one has $\beta(\tilde{\alpha}_t) = 1/2$. Already this scaling approach captures the tendency of the effective junction conductance to decrease with decreasing T due to charging effects. In order to see that one should proceed with scaling from $\omega_c \sim E_C$ to $\omega_C \sim T$ and identify the (dimensionless) junction conductance with the renormalized value $\tilde{\alpha}_t(\omega_c \sim T)$. This approach is sufficient for strong tunneling at high temperatures, namely if the final renormalized tunneling conductance still satisfies $\tilde{\alpha}_t(\omega_c \sim T) \gtrsim 1$. In general the strong tunneling approach may lead to a small renormalized conductance such that (33) ceases to be valid. For weak tunneling other scaling approaches, derived in an expansion in the tunneling conductance and equivalent to what we described in Section III, can be applied. In this situation, Falci et al. [12] suggested a 2-stage scaling procedure, where the renormalized conductance after the strong tunneling rescaling was used as an entry parameter for the weak tunneling scaling.

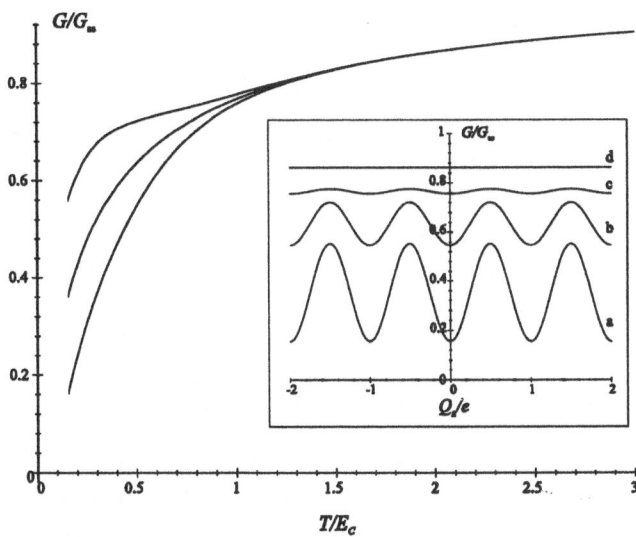

Figure 6. Maximum and minimum linear conductance of a SET transistor as a function of temperature obtained from the Langevin equation analysis (eq. (29)) for $\tilde{\alpha}_t = 2$. The intermediate curve shows the linear conductance averaged over all values of the gate charge. Inset: Conductance as a function of the gate charge for the same $\tilde{\alpha}_t$ at different temperatures $T/E_C = 0.15$ (a), 0.5 (b), 1 (c) and 2 (d).

Various theoretical approaches led to the conclusion that the strong electron tunneling $\tilde{\alpha}_t \gg 1$ reduces the charging energy, i.e. the effective capacitance is renormalized. Panyukov and Zaikin [18] treated the problem by means of instanton techniques. They concluded that electron tunneling affects both the scale and the functional dependence of the ground state energy $E(Q_g)$. At not too low temperatures $T \gtrsim w$ they find

$$E(Q_g) = -\frac{w}{2}\cos(\frac{2\pi Q_g}{e}) \tag{34}$$

with [18]

$$w = \frac{32\tilde{\alpha}_t E_C}{\pi^2}\exp(-2\tilde{\alpha}_t + \gamma). \tag{35}$$

A similar result, differing only in the numerical coefficient, has been obtained in a semiclassical analysis of the effective action [12].

With the aid of relations (34), (35) we can derive the first order correction in $1/\tilde{\alpha}_t$ in the renormalization group equation (33)

$$\beta(\tilde{\alpha}_t) = 1/2 + 1/4\tilde{\alpha}_t. \tag{36}$$

This result has been first derived in [18] by means of the instanton technique [30] and was very recently rederived by a two-loop RG calculation [29]. Note, that the RG procedure based on the equations (33,36) can be applied only at temperatures and/or energies above the typical scale $\tilde{\alpha}_t E_C \exp(-2\tilde{\alpha}_t)$. At lower energies the renormalized value of $\tilde{\alpha}_t$ becomes of order one and the expansion of the β-function in $\tilde{\alpha}_t$ breaks down. Thus – in contrast to the statement in Ref. [29] – the value of the renormalized capacitance at low T can not be derived from the RG analysis (33) and (36) [31].

At lower temperatures the form of the lowest energy band $E(Q_g)$ turns out to be even more complicated [18, 12] and the $\tilde{\alpha}_t$-dependence of the prefactor of the expression for w changes from linear in $\tilde{\alpha}_t$ for $T > w$ to quadratic in $\tilde{\alpha}_t$ for $T = 0$. Instanton techniques [18] yield

$$E_{C,\text{eff}} = 16\tilde{\alpha}_t^2 E_C \exp(-2\tilde{\alpha}_t + \gamma) . \tag{37}$$

The exponential dependence on $\tilde{\alpha}_t$ for sufficiently large $\tilde{\alpha}_t$ has been confirmed by renormalization group arguments [16, 2, 12] as well as Monte Carlo methods [12, 20, 29], see Figs. 7 and 8 where the results of instanton calculations [18, 21] were plotted together with the Monte Carlo data [20, 29]. The prefactor remains a point of controversial discussions in the literature (cf. [18] and [21], see also Figs. 7 and 8). Irrespective of this detail an important consequence of the strong capacitance renormalization for $\alpha_t \gtrsim 1$ is the exponential reduction of the temperature range where charging effects are observable.

A consequence of the renormalization group approach (33) has been pointed out in Ref. [18]. It relies on the *assumption* that the system linear conductance is determined by the renormalized value $\tilde{\alpha}_t(\omega_c \sim T)$ as

$$G = \frac{2e^2}{\pi\hbar} \tilde{\alpha}_t(\omega_c \sim T) . \tag{38}$$

Combining the above scaling approach, the high temperature expansion (32) (with $\kappa = 0$), and the expression (36) for β to first order in $1/\alpha_t$ we get for the Q_g-averaged conductance

$$\frac{G}{G_{\text{as}}} = 1 - \frac{1}{2\tilde{\alpha}_t} \left\{ \ln\left(1 + \frac{4\tilde{\alpha}_t^2}{3(1+2\tilde{\alpha}_t)} \frac{E_C}{T}\right) \right.$$
$$\left. + \ln\left[1 + \frac{1}{2\tilde{\alpha}_t} \ln\left(1 + \frac{4\tilde{\alpha}_t^2}{3(1+2\tilde{\alpha}_t)} \frac{E_C}{T}\right)\right] \right\} . \tag{39}$$

Although the above scaling approach to the conductance calculation is intuitively attractive (and the result (39) fits reasonably with the available experimental data [5, 6]) it has to be stressed that it depends on the unproven assumption (38).

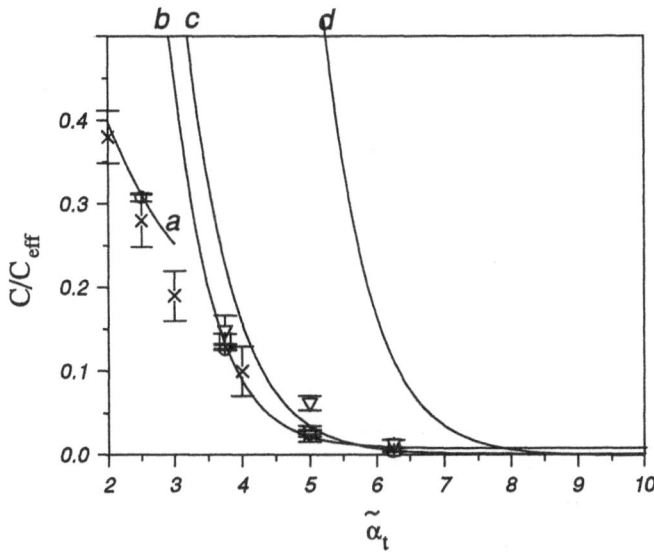

Figure 7. Comparison between various theoretical predictions and Monte Carlo data for the renormalized capacitance C_{eff} in the strong tunneling regime $\tilde{\alpha}_t \gtrsim 2$. The Monte Carlo data by Hofstetter and Zwerger [29] are denoted by crosses. The data obtained by Wang *et al.* [20] at $E_C/T = 100$, 200 and 500 are indicated respectively by circles, squares and triangles. At $\tilde{\alpha}_t = 5$ the data points [20] for $E_C/T = 100$ and 200 essentially coincide with that of [29]. The curve (a) shows the perturbative result by Grabert [11]. For all values of $\tilde{\alpha}_t < 2$ (not shown here) this result agrees well with the Monte Carlo data [20] and [29]. The curves (b), (c) and (d) indicate the results obtained by means of the instanton technique at sufficiently large $\tilde{\alpha}_t$: the results by Panyukov and Zaikin [18] obtained at sufficiently high T (the curve (b) – eqs. (34), (35)) and at $T = 0$ (the curve (c) – eq. (37) of this paper), as well as the $T = 0$ results of Wang and Grabert (the curve (d) - eq. (10) of [21]).

In contrast, the real-time path-integral techniques presented here are free from this ambiguity and allow for a direct evaluation of the *I-V* characteristics and the system conductance. We note, furthermore, that the results obtained within the real and imaginary time methods are consistent with each other. E.g. the renormalization of the effective energy difference between the two lowest charge states, derived in Ref. [12], is contained in the self-consistent solution presented in Section 3. Furthermore, comparing the expressions for w_0 (28) and the bandwidth w (35) we immediately see that these two parameters coincide up to a numerical coefficient: $w = 16w_0$. This means the requirement for the validity of the quasiclassical Langevin equation (28) roughly coincides with the requirement that the temperature (or voltage) is larger than the effective bandwidth w.

Still no quantitative theory for the conductance at lower temperatures and not too low values $\tilde{\alpha}_t \gtrsim 1$ has been provided. Although the two limiting

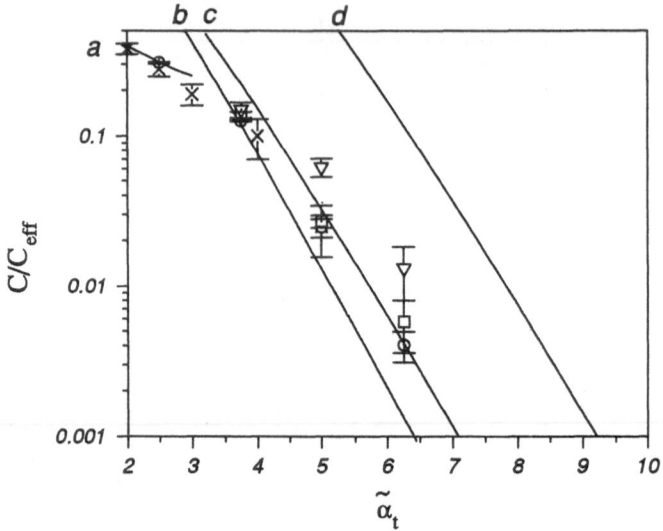

Figure 8. The data of Fig. 7 on a logarithmic scale.

descriptions presented here do not allow for a quantitative description of this parameter range it satisfactory to notice that both show the same qualitative trend in this range.

Another question of interest is the conductance at very large $\tilde{\alpha}_t \gg 1$ and very low $T \lesssim w_0$. In the limit $\tilde{\alpha}_t \gg 1$ the conductance oscillations with Q_g are exponentially small (cf. (29)). Then for all Q_g from (29) we have

$$G(T \approx w_0)/G_{as} \simeq b/\tilde{\alpha}_t, \quad b \sim 1. \tag{40}$$

Thus we can *conjecture* that the low temperature maximum conductance of a SET transistor is *universal* in the limit of large $\tilde{\alpha}_t$ being of the order of the inverse quantum resistance unit $2e^2/\pi\hbar$. This conjecture is also consistent with the scaling analysis of Refs. [16, 18, 12] combined with the results of Section 3. Starting from large $\tilde{\alpha}_t \gg 1$ we first use the renormalization group procedure (33,36) which should be cut at $\tilde{\alpha}_t(\omega_c) \sim 1$. In the second stage we expand in $\alpha_t \approx 1/\pi^2$ as described in Section 3 – starting with the renormalized value instead of the bare one. Apart from logarithmic corrections we thus arrive at the maximum conductance of order of the inverse quantum resistance, no matter how large the initial conductance is.

5. Conclusions

In this paper we have described single-electron tunneling in systems with strong charging effects beyond perturbation theory in the tunneling conductance. For this purpose we considered the real-time evolution of the reduced density matrix of the system. We presented two approximation schemes:

In the first part, valid for not too strong tunneling, $\alpha_t \lesssim 1$, we presented a systematic diagrammatic expansion, which allowed us to identify the different contributions, sequential tunneling, inelastic cotunneling and inelastic resonant tunneling. When we restricted ourselves to diagrams corresponding to maximally two-fold off-diagonal matrix elements of the density matrix we can formulate a self-consistent resummation of diagrams. At low temperatures we, furthermore, can restrict our attention to two consecutive charge states. In this limit, there exist no crossing diagrams, and we can evaluate the summation in closed form. The most important results are a renormalization of system parameters and a life-time broadening of the conductance peaks. These two approximations are justified for tunneling conductances satisfying $\alpha_t \ln\left(E_C/\max\{eV/2, 2\pi T\}\right) \lesssim 1$ and allow for a qualitative analysis of the system conductance also for larger values of α_t.

In the second part of the paper we developed an alternative approach based on quasiclassical Langevin equations for the junction phase φ. This approach assumes that fluctuations of the phase are small and that the noise can be treated perturbatively. This is a suitable approximation for large values $\tilde{\alpha}_t = \pi^2 \alpha_t$ or in the high temperature limit. For weak tunneling $\tilde{\alpha}_t \lesssim 1$ this scheme turns out to be justified only for high temperatures and/or voltages $\max(T, eV) \gg E_C$, whereas for stronger tunneling, $\tilde{\alpha}_t \gtrsim 1$, phase fluctuations are substantially suppressed. The results derived in this approach are valid, provided $\max\{T, eV\} \gg \tilde{\alpha}_t E_C \exp(-2\tilde{\alpha}_t)$. This range expands rapidly with increasing $\tilde{\alpha}_t$.

In conclusion, we found an effective action description of a single-electron transistor. We analyzed it in two limits. The charge representation, which is valid as long as $\alpha_t \lesssim 1$, provides the basis for a systematic diagrammatic description of coherent tunneling processes including resonant tunneling. The phase representation is suitable at large values of $\tilde{\alpha}_t \gtrsim 1$. In both cases we calculated the gate-voltage and temperature-dependent conductance of a single electron transistor. The dimensionless parameters in the two limits differ by a factor $\pi^2 \alpha_t = \tilde{\alpha}_t$. As a result the range of validity of the two approaches overlaps and, at least qualitatively, the two approaches cover the whole range of parameters.

Acknowledgement

The authors are grateful to D. Esteve, G. Falci, D.S. Golubev and G.T. Zimanyi for useful discussions. We thank the members of the Saclay group for sending us their data prior to publication. The project was supported by the DFG within the research program of the Sonderforschungbereich 195 and by INTAS-RFBR Grant No. 95-1305.

126

References

1. D. V. Averin and K. K. Likharev, in *Mesoscopic Phenomena in Solids*, B. L. Altshuler, P. A. Lee and R. A. Webb, eds., p. 173 (Elsevier, Amsterdam, 1991).
2. G. Schön and A.D. Zaikin, Phys. Rep. **198**, 237 (1990).
3. *Single Charge Tunneling*, NATO ASI Series, Vol. **294**, edited by H. Grabert and M.H. Devoret, (Plenum Press), 1992.
4. Proceedings of the NATO ARW *Mesoscopic Superconductivity*, Physica B **203**, Nos. 3, 4 (1994), edited by F.W.J. Hekking, G. Schön and D.V. Averin.
5. P. Joyez, V. Bouchiat, D. Estève, C. Urbina, and M.H. Devoret, submitted to Phys. Rev. Lett.
6. D. Chouvaev *et al.*, in preparation.
7. D.V. Averin and Yu.V. Nazarov, in Ref. [3].
8. K.A.Matveev, Sov. Phys. JETP 72, 892 (1991).
9. D.S. Golubev and A.D. Zaikin, Phys. Rev. B **50**, 8736 (1994); A.D. Zaikin, D.S. Golubev, and S.V. Panyukov, in Ref. [4].
10. H. Schoeller and G. Schön, Phys. Rev. B **50**, 18436 (1994), and also in Ref. [4].
11. H. Grabert, Phys. Rev. B **50**, 17364 (1994).
12. G. Falci, G. Schön and G. T. Zimanyi, Phys. Rev. Lett. **74**, 3257 (1995), and also in Ref. [4].
13. J. König, H. Schoeller, and G. Schön, Europhys. Lett. **31**, 31 (1995); J. König, H. Schoeller, G. Schön, and R. Fazio, in *Quantum Dynamics of Submicron Structures*, eds. H. A. Cerdeira *et al.*, NATO ASI Series E, Vol. **291** (Kluwer, Dordrecht) 1995, p.221.
14. J. König, H. Schoeller, and G. Schön, Phys. Rev. Lett. **78**, 4482 (1997)
15. J. König, H. Schoeller, and G. Schön, Phys. Rev. Lett. **76**, 1715 (1996); J. König, J. Schmid, H. Schoeller, and G. Schön, Phys. Rev. B **54**, 16820 (1996).
16. F. Guinea, and G. Schön, Europhys. Lett. **1**, 585 (1986); J. Low Temp. Phys. **69** 219 (1987).
17. A.D. Zaikin, and S.V. Panyukov, Zh. Eksp. Teor. Fiz. **94**, 172 (1988) [Sov. Phys. JETP **67**, 2487 (1988)]; J. Low Temp. Phys. **73**, 1 (1988).
18. S.V. Panyukov, and A.D. Zaikin, Phys. Rev. Lett. **67**, 3168 (1991).
19. A.D. Zaikin and S.V. Panyukov, Phys. Lett. A **183**, 115 (1993).
20. X. Wang, R. Egger, and H. Grabert, Europhys. Lett., **38**, 545 (1997); see also X. Wang, thesis (Freiburg, 1996).
21. X. Wang, and H. Grabert, Phys. Rev. B **53**, 12621 (1996).
22. U. Eckern, G. Schön, and V. Ambegaokar, Phys. Rev. B **30**, 6419 (1984).
23. A.O. Caldeira, and A.J. Leggett, Ann. Phys. (N.Y.) **149**, 374 (1983).
24. A. Schmid, J. Low Temp. Phys. **49**, 609 (1982).
25. D.S. Golubev, and A.D. Zaikin, Phys. Rev. B **46**, 10903 (1992); Phys. Lett. A **169**, 337 (1992).
26. D.S. Golubev, and A.D. Zaikin, Zh. Eksp. Teor. Fiz. Pis'ma Red. **63**, 953 (1996) [JETP Lett. **63**, 1007 (1996)].
27. D.S. Golubev, J. König, H. Schoeller, G. Schön, and A.D. Zaikin, submitted to Phys. Rev. B.
28. See also an earlier paper by A.A. Odintsov, Zh. Eksp. Teor. Fiz. **94**, 312 (1988) [Sov. Phys. JETP **67**, 1265 (1988)] where the analogy with the polaron problem has been exploited and a similar approximation has been made.
29. W. Hofstetter, and W. Zwerger, Phys. Rev. Lett. **78**, (1997).
30. In Ref. [18] the function β was derived for a more general case of a tunnel junction interacting with a linear Ohmic dissipative environment.
31. Also at high temperatures, the RG analysis alone appears to be insufficient to determine the renormalized capacitance unambiquously, because the corresponding RG equations do not depend on the gate charge.

SPIN TEXTURES IN QUANTUM DOTS

J. H. OAKNIN, B. PAREDES AND C. TEJEDOR
Departamento de Física Teórica de la Materia Condensada,
Universidad Autónoma de Madrid,
Cantoblanco, 28049, Madrid, Spain.

AND

L. MARTíN-MORENO
Departamento de Física de la Materia Condensada, ICMA-
CSIC, Universidad de Zaragoza,
Zaragoza 50015, Spain.

1. Abstract

We present a microscopic analysis of spin textures in quantum dots in the presence of high magnetic fields. For filling factor close to 1. our main result is a set of analytical many-body wave-functions of spin excitations which describe all of the necessary quantum numbers. These states present topological structure of different spin textures, some of which are located at the bulk, and others at the edge. Bulk charged skyrmions can be expressed as a condensate of spin excitons interacting via a two-body repulsive interaction. The size of the skyrmion is given by the number of excitons present in the condensate. Edge spin textures turn out to be non interacting bosons located at the boundary of the dot. A branch of these excitations starts with lower energy than the branch of polarized charge edge excitations. When the number of electrons is of the order of a few tenths, edge spin textures are responsible for the edge reconstruction of **the droplet.**

2. Introduction

For a long time, integer and fractional Quantum Hall effect (QHE) where consider as due to two different physical mechanisms producing the uncompressibility of a two dimensional (2D) electron gas under the action of a high magnetic field B. Fractional QHE can be only due to electron-electron

N. García et al. (eds.), Nanoscale Science and Technology, 127–136.

interaction within a Landau level. On the contrary, integer QHE was understood simply in a single particle picture due to the quantization of the kinetic energy by the magnetic field. The energy gap of a 2D electron gas at filling factor $\nu = 1$ was attributed to Zeeman and exchange energies required to produce one spin-flip transition. The fact that the experimental gap[1] resulted to be much smaller than the theoretical one given by such description was usually attributed to disorder effects. Over the last years, a complementary and interesting many-body effect reducing the energy gap has been studied by the appropriate use of field theories named classical nonlinear-σ model (CNLSM)[2, 3] and Hartree-Fock (HF) approximations [4, 5] which predict the existence of a spin texture excitation, known as skyrmion, carrying topological and real charge equal to one. Recently, the interest in skyrmions has increased due to the experimental observation of several properties of these spin-textures.[6, 7, 8, 9, 10] For a g-Lande factor equal to zero, the skyrmion is infinite in size and lowers the transport energy gap up to a factor of two. That gain is reduced for increasing g and the skyrmion becomes localized. Therefore, in real cases with non zero g-factor, skyrmions can exist in finite size systems as a quantum dot (QD). On top of the bulk spin-textures appearing in the infinite case, edge spin-textures are also possible in a QD. They can play a crucial role in lateral transport experiments[11, 12, 13, 14] due to their high overlap with electronic wavefunctions in the leads. In vertical transport experiments[15, 16] bulk and edge excitations can be competitive in importance with each other.

CNLSM and HF pictures are a reasonable starting point but they can not be good descriptions of skyrmions in a QD because they do not give wavefunctions with well defined quantum numbers, something crucial in such a small system. Due to the high symmetry, the modulus (S^2) and the third component (S_z) of the spin, and third components of both the total (M) and the center-of-mass (M_{CM}) angular momenta must be well defined. In the same spirit in which BCS wave-functions for superconductors are combined to obtain states with a well defined number of particles, a linear superposition of the mean-field wave-functions can be made so that M and S_z are well defined[17]. The other two magnitudes M_{CM} and S^2 turn out to be good quantum numbers too if some conditions are imposed to the eigenstates obtained from the unitary transformation[18].

We analyze both bulk and edge spin-textures in a system of N electrons moving in 2D, in the presence of both a high perpendicular magnetic field B and a parabolic confinement potential. Analytical many-body wavefunctions of spin excitations with all quantum numbers (M, M_{CM}, S^2 and S_z) properly described, are obtained. This microscopic description including quantum fluctuations is directly applicable up to a few thousand electrons which means rather large QD. We obtain that skyrmions can be expressed as a condensate of spin excitons interacting via a two-body repulsive interaction. The number of excitons present in the condensate controls the size of the skyrmion. We compute the skyrmion energy and spin as a function of Zeeman energy.

As far as edge spin-textures are concerned, the interesting questions are: first to know whether these edge non-polarized excitations are lower or higher in energy than the well known (spin polarized) edge charge excitations[19, 20] and second, to investigate which of them is responsible for the edge reconstruction of a QHD at $\nu = 1$ when increasing the magnetic field. For GaAs, the branch of edge spin-textures starts with lower energy than the branch of polarized excitations. If the number of electrons is not large (≤ 100) there are no crossings of the two branches and edge spin textures are responsible for the edge reconstruction of the QHD as suggested by recent experiments.[13] On the contrary, since crossings among the spin and charge branches appear for larger number of electrons, edge reconstruction seems to be due to polarized charge excitations in the large N case.

3. Spin-textures in a QD

We consider interacting electrons moving in the xy plane, confined by a parabolic potential characterized by a bare frequency ω_0. The system is in the presence of an external magnetic field. We work in the high field regime in which it is possible to project the Hamiltonian H onto the lowest Landau level (LLL) with up and down spins. In the symmetric gauge,

$$H = H^{SP} + \frac{1}{2} \sum_{m_i, \sigma_j} V_{m_1 m_2 m_3 m_4} c^\dagger_{m_1, \sigma_1} c^\dagger_{m_2, \sigma_2} c_{m_3, \sigma_2} c_{m_4, \sigma_1} \qquad (1)$$

where $c^\dagger$ and c are the electron creation and annihilation operators respectively, m_i are single-particle (SP) angular momenta and σ_j denote spins. We consider matrix elements $V_{m_1 m_2 m_3 m_4}$ for the Coulomb interaction. The single particle part H^{SP} has a spectrum

$$E^{SP}(M, S_z) = \left(\frac{N}{2} + \frac{\Omega - \omega_c}{2\Omega} M \right) \hbar\Omega + g\mu_B S_z B \qquad (2)$$

where $\omega_c = eB/m^*c$, $\Omega = [\omega_c^2 + 4\omega_0^2]^{1/2}$, and μ_B is the Bohr magneton. Due to the circular symmetry, the spectrum separates in subspaces with well defined M, M_{CM}, S^2, S_z. Since the interaction scales with the magnetic length $l_B = \sqrt{\hbar c/m^*\Omega}$, the energy ordering of states within a subspace does not depend neither on ω_0 nor on B. Only the relative energies in different subspaces change with field or confinement.

In the following paragraphs, we are going to built up trial wave-functions both for skyrmions and edge spin textures. In both cases we have checked the quality of our proposals by computing the overlap of such model states with the ones that are obtained from a numerical diagonalization of the Hamiltonian (1). We are able to perform such diagonalizations up to 20 electrons obtaining an overlap always higher than 0.99 [18]. Therefore we can consider that the trial wave-functions we present give the adequate description of spin textures of QD.

3.1. SKYRMIONS

Although wave-functions for spin textures have been previously obtained [18] in a rather formal way, we intend here to give a simpler picture of them. For that purpose we start with the well known[4] mean-field wave-functions of a skyrmion of charge 1 on top of $N-1$ electrons in a QHF state:

$$|\Psi\rangle = -c_{0\downarrow}^\dagger \prod_{j=0}^{N-2} (u_j c_{j\uparrow}^\dagger + v_j c_{j+1\downarrow}^\dagger)|\phi\rangle \qquad (3)$$

where $|\phi\rangle$ is the vacuum state. This wave-function is not an eigenstate of the spin and it has non-zero expectations values for the three components of $\vec{S}$ so that it can be associated to a classical vector field. In $|\Psi\rangle$, each spin flip implies that the electron increases its angular momentum m in 1. Therefore, in first quantization picture, the N-electrons wave-function is a Slater determinant built up from single particle states which have a spinor as a common factor:[3]

$$\begin{pmatrix} \partial_z \\ \xi e^{i\varphi}/\sqrt{2} \end{pmatrix} \frac{\sqrt{2}\phi_{j+1}}{\sqrt{j+1}} \qquad (4)$$

where $z = x + iy$ and ϕ_{j+1} is a single particle eigenstate in the LLL. The wave-function (4) involves a rotation of 2π of the xy-component of the spin when describing a closed path around the center of the skyrmion. Moreover, S_z changes its sign when going from the center to the edge of the QHD. Those two ingredients imply a topological and real charge 1.[3]

A first step to get states with well defined quantum numbers is to introduce a skyrmion wave-function $|\Psi(\varphi, \xi)\rangle$ depending on two parameters built up from (3) taking:

$$\frac{u_j}{v_j} = \frac{\xi}{\sqrt{j+1}} e^{i\varphi}. \qquad (5)$$

φ is an arbitrary parameter fixing a broken-symmetry direction of the spin vector field in the xy plane, while the parameter ξ controls the expectation value of M and S_z, i. e. the skyrmion size. $S_z^{SK}(r) \sim (r^2 - 2\xi^2)/(r^2 + 2\xi^2)$

at large distances. This is the shape of skyrmions in the CNLSM with a size $\xi/\sqrt{2}$.[4] Remarkably, such dependence of $S_z^{SK}(r)$ holds even in the case of skyrmion sizes comparable with the magnetic length l_B, regime well beyond the range of applicability of the classical model.

In order to have well defined quantum numbers, one must perform the transformation

$$|\Psi_n^{SK}\rangle = -\frac{n!}{2\pi\xi^n}\int_0^{2\pi} d\varphi e^{-in\varphi}|\Psi(\varphi,\xi)\rangle = (\Lambda_1^\dagger)^n|C_{N-1}^1\rangle. \tag{6}$$

where

$$\Lambda_1^\dagger = \sum_m \sqrt{\frac{m!}{(m+1)!}} c_{m+1,\downarrow}^\dagger c_{m,\uparrow} \tag{7}$$

and $|C_{N-1}^1\rangle$ is a compact state with $N-1$ electrons in the lowest energy single particle states ($m = 0, 1, ..., N-2$) with spin up and 1 electron in the lowest energy single particle state ($m = 0$) with spin down. In fact, the function (6) is not an exact eigenstate of S^2 but with the inclusion of an adequate projector[18, 21] it becomes a state with all the quantum numbers well defined. Skyrmions with topological and real charge larger than 1 can be obtained in a similar way [18].

3.2. EDGE SPIN TEXTURES

The procedure above described, allows to obtain an eigenstate by means of the application of the operator $\Lambda_1^\dagger$ to a compact state. A similar picture is possible for building up edge spin textures for a compact state $|C_{N-k}^k\rangle$ [18, 22]. One can obtain [22] a set of states (apart from a normalization constant)

$$(\Sigma_\Delta^\dagger)^n|C_{N-k}^k\rangle \tag{8}$$

Δ is an integer, either positive or negative, and the operators take the form:

$$\Sigma_\Delta^\dagger = \sum_m \sqrt{\frac{(m+\Delta)!}{m!}} c_{m+\Delta,\downarrow}^\dagger c_{m,\uparrow} \qquad \Delta > 0$$

$$\Sigma_\Delta^\dagger = \sum_m \sqrt{\frac{m!}{(m+\Delta)!}} c_{m,\downarrow}^\dagger c_{m+\Delta,\uparrow} \qquad \Delta < 0. \tag{9}$$

Contrary to the previous case, Σ-type operators do not have any restriction with respect to the compact state they can act on. So we can try to understand their physical significance in the simplest case of the action on the fully polarized GS at $\nu = 1$, i.e. $|C_N^0\rangle$. As in the case of skyrmions, mean field wave-functions obtained from these states have a spinor factor which implies states with a xy-spin rotation of $2\pi\Delta$ when moving along a closed path around the center of the QHD. However, the other ingredient necessary to have topological charge Δ is missing. The z-component of the spin is not changing from 1 to -1 when going from the center to the edge of the QHD. Now $S_z(r)$ is practically always positive, and it is only at the edge where it takes on a small negative value in a narrow spatial region.

4. Dispersion relations

4.1. SKYRMIONS

Energies of skyrmion-like states $|\Psi_n^{SK}\rangle$ can be computed by applying Wick's theorem to

$$E_n^{SK} = \frac{\langle \Psi_n^{SK}|H|\Psi_n^{SK}\rangle}{\langle \Psi_n^{SK}|\Psi_n^{SK}\rangle} \tag{10}$$

One obtains [18] the total energy:

$$E_n^{SK} = E^{SP}(M, S_z) + E_0 + \alpha_\Lambda n + \beta_\Lambda \begin{pmatrix} n \\ 2 \end{pmatrix} \tag{11}$$

where E_0 is the interaction energy of the compact state $|C_{N-1}^1\rangle$. α_Λ and β_Λ have analytical expressions in terms of $V_{m_1 m_2 m_3 m_4}$. The most important result is that $\alpha_\Lambda < 0$ and $\beta_\Lambda > 0$, which implies that *skyrmions in a QHD can be represented as confined bosons interacting via a two-body repulsive interaction.* Figure 1 shows the parabolic interaction energy $E_n^{SK} - E^{SP}(M, S_z) - E_0$ of the skyrmion branch for $N = 30$. Notice that we are studying excitations with a particular relation between angular momentum M and spin S_z. Therefore, Eq. (11) gives the energy of a set of excitations with varying spin.

The minimum of the parabola gives the most stable skyrmion. Its size and spin are given by n^{SK}, the integer number of applications of $\Lambda_1^\dagger$ corresponding to the state at the minimum. n^{SK} depends on the g-factor through $E^{SP}(M, S_z)$. The spin of the skyrmion is a quantity already measured by different methods in the case of 2D systems [6, 7, 8, 9, 10].

Experimental information on skyrmions in QD is already available from the studies of the transition from $|C_{N-1}^1\rangle$ to $|C_N^0\rangle$ when the magnetic field is varied. Single electron capacitance[15, 16] and transport[11] experiments in QD have been interpreted[11, 23] in terms of highly correlated ground states appearing in numerical calculations. Those states can be identified [18] as skyrmion-like (Λ-type) excitations.

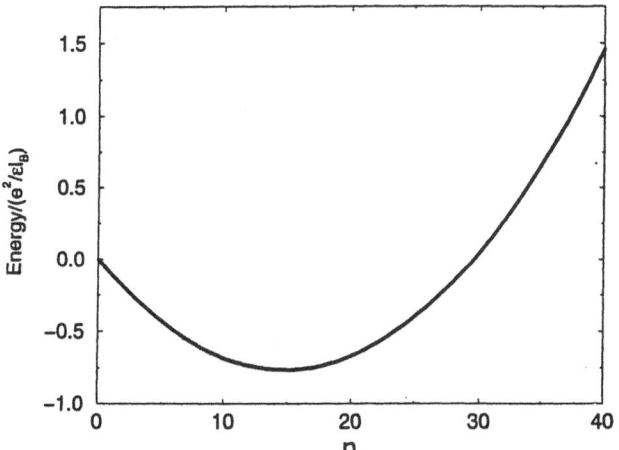

Figure 1. Interaction energy (in units of $e^2/\epsilon l_B$) as a function of n for skyrmions in a QD with 30 electrons.

It must be pointed out that to recover the 2D skyrmions as the thermodinamic limit of QD with infinite size presents some problems which have been carefully discussed [18, 24].

4.2. EDGE SPIN TEXTURES

Since edge spin excitations are similar for all the compact states, we just analyze here the QD at filling factor $\nu = 1$ described by $|C_N^0\rangle$. Wick's theorem can be also applied to calculate the energy of edge spin textures of the type $(\Sigma_\Delta^\dagger)^n$. Although everything is similar to the case of skyrmions, an important difference appears. For each value of the angular momentum Δ, one gets the same expression (11) with E_0 being now the interaction energy of the compact state $|C_N^0\rangle$, $\alpha_\Sigma > 0$ and $\beta_\Sigma \simeq 0$. This means that edge spin textures for each Δ are non interacting bosons with an energy:

$$E_{\Delta,n}^{EST} = E^{SP}(M, S_z) + E_0 + \alpha_{\Sigma_\Delta} n. \tag{12}$$

The dependence of α_{Σ_Δ} on Δ is not very large so that confinement dominates on total energy giving an almost linear dispersion relation as a function of the excess Δ of angular momentum. As shown in Figure 2, edge spin textures coexist with spin-polarized charge magnetoexcitons at the edge which are described by [20]

$$J_\Delta^\dagger |C_N^0\rangle = \sum_m \sqrt{\frac{m!}{(m+\Delta)!}} c_{m+\Delta}^\dagger c_m |C_N^0\rangle. \tag{13}$$

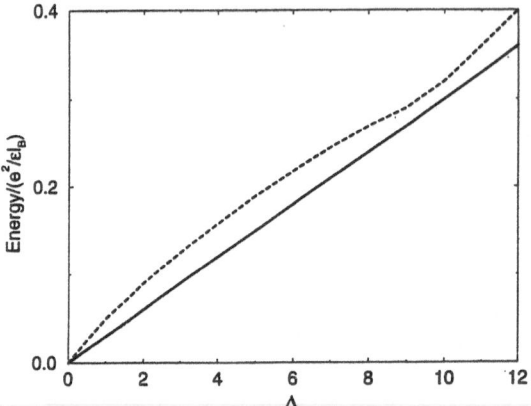

Figure 2. Energy (in units of $e^2/\epsilon l_B$) of edge spin textures (continuous line) and polarized charge excitations (dashed line) as a function of Δ.

Let us now analyze the excitations at $\nu = 1$ to see how an increase of the magnetic field produces a change of the GS. Such an increase of B simply rescales l_B in the interactions and alters the SP part of the energy through both the Zeeman term and the value of Ω in Eq. (2). QD edge reconstruction produces the new GS. The two kind of excitations which can be the new GS of the system when increasing B are: spin-polarized charge magnetoexcitons and edge spin excitations. The branch having the lowest energy depends on both N and the Zeeman energy. In the $g = 0$ limit, our calculations show that the lowest energy excitations that are localized *at the edge* (i. e. Δ small) are always non-polarized spin-textures because the initial slope of J-type excitations is always higher than α_Σ. Since the dispersion relation of magnetoexcitons is not linear, crossings of the different branches for high angular momenta are possible and spin polarized magnetoexcitons can be the new GS. We obtain that the edge spin branch has lower energy than the entire charge excitation branch for $N \leq 100$, while there is a crossing between both branches for $N \geq 100$. Obviously, the charge branch is favored on increasing Zeeman energy. However we have checked that, for GaAs ($g = 0.44$) and typical values of magnetic fields, $\alpha_\Sigma + g\mu_B B$ is still lower than the slope of J-type excitations at small Δ. Hence, non-polarized spin-textures are still the lowest energy excitations of QD at $\nu = 1$.

5. Conclusions

We have analyzed spin textures in a QD containing up to a few thousand electrons. Our main conclusions are:

1) We obtain analytical many-body wave-functions for spin excitations with all quantum numbers (M, M_{CM}, S^2 and S_z) properly described. Such a microscopic description naturally includes quantum fluctuations. Our states describe either skyrmions in QD or edge spin textures.

2) Our description reveals bulk charged skyrmions $(\Lambda_1^\dagger)^n |C_{N-k}^k\rangle$ (for $k > 0$) that can be expressed as a condensate of spin excitons interacting via a two-body repulsive interaction. The number n of excitons present in the condensate is the integer index controlling the size of the skyrmion. This framework allows the calculation of the skyrmion energy as a function of its size for both zero and finite Zeeman energy.

3) We also find branches of edge spin-textures $(\Sigma_\Delta^\dagger)^n |C_{N-k}^k\rangle$ (for $k \geq 0$ and Δ both positive and negative). For GaAs at $\nu = 1$, branches of $(\Sigma_\Delta^\dagger)^n |C_N^0\rangle$ edge spin-textures begin with energy lower than that of the branch of polarized edge excitations. For $N \leq 100$ there are no crossings of the spin textures and magnetoexciton branches so that edge spin textures are responsible for the edge reconstruction of the QD as suggested by recent experiments.[13] On the other hand, edge reconstruction is due to polarized charge excitations for $N \geq 100$.

6. Acknowledgements

This work has been supported in part by CICYT of Spain under contracts No. MAT 94-0982-C02-01 and 94-0058-C02-01.

References

1. A. Usher, R. J. Nicholas, J. J. Harris and C. T. Foxon, Phys. Rev. B **41**, 1129 (1990).
2. S. L. Sondhi, A. Karlhede, S. A. Kivelson and E. H. Rezayi, Phys. Rev. B **47**, 16419 (1993).
3. K. Moon, H. Mori, Kun Yang, S. M. Girvin, A. H. MacDonald, L. Zheng, D. Yoshioka and Shou-Cheng Zhang, Phys. Rev. B **51**, 5138 (1995).
4. H. A. Fertig, L. Brey, R. Cote and A. H. MacDonald, Phys. Rev. B **50**, 11018 (1994).
5. L. Brey, H. A. Fertig, R. Cote and A. H. MacDonald, Phys. Rev. Lett., **75**, 2562 (1995).
6. R. Tycko, S. E. Barret, G. Dabbagh, L. N. Pfeiffer and K. W. West, Science **286**, 1460 (1995); S. E. Barret, G. Dabbagh, L. N. Pfeiffer, K. W. West and R. Tycko, Phys. Rev. Lett., **74**, 5112 (1995).
7. A. Schmeller, J. P. Eisenstein, L. N. Pfeiffer and K. W. West, Phys. Rev. Lett., **75**, 4290 (1995).
8. E. H. Aifer, B. B. Goldberg and D. A. Broido, Phys. Rev. Lett., **76**, 680 (1996).
9. D. K. Maude, M. Potemski, J. C. Portal, M. Henini, L. Eaves, G. Hill and M. A. Pate, to be published.
10. V. Bayot, E. Grivei, S. Melinte, M.B. Santos and M. Shayegan, Phys. Rev. Lett. **76**, 4584 (1996).
11. T. Schmidt, M. Tewordt, R. H. Blick, R. J. Haug, D. Pfannkuche, K. v. Klitzing, A. Foster and H. Luth, Phys. Rev. B **51**, 5570 (1995).

136

12. O. Klein, C. de C. Chamon, D. Tang, D. M. Abusch-Magder, X.-G. Wen, M. A. Kastner and S. J. Wind, Phys. Rev. Lett. **74**, 785 (1995).

13. O. Klein, D. Goldhaber-Gordon, C. de C. Chamon and M. Kastner, Phys. Rev. B **53**, R4221 (1996).

14. N. C. van der Vaart, L. P. Kouwenhoven, M. P. de Ruyter van Steveninck, Y. V. Nazarov, C. J. P. M. Hartmans and C. T. Foxon, Phys. Rev. B **55**, 9746 (1997).

15. R. C. Ashoori, H. L. Stormer, J. S. Weiner, L. N. Pfeiffer, K. W. Balswin and K. W. West, Phys. Rev. Lett. **71**, 613 (1993).

16. R. C. Ashoori, Nature **379**, 413 (1996).

17. C. Nayak and F. Wilczek, Phys. Rev. Lett., **77**, 4418 (1996).

18. J. H. Oaknin, L. Martin-Moreno and C. Tejedor, Phys. Rev. B, **54**, 16850 (1996); erratum **55**, 15943 (1997).

19. C. de C. Chamon and X. G. Wen, Phys. Rev. B **49**, 8227 (1994).

20. J. H. Oaknin, L. Martin-Moreno, J. J. Palacios and C. Tejedor, Phys. Rev. Lett. **74**, 5120 (1995).

21. The effect of such projector reduces with increasing N.

22. J. H. Oaknin, B. Paredes, L. Martin-Moreno and C. Tejedor, to be published.

23. J. J. Palacios, L. Martin-Moreno, G. Chiappe, E. Louis and C. Tejedor, Phys. Rev. B **50**, 5760 (1994).

24. J. J. Palacios, D. Yoshioka and A. H. MacDonald, Phys. Rev. B **54**, R2296 (1996).

THEORY OF NEAR FIELD OPTICS

Surface Imaging by a Levitating Particle

A. MADRAZO AND M. NIETO-VESPERINAS

Instituto de Ciencia de Materiales de Madrid,
Consejo Superior de Investigaciones Científicas
Cantoblanco. Madrid 28049. Spain.

Abstract. We address the scattering of light by a system consisting of a dielectric corrugated object surface in presence of a metallic nanoparticle suspended at subwavelength distance from it. The sample is illuminated from its dielectric side by a focused beam either at normal or total internal reflection incidence. The near field distribution projected into the far zone by the particle, as well as its imaging characteristics, is analyzed. A filtering of the far field intensity is imposed so that the signal coming from the nanoparticle is extracted out of the component due to the surface. The superresolution dependence on nanocylinder size and its distance to the sample is discussed. Also, the image contrast versus the path followed by the particle on scanning the sample is analyzed.

1. Introduction

Near-field optics (NFO) apertureless techniques have received much attention in the last years [1, 2, 3, 4, 5]. Also recently, topographic[6] and molecular spectroscopy[7] with 1 nm resolution has been claimed by using these procedures. This possibility was already pointed out in earlier NFO studies, (see the review of Ref. 8).

The theory of this process should explain how the subwavelength surface structure information is encoded in the far-field produced by the system tip-sample, and the dependence of the superresolution achieved on the coupling of the evanescent waves into propagating components travelling to the far zone through scattering by the tip. Theoretical works initially studied this question even though based on either the electric dipole [9, 10] or the passive probe approximations [11, 12, 13].

N. García et al. (eds.), Nanoscale Science and Technology, 137–154.

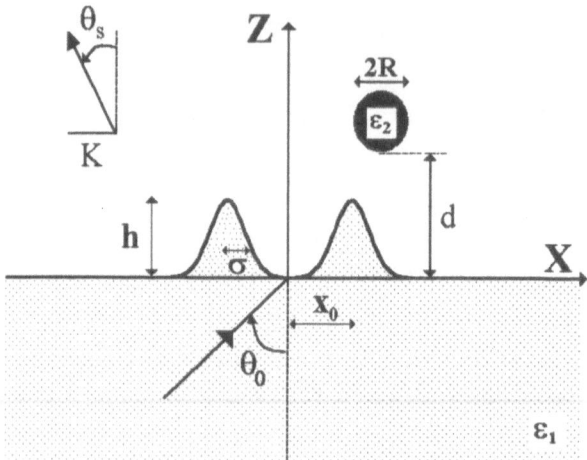

Figure 1. Scattering geometry.

In the next pages we shall put forward a model based on the scattering of an *S*-polarized electromagnetic wave by a system consisting of a metallic cylinder of nanometric dimensions and a corrugated dielectric surface, constant along the direction of the cylinder axis, (see Fig.1). This nanocylinder constitutes a 2-D representation of a detecting nanoparticle levitating above the object surface, as done in recent experiments [4]. The active part of a tip could also be modelled by such a system, [9, 10].

The mathematical framework is based on a generalization of the extinction theorem to multiply connected domains[14], already used in other scattering configurations and phenomena in NFO microscopy[15, 16]. However, it is worth stressing, that the problem of calculating the real active part of an extended tip for several configurations is interesting *per se*. In this connection, pictures of a extended tip illuminated with the evanescent waves excited by incidence from the dielectric side under total internal reflection[3], suggest that the scattering surface of the tip is like a sphere whose radius is actually smaller than that of the tip. Our study, although being 2-*D*, permits an analysis of the optical coupling between the probe and the sample, and hence, apart from some quantitative details pertaining to the response of the system to the polarization characteristics of the incident wave which would certainly require a full 3-*D* treatment (see Refs. 17, 18, and 19 for 3-D calculations), it allows an understanding and interpretation of several physical processes involved in apertureless NFO measurements. Specifically, it is shown how the surface subwavelength information contained in the near field is transferred and encoded into the far-field on scattering by the probing cylinder. This requires, however, a filtering of the far zone intensity in order to separate the component due to the tip from that coming from the surface. Also, we investigate the dependence of the optical signal on tip size, (namely, cylinder diameter), as well as on tip-sample distance, angle of incidence, and on the path followed by the cylinder as it scans close to the sample.

2. Theoretical Model

The physical system is shown in Fig. 1. The geometry is $2 - D$, and it contains a corrugated interface, constant along the OY axis, with profile $z = D(x) = h[exp(-(x - x_0)^2/\sigma^2) + exp(-(x + x_0)^2/\sigma^2)]$ separating air $(z > D(x))$ from a dielectric half-space $(z < D(x))$ with permittivity $\epsilon_1 = 2.126$ (glass). The nanoparticle is simulated by a cylinder of radius R whose axis is along OY, at a distance $d + R$ of the plane $z = 0$, and dielectric permittivity $\epsilon_2 = (-9.89, 1.05)$ (corresponding to gold at a wavelength in vacuum $\lambda = 652$ nm). An s-polarized focused Gaussian beam with half-width $W = 8\lambda$ illuminates the system from the dielectric half-space at an angle of incidence θ_0. The complex amplitude of the scattered electric field $\mathbf{E}(\mathbf{r}) = (0, E(\mathbf{r}), 0)$ at a point $\mathbf{r}$ in the far zone in air $(kr \to \infty, k = 2\pi/\lambda)$ has the expression[20]:

$$E(r, K) = iq[T_1(K) + T_2(K)]\frac{exp(ikr)}{\sqrt{r}}, \qquad (1)$$

where $K = -ksin\theta_s$, θ_s being the scattering angle (see Fig. 1) and $q = \sqrt{k^2 - K^2}$, $(K \leq k)$. The functions $T_1(K)$ and $T_2(K)$, are the scattering amplitudes generated by the corrugated dielectric surface and by the nanocylinder, respectively, and they *only* contain propagating plane waves towards $z > 0$. It should be noted that both functions are calculated without any approximation on the order of the interaction between the particle and the surface, using the expressions[14]:

$$T_1(K) = \frac{1}{4\pi q} \int_D ds' \left[(\mathbf{n}'.\mathbf{k})E(\mathbf{r}') - i\frac{\partial E(\mathbf{r}')}{\partial n'} \right] \exp(-i(Kx' + qz')), \qquad (2)$$

$$T_2(K) = \frac{1}{4\pi q} \int_C ds' \left[(\mathbf{n}'.\mathbf{k})E(\mathbf{r}') - i\frac{\partial E(\mathbf{r}')}{\partial n'} \right] \exp(-i(Kx' + qz')), \qquad (3)$$

where $\mathbf{k} = (K, q)$, and D and C stand for the surface of the sample and the cylinder, respectively. $\mathbf{n}'$ denotes the unit outward normal to each surface. Eqs. (2) and (3) express $T_1(K)$ and $T_2(K)$ in terms of the limiting values of the electric field, $E(\mathbf{r}')$ and its normal derivative, $\partial E(\mathbf{r}'/\partial n')$ on the surfaces D and C, respectively. These two functions are calculated from the extinction theorem boundary condition taking into account the coupling of the surface and the nanoparticle[15, 16]. Nevertheless, from an interpreting point of view, it will be useful to express $T_2(K)$ in terms of its equivalent expression given by a volume integral [21]:

$$T_2(K) = \frac{ik^2}{4\pi q} \int_{V_{tip}} d\mathbf{r}' \left(\epsilon_2(\mathbf{r}') - 1\right) E_{in}(\mathbf{r}') \exp(-i(Kx' + qz')), \qquad (4)$$

where V_{tip} is the cylinder domain and $E_{in}(\mathbf{r}')$ is the electric field *inside* the cylinder. The factor $(\epsilon_2(\mathbf{r}')-1)E_{in}(\mathbf{r}')$ is proportional to the polarization field, $P_{in}(\mathbf{r}')$ inside the cylinder. In the electric-dipole approximation one assumes that the polarization field $P_{in}(\mathbf{r}')$ is constant across the particle volume V_{tip} and then $T_2(K)$ can be expressed as:

$$T_2(K) \simeq \frac{ik^2}{q} \alpha(R, \epsilon_2) E(\mathbf{r_{tip}}), \qquad (5)$$

where $\alpha(R, \epsilon_2)$ is the electric-dipole polarizability and $E(\mathbf{r_{tip}})$ is the electric field *in absence of cylinder* at the point $\mathbf{r_{tip}} = (x_{tip}, d + R)$. Eq. (5) shows that the far-field is then proportional to the near-field.

It should be pointed out that, as it will be shown in the next section, the resolution limit given by the scattering probe is imposed by the tip size, in addition to the distance d (in fact, this is the same resolution limit yielded by NFO techniques with an aperture probe[22]).

Also, it is important to remark that Eq. (5) is the limit of Eq. (4) when the tip degenerates into a point at $\mathbf{r} = \mathbf{r_{tip}}$, namely, when $\epsilon_2(\mathbf{r}) = \epsilon_2\delta(\mathbf{r} - \mathbf{r_{tip}})$ and no coupling is assumed between particle and surface. Otherwise, Eq. (4) exhibits the integration (convolution) of the near-field inside the particle domain V_{tip} with the potential function $\epsilon_2(\mathbf{r} - \mathbf{r_{tip}}) - 1$; this underlines the loss of resolution as R increases.

Let us note that there exist several non-independent assumptions in Eq. (5). First, the dielectric function $\epsilon_2(\mathbf{r}')$ is constantly equal to its bulk value ϵ_2 across the cylinder domain V_{tip}. Second, the electrostatic approximation (i.e., the amplitude and phase variations of the electric field inside the tip are negligible) is valid. These approximations are mainly controlled by the nanoparticle size. It is generally accepted that for small enough particles both assumptions are valid. Nevertheless, for very small metallic particles (for instance a gold sphere with $R \sim 10^{-3}\lambda$) it has been shown[23] that quantum-size effects become important and they must be taken into account. This can be done in two steps. First, one can use quantum mechanics to calculate the position-dependent dielectric function of the nanoparticle; then, once this function is known, it is possible to solve the scattering problem by using Maxwell's equations. In the numerical calculations presented in this work we have used bulk values for the dielectric permittivities, limiting the lower size of the particle to $R = 0.01\lambda$, for which quantum-size effects are negligible. Finally, note that in Eq. (5), the coupling between the cylinder and the semi- infinite medium is either negligible[11, 13], or otherwise, the electric-dipole polarizability is described by an effective anisotropic tensor $\alpha_{eff}(R, \epsilon_2, z_{tip})$[24] that takes this interaction into account.

Although, for the sake of brevity, it is not shown here, when the scattering volume V_{tip} is small the factor $|qT_2(K)|^2$ does not strongly depend on the variable K for the present incident polarization (see Eq. (4)). Then, we have fixed somewhat arbitrarily the scattering angle to $\theta_s = 50^0$.

3. Illustrations

In the numerical experiments presented in this section, the integration over the surface D has been restricted to a finite length $L = 50\lambda$. The choice of such a sufficiently long segment, together with the beam tapered shape of the incident wave prevents spurious effects arising from scattering at the surface edges. We have used a density of sampling points in the discretization of the surface D given by an interval $\Delta x = 0.001\lambda$ if $|x| \leq 0.5\lambda$ (corrugated profile region) and $\Delta x = 0.05\lambda$ otherwise (flat profile region). A sampling interval $\Delta s = 0.001\lambda$ has been taken on the cylinder. The numerical consistency of the results is checked on controlling both convergence with the number of sampling points and their unitarity (energy balance).

We consider the surface profile $z = D(x)$, given in Section 2, with parameters: $h = 0.05\lambda$, $\sigma = 0.06\lambda$, and $x_0 = 0.15\lambda$. Figs. 2(a) and 2(b) show a comparison of the electric field intensity distribution at a constant line $z_0 = 0.065\lambda$, obtained in the absence of cylinder (solid line), with the electric field intensity *inside* a cylinder of radius $R = 0.01\lambda$ (dotted line) when it scans the surface with its center along the constant height $z_{tip} = 0.065\lambda$. The angles of incidence are $\theta_0 = 0^0$ and 50^0, respectively. Obviously, due to absorption in the metallic volume, the intensity distribution inside the cylinder is weaker than that transmitted by the surface in absence of cylinder. Nevertheless, it is worth noting that both intensity distributions are proportional to each other. Also, the image contrast is larger in Fig. 2(b) namely, for incidence under total internal reflection, $\theta_0 = 50^0$. It should be remarked that, although the field intensity inside the cylinder is proportional to the intensity distribution in absence of it, this cylinder strongly distorts the exterior intensity distribution around it. This is shown in Figs. 3(a) and 3(b) that display the field intensity distribution along the line $z_0 = 0.075\lambda$ just between the surface profile and a cylinder of radius

142

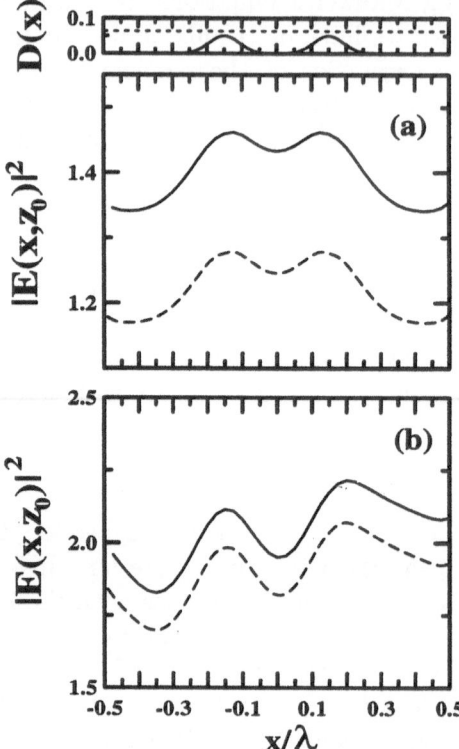

Figure 2. Comparison between the electric field intensity distribution at $z_0 = 0.065\lambda$ in the absence of nanoparticle (solid line) and the electric field intensity inside a gold cylinder (dotted line) with radius $R = 0.01\lambda$ when it scans the surface with its center along the line $z_{tip} = 0.065\lambda$. Surface parameters: $h = 0.05\lambda$, $x_0 = 0.15\lambda$, and $\sigma = 0.06\lambda$. 2(a): $\theta_0 = 0^0$. 2(b): $\theta_0 = 50^0$. The figure on the top shows the sample profile (thick line), as well as the height z_0 (dotted line).

$R = 0.05\lambda$ with its center at the point $\mathbf{r_{tip}} = (x_{tip}, 0.15\lambda)$ for $\theta_0 = 0^0$ and $\theta_0 = 50^0$, respectively. It is shown, from top to bottom, the intensity distribution for differents lateral positions of the nanoparticle, namely, $x_{tip} = 0.5\lambda$, 0.15λ, 0, -0.15λ, and -0.5λ, respectively. Also, it is shown for the sake of comparison the intensity distribution transmitted by the sample in the absence of cylinder (dotted line) at the same height z_0. The presence of the metallic particle clearly modifies the intensity distribution over a range of several wavelengths around it. The intensity distributions exhibit a pronounced dip just at the position of the cylinder due to a cavity effect produced by the presence of the metallic nanoparticle at subwavelength distance from the sample. However, although not shown here, the field intensity in the cylinder is again proportional to the intensity distribution in its absence. This is the fact meant when the particle is said to be *passive*

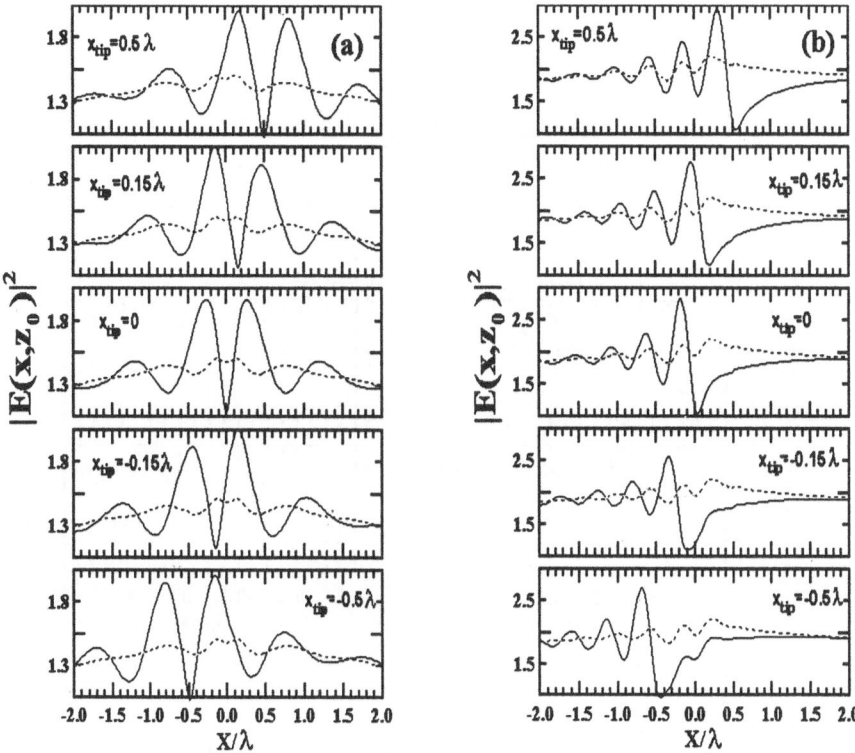

Figure 3. Electric field intensity distribution along the line $z_0 = 0.075\lambda$ when a cylinder of radius $R = 0.05\lambda$ is placed with its center at the point $r_{tip} = (x_{tip}, 0.15\lambda)$. From top to bottom it is shown; $x_{tip} = 0.5\lambda$, $x_{tip} = 0.15\lambda$, $x_{tip} = 0$, $x_{tip} = -0.15\lambda$, and $x_{tip} = -0.5\lambda$, respectively. Dotted line: intensity distribution at the same height z_0 in absence of cylinder. 3(a): $\theta_0 = 0^0$. 3(b): $\theta_0 = 50^0$.

[25, 26, 27].

Next, we analyze the imaging properties of the nanoparticle, namely, of the nanocylinder, close to the dielectric surface. Fig. 4 shows a set of numerical calculations when the nanoparticle scans the surface at a constant height $d = 0.055\lambda$ from $z = 0$. It is displayed, from top to bottom: the far-field intensity (FFI), $|E(r, K)|^2$ scattered by the whole system: particle plus surface (Figs. 2(a) and 2(b)); the factor $|qT_1(K)|^2$, namely, the FFI scattered by the dielectric surface (Figs. 4(c) and 4(d)); the FFI scattered by the nanocylinder, $|qT_2(K)|^2$ (Figs. 4(e) and 4(f)); and the electric field intensity *inside* the cylinder, $|E(x, d+R)|^2$ (Figs. 4(g) and 4(h)). Two particle sizes have been used: $R = 0.05\lambda$ (broken line) and $R = 0.01\lambda$ (solid line), as well as two angles of incidence: $\theta_0 = 0^0$ (left column), and $\theta_0 = 50^0$ (right column) for which total internal reflection occurs. First, let us focus our attention on Figs. 4(a), 4(b) and Figs. 4(c), 4(d), i.e., on the

144

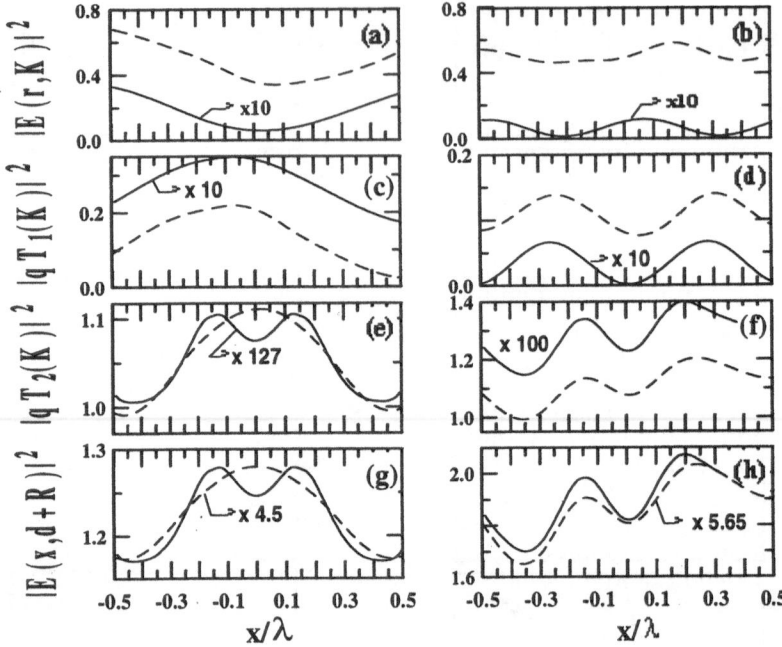

Figure 4. Numerical calculations as the nanoparticle scans the surface at a constant height $d = 0.055\lambda$. From top to bottom it is shown: the far field intensity scattered by the whole system (4(a) and 4(b)), by the surface (4(c) and 4(d)), and by the cylinder (4(e) and 4(f)), $|E(r, K)|^2$, $|qT_1(K)|^2$, and $|qT_2(K)|^2$, respectively. $K = -ksin\theta_s$, $\theta_s = 50^0$. The electric field intensity inside the nanocylinder, $|E(z = d + R)|^2$, is also shown (4(g) and 4(h)). The surface parameters are: $h = 0.05\lambda$, $x_0 = 0.15\lambda$, and $\sigma = 0.06\lambda$. Solid line: $R = 0.01\lambda$. Broken line: $R = 0.05\lambda$. Left column: $\theta_0 = 0^0$. Right column: $\theta_0 = 50^0$.

FFI scattered from the total system: cylinder plus sample, and on the FFI produced by the dielectric surface. It is observed that $|qT_1(K)|^2$ depends on the nanoparticle size. This is due to the electromagnetic coupling between the nanoparticle and the sample. Nevertheless, this coupling stems mainly from electromagnetic radiation reflected from the cylinder towards the sample which, in turn, reemits it to the far zone. Multiple reflections between the cylinder and the sample are negligible as we have shown in Fig. 2. The proportionality between the near-field inside the cylinder and the corresponding one in the absence of probe is a consequence of the low bouncing degree in the cavity formed by the nanocylinder and sample.

Also, it is worth noting that these magnitudes do not present any sub-wavelength feature of the surface. Furthermore, they show a smooth oscillation with period of the order of the incident wavelength, λ. However, the FFI scattered by the nanoparticle, $|qT_2(K)|^2$, contains subwavelength features that can be easily correlated with those present in the surface profile. Also, on comparing $|qT_2(K)|^2$ with the field intensity inside the nanocylinder, $|E(x, d + R)|^2$, one notes that both quantities are almost proportional to each other. In consequence, as it has been shown in Fig. 2, the near-field intensity above the surface *in the absence of cylinder* is also proportional to both $|qT_2(K)|^2$ and $|E(x, d + R)|^2$. This fact corroborates the assumption that the cylinder can be considered a passive probe[25, 26], namely, it scatterers to the far zone a signal proportional to that produced in the near- field by diffraction from the sample in the absence of nanoparticle, even though the particle perturbs the field intensity distribution around it. Nevertheless, it is noticeable that, for the metallic probe addressed in this work, the cylinder size condition $R \leq 0.05\lambda$, which supports the 'passive probe' assumption, is more restrictive than the one: $R \lesssim 0.15\lambda$ for the case of a dielectric cylinder[15]. Furthermore, the particle size fixes the lateral resolution that it is possible to achieve [1, 3, 4, 5, 6, 7] when the cylinder-sample distance is keep constant. It is worth noting from Fig. 4(e) that the signal scattered from the larger cylinder, ($R = 0.05\lambda$, size similar to the half-width of the profile bumps), presents an integration on the cylinder domain. In addition, it exhibits a loss of resolution due to the larger distance from the center of the cylinder to the plane $z = 0$. Hence. it does not resolve any feature of the surface structure. An extended study of the probe-sample interaction versus d will be published elsewhere[28]. The same conclusion can be obtained from Fig. 4(f). Although in this case an incidence under total internal reflection, enhances the image contrast[7], the position of the second maximun in the image is shifted about 0.1λ (namely, a distance equal to the nanocylinder diameter), with respect to those in the surface profile.

Fig. 5 shows $|qT_2(K)|^2$ for a surface profile of even smaller details. Now, the surface parameters are: $h = 0.05\lambda$, $x_0 = 0.025\lambda$, and $\sigma = 0.01\lambda$. The particle has a radius $R = 0.01\lambda$ and it scans the surface at a constant height $d = 0.052\lambda$. The angle of incidence is $\theta_0 = 0^0$ and $\theta_0 = 50^0$ in Fig. 5(a) and 5(b), respectively. It is remarkable that, although the surface profile contains extremely subwavelength localized objects, the positions of the profile bumps clearly appear as two dips in the intensity distributions, thus, indicating a contrast reversal with respect to the surface profile for both angles of incidence. Although, for $\theta_0 = 50^0$ the curve does not exhibit the surface profile symmetry, the image shows an enhanced contrast with respect to that obtained at $\theta_0 = 0^0$. Again, these distributions are proportional to the near-field intensity distribution obtained inside the cylinder as this scans with its center along the plane $z = 0.052\lambda$. The lack of resemblance of

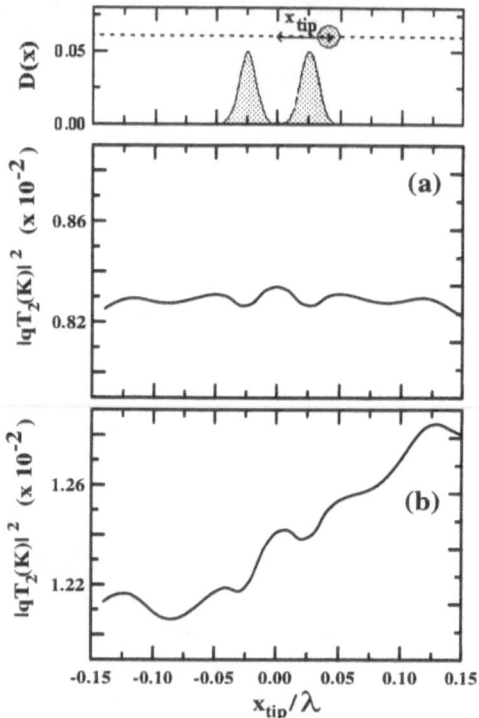

Figure 5. Far field intensity scattered by the nanocylinder $|qT_2(K)|^2$ when it scans the surface at a constant height $d = 0.052\lambda$. $K = -ksin\theta_s$, $\theta_s = 50^0$. The surface parameters are $h = 0.05\lambda$, $x_0 = 0.025\lambda$, and $\sigma = 0.01\lambda$. $R = 0.01\lambda$. 5(a): $\theta_0 = 0^0$. 5(b): $\theta_0 = 50^0$. The top figure shows the sample profile (thick line), as well as the path followed by the cylinder (dotted line).

this field distribution to the surface profile is a well known fact[29] due to multiple scattering.

Up till now, we have considered that the cylinder scans the surface at a constant height $z_{tip} = R + d$ (constant-height mode). However, it has recently been shown[30] that this operation mode is completely equivalent to the constant-intensity mode, i.e., the detected intensity is kept constant and then the resulting vertical displacements of the nanoparticle are those recorded. Furthermore, it has been shown that the images obtained in both operation modes are almost proportional to each other[30]. In several experimental configurations it is even possible that the particle scans the surface following a line parallel to its topography by using, for instance, an atomic force microscope. Hence, we next discuss the imaging properties of the particle in this operation mode.

Fig. 6 shows the FFI scattered by the cylinder at $K = -ksin\theta_s$, $\theta_s = 50^0$

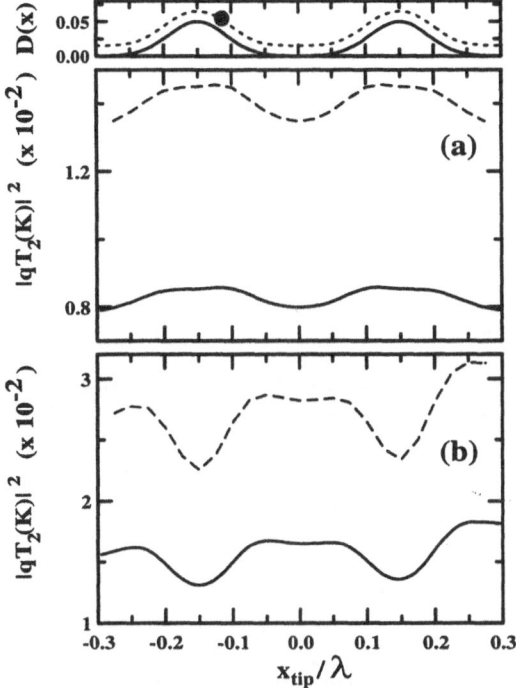

Figure 6. Far Field intensity scattered by the particle $|qT_2(K)|^2$ when it scans the surface following its topography. $K = -k\sin\theta_s$, $\theta_s = 50^0$. The surface parameters are $h = 0.05\lambda$, $x_0 = 0.15\lambda$, and $\sigma = 0.06\lambda$. $R = 0.01\lambda$. The constant distance between the cylinder and the surface profile is $R + 0.005\lambda$. Solid line: $|qT_2(K)|^2$. Broken line: Near field intensity in the absence of cylinder along the path followed by the center of the nanocylinder. The top figure shows the sample profile (thick line), as well as the path followed by the cylinder (dotted line). 6(a): $\theta_0 = 0^0$. 6(b): $\theta_0 = 50^0$.

when it scans the surface (same surface profile parameters as in Fig. 4) with its center at a constant distance to the surface profile $R + 0.005\lambda$. The cylinder radius being $R = 0.01\lambda$. The angles of incidence are $\theta_0 = 0^0$, and $\theta_0 = 50^0$ in Figs. 6(a), and 6(b), respectively. The electric field intensity *in the absence of cylinder* (broken line) along the path followed by the cylinder (dotted line in the top pannel) is also shown. Again, the FFI scattered by the nanoparticle and the field intensity in the absence of probe are proportional to each other, and, then, the passive probe assumption remains valid in the constant-distance mode of operation. Although there are no important differences between the FFI scattered by the nanocylinder in the constant height mode (Fig. 4(e)) and in the constant distance mode (Fig. 6(a)) for normal incidence, it is observed that, for incidence under total internal

148

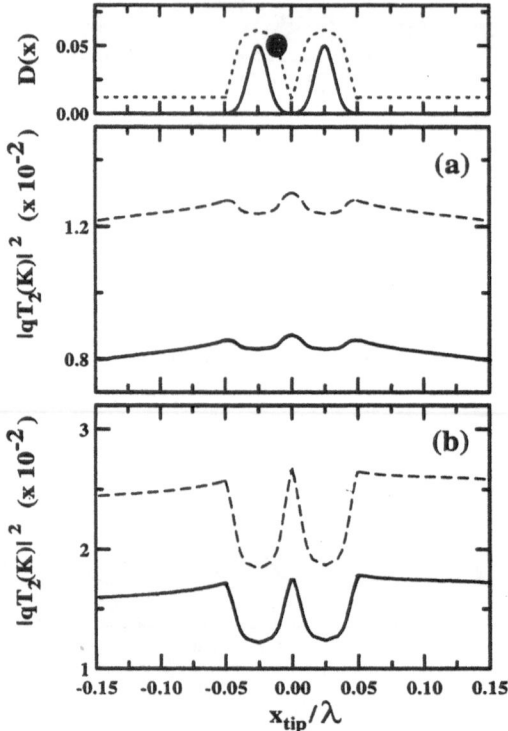

Figure 7. Same as Fig. 6 for the surface parameters: $h = 0.05\lambda$, $x_0 = 0.025\lambda$, and $\sigma = 0.01\lambda$. The distance between the cylinder and the surface profile is $R + 0.002\lambda$. 7(a): $\theta_0 = 0^0$. 7(b): $\theta_0 = 50^0$.

reflection, the image obtained at constant distance (Fig. 6(b)) is better than the one obtained when the cylinder scans the surface at a constant height (Fig. 4(f)). First, the constant-distance mode partially removes the lack of symmetry displayed in the images obtained in the constant-height mode (Fig. 4(f)). Second, Fig. 6(b) exhibits a constrast reversal with respect to the surface profile due to the fact that, as the separation between the nanocylinder and the mean plane $z = 0$ increases, while the nanocylinder scans the profile Gaussian bumps, the coupling between the nanoparticle and the evanescent wave created on incidence at total internal reflection decreases.

For the sake of comparison, we show in Fig. 7 the constant distance images obtained from a cylinder with $R = 0.01\lambda$ when it scans the surface profile addressed in Fig. 5. The distance between the cylinder and the surface profile is now $R + 0.002\lambda$. Once again, the images obtained by keeping the particle-sample distance constant, show much better quality than those

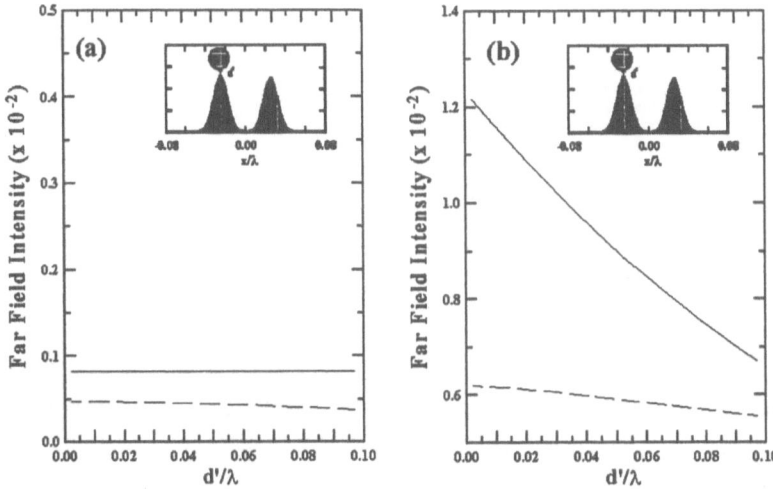

Figure 8. Far field intensity dependence on nanocylinder-sample distance. Same surface parameters as in Fig. 5. The lateral position of the cylinder is $x_{tip} = -x_0$. $R = 0.01\lambda$. Solid line: $|qT_2(K)|^2$. Broken line: $|qT_1(K)|^2$. $K = -ksin\theta_s$, $\theta_s = 50^0$. 8(a): $\theta_0 = 0^0$. 8(b): $\theta_0 = 50^0$.

corresponding to the constant-height mode, (Fig.5). Namely, they show a symmetry quite close to that presented in the surface profile. Also, it is remarkable that the interference pattern presented in the curves of Fig. 5, outside the bump location, is washed out when the cylinder follows a path parallel to the surface profile.

4. Filtering of the Nanoparticle Signal

It was shown that *only* the FFI scattered by the particle translates into the far zone the near-field of the surface, and hence, it contains the subwalength information about the surface profile. Furthermore, it has been shown that the cylinder radius determines the lateral resolution that can be obtained. Hence, a filtering data processing of the detected signal, $|E(r, K)|^2$, (cf. Eq.(1)), is necessary in order to extract the far-field signal $|qT_2(K)|^2$, which stems from the nanocylinder, from the one $|E(r, K)|^2$ of the whole system: particle plus surface. However, note that this filtering in the detected signal is not needed in the collection-mode techniques (PSTM, STOM) due to the fact that, in these cases, the detected signal is directly the field intensity diffracted by the probe, i.e., $|qT_2(K)|^2$, or an integration of $|qT_2(K)|^2$ over an interval of scattering angles[10, 11, 12]. We next analyze how this filtering can be done by exploiting the dependence of $|qT_2(K)|^2$ on the particle-sample distance.

Figs. 8(a) and 8(b) show this dependence of the FFI on the cylinder-

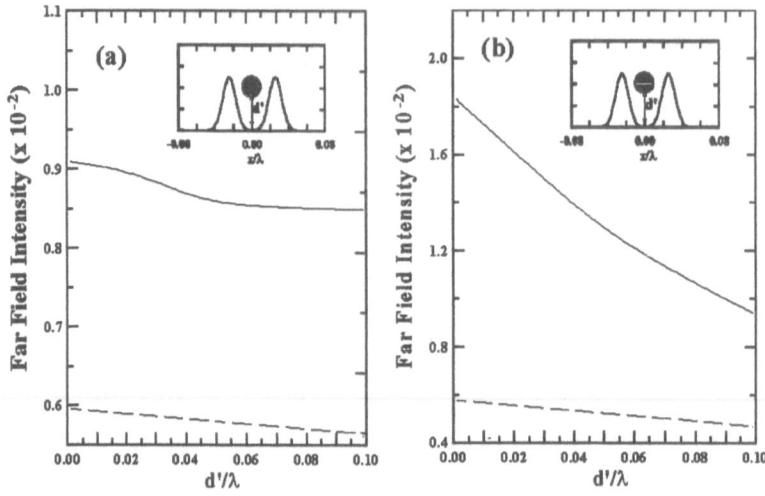

Figure 9. Same as Fig. 8 for a lateral position of the nanocylinder $x_{tip} = 0$. 9(a): $\theta_0 = 0^0$. 9(b): $\theta_0 = 50^0$.

sample distance for $\theta_0 = 0^0$ and 50^0, respectively. The lateral position of the nanoparticle is fixed just above the left bump on the surface profile, $x_{tip} = -x_0$. We consider the same profile parameters as those addressed in Fig. 5, then the cylinder is moved away from the sample along OZ. The solid and the broken curves show the cylinder-sample distance d dependence of the FFI scattered by the cylinder and by the surface, respectively. Figs. 9(a) and 9(b) show the same dependence as in Figs. 8(a) and 8(b) but now the lateral position of the cylinder is fixed at the center of the surface profile, $x_{tip} = 0$.

It is remarkable that, for $\theta_0 = 50^0$ (Figs. 8(b) and 9(b)), $|qT_2(K)|^2$ exhibits a much faster variation with d than $|qT_1(K)|^2$. Furthermore, $|qT_2(K)|^2$ exponentially decreases as d increases, due to the coupling of the near-field evanescent components transmitted by the surface with propagating waves travelling into $z > 0$ through scattering with the particle, (this, in turn, confirms the main premise supporting the technique reported in Ref. [3]). $|qT_2(K)|^2$ has a intensity decay length of 0.159λ, which is close to the theoretical value: 0.16λ for a flat interface. However, Fig. 8(a) shows that both $|qT_2(K)|^2$ and $|qT_1(K)|^2$ have a weakly dependence on d. This is due to the large distance between the particle center and the mean plane of the surface, which prevents the detection of the evanescent waves created on scattering from the surface corrugation. These evenescent waves are, of course, more weakly excited than those due to total internal reflection, as a comparison of Fig. 8(a) and 8(b) evidences. On the other hand, when the cylinder enters in the selvedge region, and thus approaches the surface mean plane, these scattering evanescent waves are detected, as shown in Fig. 9(a) where, again, $|qT_2(K)|^2$ exhibits a much faster variation than $|qT_1(K)|^2$ for distances $d < 0.05\lambda$, even though, this variation is not so pronounced as in the case of total internal reflection. This, again, is because the scattering evanescent waves are weaker than those from total internal reflection.

The fact that $|qT_2(K)|^2$ shows a cylinder-sample distance dependence quite different to that of $|qT_1(K)|^2$, could yield an interpretation and supports the procedure used in several experimental configurations [3, 7] in order to extract the signal scattered by the nanoparticle from that scattered by the whole system: particle plus sample. To accomplish this, a tip is periodically vibrated in the z-direction at a frequency Ω and then, the time dependent part of the FFI is filtered, thus rejecting in the detection any signal with frequency different to Ω.

5. CONCLUSIONS

A theory has been established showing how near field information, and hence, surface topographic features, are both encoded in the far-field intensity scattered by a system consisting of a nanoparticle raster scanning a corrugated surface. We note that preliminar experimental work has been done using a levitating nanoparticle as a scattering probe[4].

Also, we have shown the mechanisms leading to a filtering of the far field intensity so that the signal coming from the particle, (which contains the desired information), is separated from that due to the surface, (which acts as a background). The illustration has been made with an s- polarized wave, incident on a dielectric interface from the denser side. It has been found that, for the present geometrical configuration and polarization of the incident beam, metallic particles with radii not larger than 0.05λ can be considered passive (i.e. in the sense quoted before, namely, the field inside the paricle equals that produced in its absence, even if this particle strongly modifies the field distribution around it). This means that these nanoparticles scatter a signal into the far zone, and this signal is proportional to the near-field transmitted by the sample in the absence of probe. Although those specific values of the particle size used in our 2-D model, can somewhat differ in a 3-D system, this study illustrates the essential fact of how the particle radius governs the lateral resolution that can be achieved. As a consequence, calculations for smaller partile radii than those addressed in this paper should show better resolutions. However, these calculations must take into account quantum size effects on the particle dielectric function.

It has been pointed out that, for incidence under total internal reflection, the images obtained when the nanoparticle scans parallel to the surface profile, resemble better the actual topography of the sample than those

152

obtained when the particle scans at a constant height, even though the contrast can be inverted. In this configuration, artifacts that reproduce the scanning path rather than the actual topography, may exist [32, 33]. Another observed phenomenon is the enhanced image contrast obtained for angles of incidence larger than the critical one.

3-D calculations [17, 18, 19, 31] can be straightforwardly done with much higher computing times. However, the main characteristics concerning the encoding of near field information in the far field, as well as the dependence of resolution on nanoparticle size and distance to the surface, will remain qualitatively similar to those shown in this 2-D study.

Acknowledgments

This research has been supported by the Dirección General de Investigación Científica y Técnica under grant PB 95-0061 and by the CE. A.M. acknowledges a scholarship from Comunidad Autonoma de Madrid.

References

1. Zenhausern F., O'Boyle M. P., and Wickramasinghe K. H. (1994) Apertureless Near-Field Optical Microscope, *Appl. Phys. Lett.* **65**, 1623-1625.
2. Garcia N. and Nieto-Vesperinas M. (1995) Theory for the Apertureless Near Field Optical Microscope: Image Resolution, *Appl. Phys. Lett.* **65**, 3399-3400.
3. Inouye Y., and Kawata S. (1994) Near-Field Scanning Optical Microscope with a Metallic Probe Tip, *Opt. Lett.* **19**, 159-161.
4. Kawata S., Inouye Y., and Sugiura T. (1994) Near-field scanning optical microscope with a laser traped probe, *Jpn. J. Appl. Phys.* **33**, 1725-1727.
5. Gleyzes P., Boccara A. C., and Bachelot R. (1995) Near Field Optical Microscopy using a Metallic Vibrating Tip, *Ultramicroscopy* **57**, 318-322.
6. Zenhausern F., Martin Y., and Wickramasinghe H. K. (1995) Scanning Interferometric Apertureless Microscopy, *Science* **269**, 1083-1085.
7. Martin Y., Zenhausern F., and Wickramasinghe H. K. (1996) Scattering Spectroscopy of Molecules at Nanometer Resolution, *Appl. Phys. Lett.* **68**, 2475-2477.
8. Pohl D. W. (1990) Scanning Near-Field Optical Microscopy, in Sheppard J. R., and Mulvey T. (eds.), *Advances in Optical and Electron Microscopy*, Academic Press, New York, pp. 243-311.
9. Vigoureux J. M., Girard C. and Courjon D. (1989) General principles of scanning tunneling optical microscopy, *Opt. Lett.* **14**, 1039-1041.
10. Van Labeke D. and Barchiesi D. (1992) Scanning-tunneling optical microscopy: a theoretical macroscopic approach, *J. Opt. Soc. Am. A* **9**, 732-739.
11. Van Labeke D. and Barchiesi D. (1993) Probes for scanning tunneling optical microscopy: a theoretical comparison, *J. Opt. Soc. Am. A* **10**, 2193-2201.
12. Barchiesi D. and Van Labeke D. (1993) Application of Mie scattering of evanescent waves to Scanning Tunneling Optical Microscopy theory, *J. Mod. Opt.* **40**, 1239-1254.
13. Xiao M. and Bozhevolnyi S. (1996) Imaging with reflection near-field optical microscope: contributions of middle and far fields, *Opt. Comm.* **130**, 337-347.
14. Madrazo A. and Nieto-Vesperinas M. (1995) Scattering of Electromagnetics Waves from a Cylinder in Front of a Conducting Plane, *J. Opt. Soc. Am. A* **12**, 1298-1309.

15. Madrazo A. and Nieto-Vesperinas M. (1996) Surface Structure and Polariton Interactions in the Scattering of Electromagnetics Waves from a Cylinder in front of a Conducting Grating: Theory for the Reflection Photon Scanning Tunneling Microscope, *J. Opt. Soc. Am. A* **13**, 785-795; ib. (1997) Reconstruction of Corrugated Dielectric Surfaces with a Model of Photon Scanning Tunneling Microscope. Influence of the Tip on the Near Field, *J. Opt. Soc. Am. A* **14**, 618-628.

16. Carminati R., Madrazo A. and Nieto-Vesperinas M. (1994) Electromagnetic Wave Scattering from a Cylinder in front of a Conducting Surface-Relief Grating, *Opt. Comm.* **111**, 26-33; Madrazo A. and Nieto-Vesperinas M. (1995) Detection of Subwavelength Goos-Hänchen shifts from Near-Field Intensities: a Numerical Simulation, *Opt. Lett.* **20**, 2445-2447; Madrazo A., Garcia N., and Nieto-Vesperinas M. (1996) Exact Calculation of Maxwell Equations for a Tip-Metallic Interface Configuration: Application to Atomic Resolution by Photon Emission, *Phys. Rev. B* **53**, 3654-3657.

17. Girard C., and Dereux A. (1994) Optical Spectroscopy of a Surface at the Nanometer Scale: A Theoretical Study in Real Space, *Phys. Rev. B* **49**, 11344-11351.

18. Girard C., Dereux A., Martin O. J. F., and Devel M. (1994) Importance of Confined Fields in Near-Field Optical Imaging of Subwavelength Objets, *Phys. Rev. B* **50**, 14467-14473.

19. Martin O. J. F., Girard C., and Dereux A. (1995) Generalized Field Propagator for Electromagnetic Scattering and Light Confinement, *Phys. Rev. Lett.* **74**, 526-529.

20. Nieto-Vesperinas M. (1991) *Scattering and Diffraction in Physical Optics*, J. Wiley, New York. Ch. 2, pag. 52.

21. It should be remarked that the scattering equations can be expressed in two completely equivalent representations, namely, by using either surface or volume integrals. See, for instance, Wolf E. (1973) in Mandel L. and Wolf E. (eds.) *Coherence and Quantum Optics*, Plenum Press, New York, pp. 339-357; or Ref. 20, Ch. 1.

22. Novotny L., Pohl D. W., and Regli P. (1994) Light Propagation through Nanometer-Sized Structures: the Two-Dimensional-Aperture Scanning Near-Field Optical Microscope, *J. Opt. Soc. Am. A* **11**, 1768-1779.

23. Keller O. (1996) Short and Long Range Interactions in Near Field Optics, in Nieto-Vesperinas M. and Garcia N. (eds.), *Optics at the Nanometer Scale*, Kluwer, the Netherlands, pp. 63-93.

24. Efrima S. and Metiu H. (1979) Classical theory of light scattering by an adsorbed molecule. I. Theory, *J. Chem. Phys.* **70**, 1602-1613.

25. Greffet J.-J., Sentenac A., and Carminati R. (1995) Surface Profile Reconstruction Using Near-Field Data, *Opt. Comm.* **116**, 20-24.

26. Carminati R., Greffet J.-J. (1995) Two-Dimensional Numerical Simulation of the Photon Scanning Tunneling Microscope: Concept of the Transfer Function, *Opt. Comm.* **116**, 316-321.

27. Weeber J. C., de Fornel F. and Goudonnet J. P. (1996) Numerical study of the tip-sample interaction in the photon scanning tunneling microscope, *Opt. Comm.* **126**, 285-292.

28. Madrazo A., Carminati R., Nieto-Vesperinas M., and Greffet J.-J. (in press) Polarization effects in the optical interaction between a nanoparticle and a corrugated surface: implications for apertureless near-field microscopy, *J. Opt. Soc. Am. A*.

29. Garcia N. and Nieto-Vesperinas M. (1993) Near-Field Optics Inverse-Scattering Reconstruction of Reflective Surfaces, *Opt. Lett.* **24**, 2090-2092; Garcia N. and Nieto-Vesperinas M. (1995) A Direct Solution to the Inverse Scattering Problem for Surfaces from Near Field Intensities without Phase Retrieval, *Opt. Lett.* **20**, 949-951; Carminati R., Greffet J.-J., Garcia N. and Nieto-Vesperinas M. (1996) Direct Reconstruction of Surfaces from Near-Field Intensity under Spatially Incoherent Illumination, *Opt. Lett.* **21**, 501-503.

30. Carminati R., and Greffet J.-J. (1996) The Equivalence between Constant-Height and Constant-Intensity Images in Detection-Mode Scanning Near-Field Optical

Microscopy", *Opt. Lett.* **21**, 1208-1210.

31. Girard C., and Dereux A. (1996) Near Field Optics Theories, *Rep. Prog. Phys.* **59**, 657-699.

32. Hecht B., Bielefeldt H., Ynouye I., Pohl D. W. and Novotny L. (1997) Facts and Artifacts in Near Field Optical Microscopy, *J. Appl. Phys.* **81**, 2492-2498.

33. Carminati R., Madrazo A., Nieto-Vesperinas M., Greffet J.-J. (1997) Optical content and resolution of near-field optical images: influence of the operating mode, *J. Appl. Phys.* **82**, 501-509.

PHOTONS AND LOCAL PROBES

OTHMAR MARTI, JOACHIM BARENZ, ROBERT BRUNNER, MICHAEL HIPP[1],
OLAF HOLLRICHER, INGOLF HÖRSCH, AND JÜRGEN MLYNEK[2]

Abteilung Experimentelle Physik, Universität Ulm, D-89069 Ulm, Germany
[2]Fakultät für Physik, Universität Konstanz, D-78434 Konstanz, Germany
[1]Robert Bosch GmbH, Reutlingen, Germany

ABSTRACT. Experiments using photons as a probe provide an excellent energy resolution. The wavelength of the light, however, sets a lower limit to the probed area perpendicular to the direction of propagation. This limit is of the order of the wavelength of the light. Scanning probe microscopes, on the other hand, have a superb spatial resolution. The energy of the interaction can only be acquired with modest sensitivity and resolution. Therefore, the combination of photons and local probes might provide the best of the two worlds. In this article we explore the possibilities of combining optical and SPM techniques. Near field optical microscopy using the aperture probe is used to investigate the emission properties of vertical cavity surface emitting laser diodes (VCSEL). Mode patterns put into relation to the topography of the emission area serve to diagnose the performance of the device. In contrast to classical optical microscopy techniques, our method is able to simultaneously decompose lasing transversal modes by their wavelength with lateral superresolution. Similarly, the comparison of the emission location of nanometer sized thin luminescent layers with the shape of the overgrown structure reveals variations of the bandgap and its position. Subtle differences in images obtained with internal and external collection modes provide clues to the diffusion of the charge carriers. Finally scanning force microscope (SFM) is used to detect near field light by a mechanism based on optical modulation of the image force between a semiconducting probe tip and a glass surface due to the surface photo-voltage (SPV). This technique, which has a lateral resolution of better than 70 nm, allows the simultaneous detection of minute optical powers as small as $0.1\,\mathrm{pW}/\sqrt{\mathrm{Hz}}$ in air and of charges. The technique is applied to the measurement of optical and charge gratings of a photorefractive material.

1   Introduction and Concepts

Optical spectroscopy[1] is one of the most powerful tools to investigate the intricate interactions in atoms, molecules, and matter at larger scale. Light provides the excellent energy resolution, which allows the measurement of such subtle influences as that of parity non-

N. García et al. (eds.), Nanoscale Science and Technology, 155–174.
© 1998 *Kluwer Academic Publishers.*

conservation. Scanning probe microscopes[2] on the other hand have an energy resolution which is of the other of several tens of meV. In many cases, a better energy resolution would help to clarify interaction mechanisms and give means to identify materials. Hence, near field optical microscopy[3-10] was invented. There a fine aperture or a fine scattering center is brought close to the sample surface. The distance is kept constant by a force detection mechanism. Because of the size of the aperture and of the scattering centers it was not yet possible to get atomic resolution. By using the photoeffect, the inverse photoeffect, or by using an STM tip and the non-linearity of its tunneling junction as a mixer it is possible to obtain atomic resolution.

Aperture near field optical microscopes[7-10] are successfully applied to a broad range of samples, ranging from biological specimens[11] to semiconductor microscopic[12] and mesoscopic structures[13]. The main advantage of near field optical microscopy besides the improved lateral resolution compared to classical optical microscopy is the simultaneously acquired topography. This allows the precise location of luminescent structures with respect to the sample topography[14], a task that in classical microscopy sometimes is difficult. Since the position of emission maximum can be determined with much better resolution than the smallest separation of two nearby objects a precision in the 10 nm range is obtained. The emission pattern of laser diodes[15] is a second example where the relative position of the light intensity and the topography of the emission facets is important.

The deficiencies of the aperture probe in near field optics have also triggered the search for new probe concepts. One of these relies on the quenching of the fluorescence or luminescence in the tip region[16]. Alternatively, nanometric scattering centers close to the sample surface can resolve optically active structure separated as little as 1 nm[17,18].

Another concept for near field optical microscopy is based on induced forces acting on a silicon SFM cantilever[19,20]. Genuine optical forces such as radiation pressure and dipole forces have been observed on micrometer and nanometer-sized particles and are nowadays used in "optical tweezers" to trap and move particles. Light forces on atoms are a standard means in the field of atom optics to trap atoms and realize optical elements for atoms. Far-field radiation pressure forces on micro-cantilevers have been measured with a high sensitivity. On nanometer-sized objects like probe tips, these forces require power densities, which are too high to be of practical interest for use in near field imaging. Indirect i.e. light induced forces may have a higher sensitivity for light detection purposes. A promising candidate is the surface band bending in a semiconducting probe tip. The local light distribution and light induced processes on the sample can be measured simultaneously. An example is the investigation of light induced space charge gratings in photorefractive materials.

The main idea behind any near field optical measurement is the following: when a source or detector of sub-wavelength size is close to the sample the resolution is determined by the geometry of both of them, but not by the wavelength of the light. Similar concepts are used to describe the leaking out of microwave radiation of equipment. Since the separation of the source (detector) and the sample must be small compared to the wavelength of the light this implies that it is of the order of several nanometers. Such distances can only be stabilized by using an additional control interaction. Such an interaction might be tunneling [7], force microscopy using normal forces[21] or shear force microscopy[22,23]. Samples or probes are usually scanned using piezo ceramics[24,25]. Hence, near field optical microscopes are similar to other scanning probe devices.

2 Near Field Optics using Apertures

In this chapter we describe experiments with laser diodes and semiconductor samples. The piezoelectric shear force detection described in the first part was indispensable for the experiments.

2.1 INSTRUMENTATION

Most aspects of near field optical instrumentation are well described in the literature[26,27]. Hence, we limit our description to the piezoelectric shear force detection. A tapered glass fiber is glued into a metal tube and both are integrated in a mounting, sandwiched between two piezo segments. One of the piezo elements excites the fiber tip at mechanical resonance while the other one is used for detection. The main attractions of this setup are its simplicity, its compactness and the lack of disturbing light sources. The fiber is easy accessible and tip exchange is simple. The geometry also allows the measurement of samples covered with a few millimeters of liquid, which is important for applications in biology and medicine.

2.1.1 PIEZOELECTRICAL SHEAR FORCE SETUP

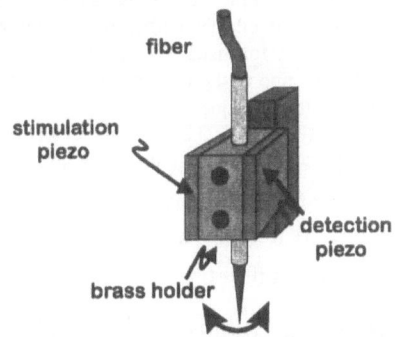

Figure 1: Illustration of the piezo electrical shear-force detection setup. The tapered glass fiber is glued into a metal tube and fixed with two setscrews in a tiny brass holder, sandwiched between two piezo plates. The stimulation piezo excites the tip at its mechanical resonance. The response signal is registered by the detection piezo and measured with a lock-in amplifier. The change in amplitude and phase during the surface approach is used for distance control.

The glass fibers for the aperture probe fabricated with a Sutter Micropipette Puller P-2000. Immediately after the pulling process, the fibers were fixed into a metal tube with cyanoacrylate glue, such that the fiber sticks 2 mm out of the metal tube. The tubes have an inner diameter of 0.26 mm, an outer diameter of 0.51 mm and a length of 19.0 mm. The resonance frequency of the fiber probe can be controlled by the tip length and is between 20 kHz and 100 kHz. The aperture at the end of the fiber is formed by angled deposition of a 100 nm layer of aluminum. The fiber is held by the piezoelectric detection head (see Figure 1). The holder (4 mm x 2 mm x 3.5 mm) is made of brass with a cylindrical hole of 1.2 mm diameter. The metal jacket of the fiber is held by two M1 set screws, allowing a fast tip exchange. Piezo plates (mass: 60 mg, dimensions: 4.5 mm x 3.5 mm x 0.5 mm) glued to the brass block form the excitation and detection system.

Figure 2 shows a typical resonance curve obtained with an excitation of 40 mV RMS. The fiber was damped by dipping it into a drop of oil. The piezoelectrical signal shows many and dispersive like resonances. In addition to the fiber resonances there are also spurious resonances of the brass holder or the piezos themselves.

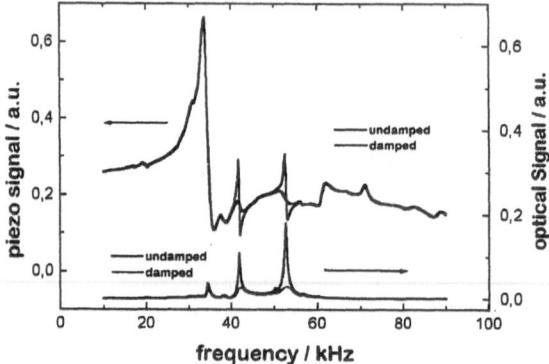

Figure 2: Shown is the signal at the detection piezo versus excitation frequency. The solid line represents the undamped system, while the dashed line is obtained by damping the oscillation by dipping the fiber tip into a drop of oil. Two fiber resonances are observed. The change in amplitude is an appropriate quantity for distance control. The inset illustrates the cross section of the detection system. The direction of the fiber oscillation is determined by the fixing of the metal tube and is not perpendicular to the piezo plates. The lower part shows the same resonance spectra, observed with a conventional optical shear-force detection system. The two dominating maxima correspond with the piezo electrical signals in the upper part.

The dispersive resonance curve of the fiber is the result of the interference of signal $A = a \cdot \cos(\omega \cdot t)$ transmitted directly from the stimulation piezo and the phase shifted signal $B = b \cdot \cos(\omega \cdot t - \delta)$ from the oscillating fiber tip.

$$S = s(\omega) \cdot \cos(\omega \cdot t - \vartheta(\omega)) = a \cdot \cos(\omega \cdot t) + b(\omega) \cdot \cos(\omega \cdot t - \delta(\omega)) \tag{1}$$

where a and $b\ (\omega)$ are the amplitudes of the signals from the stimulation piezo and the fiber tip, ω is excitation frequency and $\delta(\omega)$ the phase shift. Using the expressions for amplitude and phase of a driven harmonic oscillator

$$b(\omega) = \frac{b_0}{\sqrt{\left(1 - \dfrac{\omega^2}{\omega_0^2}\right)^2 + \dfrac{\omega^2}{Q \cdot \omega_0^2}}} \tag{2}$$

$$\tan \vartheta(\omega) = \frac{1}{Q} \cdot \frac{\omega \cdot \omega_0}{\omega_0^2 - \omega^2} \tag{3}$$

where Q is the quality factor of the system, we can calculate the amplitude s(ω) and the phase shift $\vartheta(\omega)$ of the detected signal S. The measured resonance curve agree quite reasonably with this model. (See Figure 3). Fitting the data we typically obtain $Q \approx 100$ for the fiber resonance.

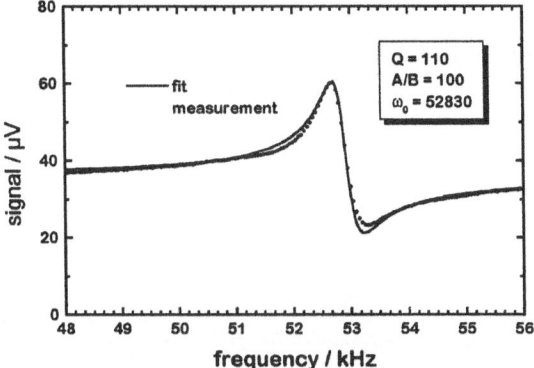

Figure 3: The amplitude of the resonance signal measured with the piezo electrical setup shows a dispersive like shape (dots). The signal at the detection piezo is assumed to be the interference of the signal transmitted directly from the stimulation piezo and a phase shifted signal from the resonating fiber tip. From this assumption we calculate a resonance behavior, represented as the solid line in the figure, with a Q-factor of 110, a relation of $a/b=100$ and a resonance frequency of $\omega= 52830$ Hz.

2.1.2 FIBER OSCILLATION AMPLITUDE

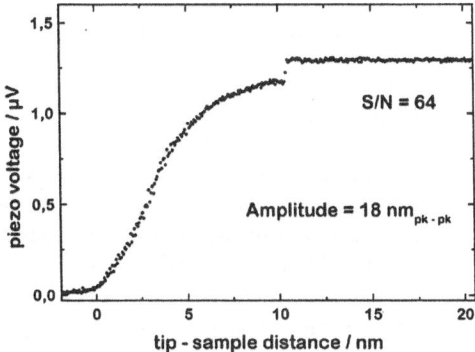

Figure 4: A typical tip-sample approach curve. An amplitude of 40 mV at the stimulation piezo corresponds to a fiber amplitude of 18 nm (peak to peak). The curve was measured with a lock-in time constant of $\tau = 1$ ms. The signal to noise ratio of 64 leads to a topographical resolution of $5\text{pm}/\sqrt{\text{Hz}}$. The step in the curve at a distance of about 10 nm from the sample surface is reproducible in various runs, but the origin is unclear.

Small and well-defined oscillation amplitudes of the fiber tip are crucial to preserve the optical resolution given by the aperture. The calibration of the fiber amplitude was done in two steps.

First the piezoelectrical signal was compared to the output of a classical optical shear force detection system[22,23]. The optical system itself was calibrated using a calibrated translation stage with 0.5 nm precision. A 40 mV RMS drive signal at the fiber resonance frequency resulted in an undamped fiber amplitude of 18 nm (peak to peak). Feedback regulation is done a few hundred Hz off the resonance. Together with the amplitude set-point of 50% this results in a fiber amplitude of 5-10 nm (peak to peak). From the approach curve (Figure 4) a topographical resolution of $5pm/\sqrt{Hz}$ is derived.

2.1.3 MEASUREMENTS IN WATER

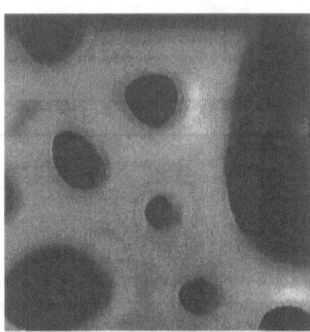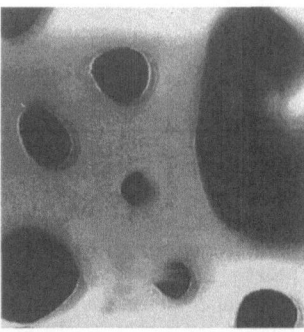

Figure 5: Shear force images of a spin coated PMMA-film on a glass substrate. The holes in the structure have a depth of about 230 nm. The sample was covered with 4 mm of water. The dither amplitude of the fiber during the scan was about 20 nm (peak to peak), which required a driving voltage of 500 mV.

Using the nearfield microscope in transmission mode, the sample was illuminated with the fiber tip and the transmitted (or fluorescent) light was focused onto an avalanche photo diode (APD) in a confocal arrangement[28]. The very small sensitive area of the APD forms the detection aperture. Since the relative positions of tip and sample must be fixed for a confocal microscope, the sample was scanned[29]. The microscope was tested using a glass slide spin coated with PMMA. After evaporation of the solvent, the sample forms a network like structure of 250 nm height with holes, 1-5 µm in diameter (see Figure 5). The piezoelectrical distance regulation works well on samples covered with water, a big advantage over standard optical distance regulation setups. Typically, the amplitude of the fiber resonance is reduced by a factor of 5 due to damping by the water. It is crucial to keep the water height at a fixed level, because the falling water level due to evaporation alters the resonance frequency of the fiber.

2.2 LASER DIODES

2.2.1 SAMPLES

The VCSEL's consist of a multilayer structure grown by molecular beam epitaxy on an n-GaAs substrate (see Figure 6). The active zone consists of three strained 8 nm InGaAs quantum wells embedded in 10 nm GaAs barriers. Surrounding AlAs and AlGaAs layers improve the longitudinal carrier confinement to the quantum wells. Details of the laser construction are published elsewhere[30,31].

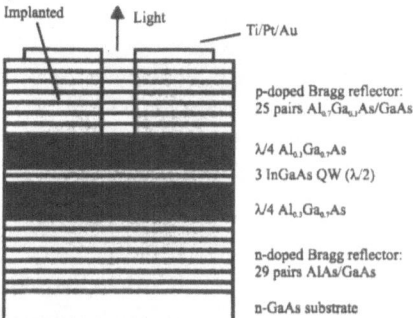

Figure 6: Structure of a VCSEL.

Current is supplied by a TiPtAu ring contact on the laser facet and a GeNiAu contact on the backside of the n-GaAs substrate. Lasers with an active diameter of 25 µm were investigated both in the far field (Figure 7). The threshold current was 6.5 mA at a voltage of 1.8 V. The output power saturates at 4 mW (driving current of 42 mA). Thermal heating induces a wavelength shift of 0.07 nm/K and a roll over of the output power due to the increasing mismatch of the mode and gain maxima. The emission wavelength is around 990 nm. Gain guiding is dominant for lower driving currents. Thermal lensing and spatial hole burning effects favors index guiding for higher drive currents[*].

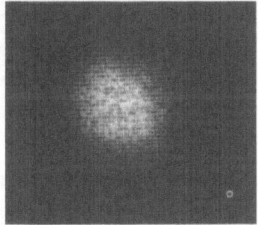 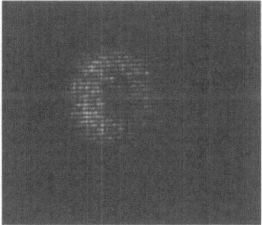

Figure 7: Far field pattern of a laser diode. Emission images of the laser device were taken at different injection currents (left: 12 mA, right: 22 mA) using a 20XN.A.0.40 microscope objective projecting the light on a video camera. The total light intensity appears Gaussian for low currents and doughnut shaped for higher injection currents.

2.2.2 INSTRUMENTATION DETAILS

The near field optical microscope was used in the collection mode. Measurements were performed using a 980 nm single mode Al-coated fiber tip. The fiber was coupled directly into a 275 mm monochromator with a 1200 lines/mm grating, giving rise to a spectral resolution of

---

[*] Note that in literature, those images are often referred to as "near-field" images, which is slightly misleading in this respect, since only the propagating near-field modes are being detected with far field microscopy.

0.08 nm. The scanning control program provided a TTL signal for each pixel, triggering a Peltier-cooled intensified CCD camera mounted at the exit of the spectrometer. It was possible to measure a complete spectrum every 200 ms (20 s per line or 20 min per 64x64 pixel image). The integral over selected regions in each spectrum was calculated for every pixel, forming an intensity map at a specific wavelength range. Furthermore, the wavelength position of the highest peak of each spectrum was determined. The VCSEL were powered by a low noise constant current source (stability: 20 pA, noise < 2pA). The temperature of the laser heat sink was stable within 100 mK

2.2.3 RESULTS

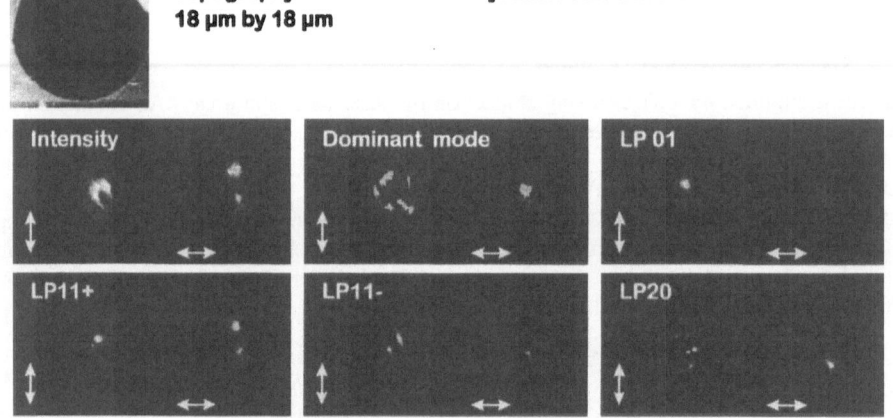

Figure 8: Spectral maps of a VCSEL Laser Diode. The top row shows the topography. The middle and the bottom rows show the different modes, polarization resolved.

Near-field images deduced from a spectral data set are shown in Figure 8 for an injection current of 3.0 mA. The scan range was 10x 10 μm for all images. Figure 8a) shows the topography. The Au contact ring for the current injection is centered on the top laser facet.

Laser modes from LP_{01} to LP_{41} are concurrently lasing. The LP_{01}, which in the far field is of Gaussian shaped, is in this near field measurement crescent-shaped. The LP^*_{11} mode together with the LP_{21} mode are dominating the pattern. Note the highly symmetric shape of the LP_{41}, mode showing the high lateral resolution of the measurement. The lateral resolution was estimated from the smallest structures imaged in the LP_{41} mode to about $\lambda/2=500$ nm which seems to be a rather arbitrary upper limit due to the intrinsic smoothness of the modes. On the other hand, from the regular shape of the LP_{41} mode, we can clearly exclude tip artifacts for an explanation of the anomalous shaped LP_{11} mode.

2.3 III-V QUARTERNARY SAMPLES

GaInAsP alloys with their wide tunable bandgap from 920 nm to 1630 nm grown lattice matched on InP are promising candidates for a broad range of optoelectronic devices[14]. Confined semiconductor structures in these material systems are important building blocks. To characterize the structures it is often needed to image them at a cleavage plane. We discuss experiments with these samples and show that the localization of the emission with respect to the sample structure is an important information.

2.3.1 SHEAR FORCE DETECTION AND „FALLING DOWN THE SAMPLE"

When imaging cleavage planes of a semiconductor wafer one often is confronted with the problem of taking images close to the border of the sample. The conventional scanning force microscope needs additional hardware or a very careful design of the experiment to avoid damage to the force-sensing cantilever. After the cantilever has fallen down from the sample its possible deflections when coming back to the edge are such that the feedback loop gets no signal indicating that the sample has a rising slope. Most often, the cantilever is bent sideways until it breaks. When the maximum extension of the z-translator allows only a tiny ($<1\mu m$) movement below the sample surface then the slope of the pyramidal or conical tip can save it.

The shear force detection scheme is well adapted to the problem of imaging samples with limited extension. The shear force mechanism reacts on the damping of the oscillation of the fiber tip. After falling down from the sample surface, the oscillating fiber tip knocks on the side wall of the sample. The operation is much like that of a tapping mode scanning force microscope[32-34]. Because of the interaction, the amplitude is decreased. Since this is exactly the same behavior as a sample' surface too close to the tip would induce, the feedback loop reacts correctly and withdraws the tip. The piezoelectrical shear force detection described above is especially well suited to the problem because it has well defined oscillation directions[35,36].

2.3.2 INVESTIGATED STRUCTURES

The samples are grown on a [100]-InP wafer partially covered by SiO_2. First, a 1 μm thick InP buffer was grown. By selective area epitaxy in MOMBE the III-V alloys grow only on the substrate but not on the oxide leading to sharp growth to non-growth transitions with various side facets. InGaAsP with a subsequent InP cap have been grown over the InP buffer. Samples investigated include one consisting of a single 400 nm thick $In_{0.9}Ga_{0.1}As_{0.24}P_{0.76}$ layer with a luminescence maximum at 1050 nm and one consisting of five 10 nm $In_{0.9}Ga_{0.1}As_{0.24}P_{0.76}$ layers each separated by 30 nm InP barriers. The structure formation is intimately connected with adsorption, diffusion and desorption of the reactants on the surface. Details of the process are published elsewhere[14].

2.3.3 SETUP OF THE REFLECTION MODE NEAR FIELD MICROSCOPE

Two setups for the reflection mode spectroscopy have been used. In the illumination mode (see Figure 9) the sample is illuminated through the fiber aperture and the luminescence is collected in the far field. We use a multimode fiber with a numerical aperture of N.A.=0.48 for collection. Compared to the collection with a microscope objective this setup is more compact and allows an easy coupling of the luminescence into a spectrometer. In addition, fiber arrays could collect a wider solid angle in a very compact arrangement.

164

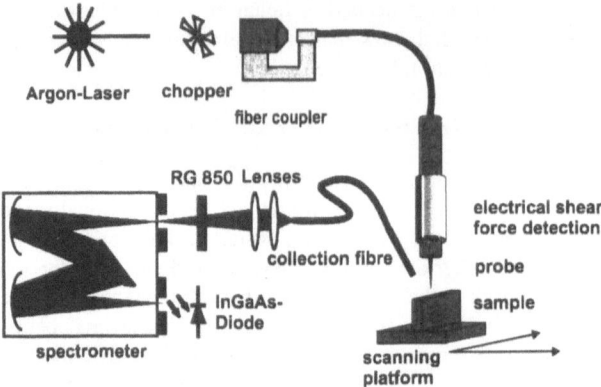

Figure 9: In the illumination mode setup, an Argon Ion Laser beam is coupled into a glass fiber with a small aperture at its end. This small aperture illuminates the sample in near field. The luminescence light is collected in far field by a multimode IR-fiber that has a numerical aperture N.A. of 0.48. The light, collected by the multimode fiber, is coupled into a spectrometer and is detected by an InGaAs-photodiode.

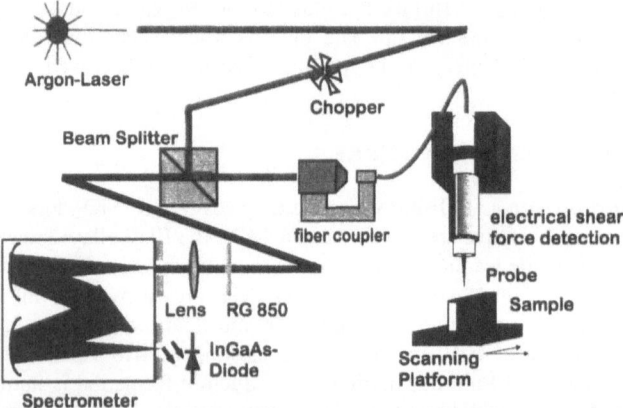

Figure 10: In the collection mode setup the Argon Ion laser beam is coupled into the fiber as in illumination mode. The fiber tip illuminates the sample in near field. The luminescence light is collected through the same aperture. A beam splitter cube separates the luminescence light from the incident beam. The luminescence light is coupled into the spectrometer and is detected by an InGaAs diode.

In collection mode, the sample is again illuminated through the fiber but the luminescence is now collected through the same aperture (Figure 10). A beam splitter cube and subsequent filters separates the luminescence light from the incident beam. The luminescence is then coupled into the spectrometer and is detected in the same manner as described in the section before. Because of the low transmission of a coated fiber ($\approx 10^{-4} - 10^{-5}$), this setup only works with uncoated fibers[37-39]. Even with uncoated tips, the illumination density has to be 1-5 MW/cm^2 to get a reasonable luminescence signal. Typically, the scan time per line is 25 seconds due to the integration time of the detector system.

2.3.4 ILLUMINATION MODE RESULTS

A double heterostructure ridge with a quaternary layer of 400 nm thickness was investigated. Near field optical images at wavelengths of 1075 nm and 1175 nm (4 nm spectral width) were measured[14] (See Figure 11: a: topography, b: intensity at 1075nm c: intensity at 1175 nm). In b) the decrease of luminescence at 1075 nm towards the edge is clearly seen while peak intensity at 1175 nm in c) seems to be 1 µm away from the edge. Hence, the vertical grown quaternary layer on the sidewall most probably is responsible for the luminescence. Luminescence occurs even if the tip is on the side of the structure (dark areas in the topographic image). This is an artifact due to the illumination of the sidewall of our structure. This effect disappears when the sidewall is coated with metal.

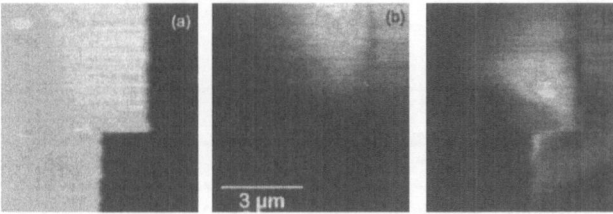

Figure 11: Figure a shows the topography of the investigated structure. In the luminescence map at 1075 nm the decrease of luminescence towards the edge is visible. At the edge the second recombination channel appears as shown in figure e. From this picture one might suppose that this luminescence is caused by the vertical grown quaternary layer on the sidewall. In these spectral maps, luminescence occurs if the tip is next to the structure (dark areas in the topographic image). As figure b,c shows the resolution is limited to about 4 µm. This is a result of charge carrier diffusion.

The diffusion length of the carriers limit the resolution to 4 µm. Electrons and holes generated outside the quarternary layer will contribute to the luminescence, as long as they can reach the low bandgap region. The luminescence data gives a tool to infer the diffusion length of the electrons and holes in InP. Using the continuity equation[14] a diffusion length L to (4.4 ± 0.5) µm was found, in good agreement to other diffusion length measurements done on comparable materials.

2.3.5 COLLECTION MODE RESULTS

Since the resolution in illumination mode is limited by the migration of the charge carriers collection mode experiments were carried out on the multiple quantum well structure. Spectral maps were acquired at different wavelengths (Figure 12). A decrease of the luminescence towards the edge is visible at 1016 nm. It is the result of a luminescence shift towards higher wavelengths. Furthermore, the charge carriers are trapped by the vertical sidewall that forms a region of lower energy. The map at 1060 nm shows an increasing luminescence approaching the edge. This is better visible in the map at 1115 nm where the luminescence of the horizontal superlattice has vanished. At this wavelength, only the side facets have a recombination channel. This is caused by the different material composition due to different growth condition on these vertical sidewalls.

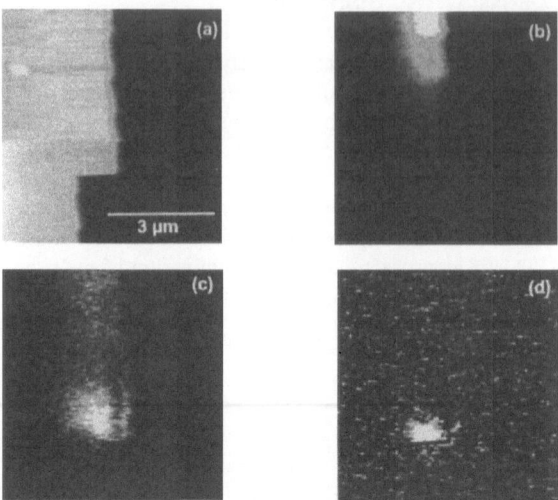

Figure 12: Topography (a) and luminescence maps at 1016 nm, 1060 nm and 1115 nm. At 1016 nm (b) a decrease of the luminescence towards the edge is visible. The decrease at this wavelength is a result of luminescence shift towards higher wavelength. Furthermore, the charge carriers are trapped by the vertical facets that form a region of lower energy as the next picture shows. The map at 1060 nm (c) shows an increasing luminescence at the edge. This is more clearly shown in the map at 1115 nm (d) while the luminescence of the horizontal layer has vanished.

3 Photons and Scanning Force Microscopy

Normally near field optical microscopes detect light by either guiding or scattering it into a detector. The following chapter demonstrates that the silicon tip of a scanning force microscope cantilever can act as a light detector itself.

3.1 LIGHT-INDUCED ELECTROSTATIC FORCES

Semiconducting probe tips are subject to the surface photo-voltage (SPV), a mechanism that can charge the tips and consequently apply a force to the cantilever[19,20]. A light beam is totally reflected internally (TIR) (Figure 13). The evanescent field close to the prism surface is probed by the tip mounted at the end of a micro-cantilever. Standard optical beam deflection is used to measure the force[40,41]. The charges on the tip surface generated by the SPV[42-44] locally induce surface charges on the prism, yielding a net attractive image force. The force on the tip is a function of the illumination. Qualitatively the force can be understood by using the theory of metal-insulator-semiconductor junctions[19,20], where the free charges in the metal are replaced by the bound charges in glass.

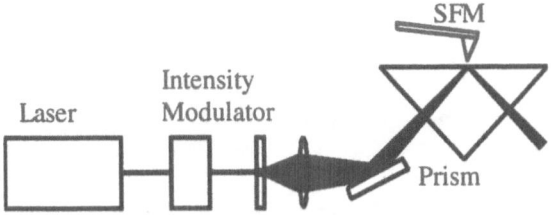

Figure 13 Experimental layout. An intensity modulated laser beam undergoes total internal reflection at the glass-air interface. Here it is imaged by a scanning force microscope (SFM).

Illumination results in a reduction in the surface photo-voltage $|\psi|$ (SPV). The response time of the SPV is given by the carrier diffusion rates through the depletion layer potential, and as such the band bending is effectively unpinned by the light. In the regime of modest light intensity the SPV may be approximated by[45,46]

$$|\delta\psi| \cong A \ln\left(1 + \frac{I}{I_s}\right) \tag{4}$$

where I is the light intensity incident on the tip, and A and I_s are parameters independent of I. In the regime of large light intensity the bands are driven flat and the SPV saturates at a value equal to the band bending with no illumination. The ac force resulting from the intensity modulation $I(\omega) = I_0(1 + M \cos(\omega t))$ is then

$$|\delta F| \cong A M (V - \psi - \phi_s)\left(\frac{I_0}{I_0 + I_s}\right)\frac{dC_i}{dz} \tag{5}$$

ω is set to the mechanical resonance of the cantilever spring. Its deflection is monitored by a lock-in amplifier. The probe is made of n-doped silicon (resistivity 0.02 Ωcm) with a tip axis in the (100) direction. The tip geometry is specified by the manufacturer to a length of 10-20 μm, an apex angle 45°, and a radius of curvature 30 nm. The experiment was performed in ambient The spring constant, resonance frequency, and the quality factor of the cantilever are about $k = 30$ N/m, ($\omega_0 = 160$ kHz, and $Q = 300$ respectively. The spot radius of the light inducing the SPV under the tip is about 100 μm.

3.2   OPTICAL NEAR-FIELD DETECTION AND IMAGING

For a fixed tip-prism distance, the ac force is found to vary almost linearly with the applied voltage (Figure 14). The force is exactly in-phase or out-of-phase with the light. The sign changes at a voltage in the range 2-5 volts depending on the tip work function. No additional phase lag is observed, allowing us to exclude photo-thermal effects[47]. The ac force δF is linear in the modulation index M, and nonlinear in the mean intensity I_0, as shown in (Figure 15) for $V = -5$ volts at a distance of 30 nm. The best fit is $I_s = 25$ pW. Hence, the ac force saturates near 10 pN. The thermally limited minimum detectable power level in air is typically $0.1 \, pW/\sqrt{Hz}$ ($V = -5$ volts).

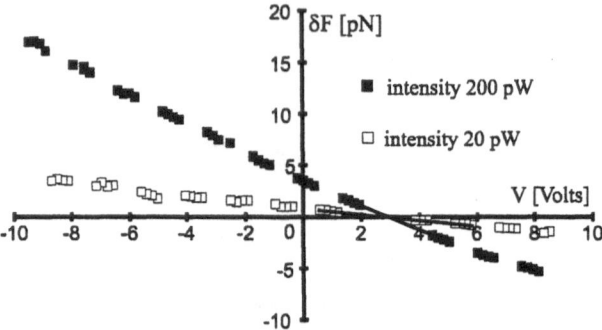

Figure 14. AC force exerted on the probe tip at the cantilever resonance frequency as a function of the voltage V applied to the tip for two light intensities incident on the tip: o = 20 pW, • = 200 pW. Negative values of the ac force correspond to a phase change of 180° relative to the light modulation. The measurement was performed in vacuum with an Ar+ laser (514 nm) and an evanescent field decay length of 230 nm. The tip-sample distance was approximately 30 nm.

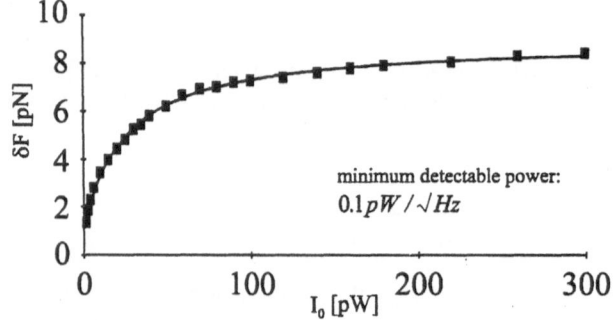

Figure 15. AC force exerted on the probe tip at the cantilever resonance frequency (160 kHz) as a function of DC light intensity incident on the tip. Measurement performed in vacuum with an Ar+ laser (514 nm) and an evanescent field decay length of 230 nm. The tip-sample distance was approximately 30 nm. The solid line is a best fit according to equation (5).

Increasing the voltage applied to the tip allows improves the sensitivity somewhat. Shifts in the cantilever resonance frequency are used to infer the gradient of the dc force on the tip. The DC force is a parabolic function of V. A semiquantitative analysis can be done by modeling the tip as a sphere[19,20].

The optical resolution of the microscope is evaluated by imaging a standing wave fringe pattern. The pattern is formed by back-reflecting the TIR beam in with a piezo mounted mirror. A scan of the tip over the immobile fringes, or a scan of the fringes under an immobile tip, yields the same result. The latter measurement ensures that there is no interference by topographical effects. A line cut of this image is shown in Figure 16. We observe a fringe contrast of 50% and the signal to noise ratio of 30 dB. The lateral resolution is estimated by the FWHM of a Gaussian window function. With the known fringe period, a deconvolution of the data yields a lateral resolution of better than 110 nm. The visibility in the fringe pattern for a Dirac delta detector is assumed to be 100%.

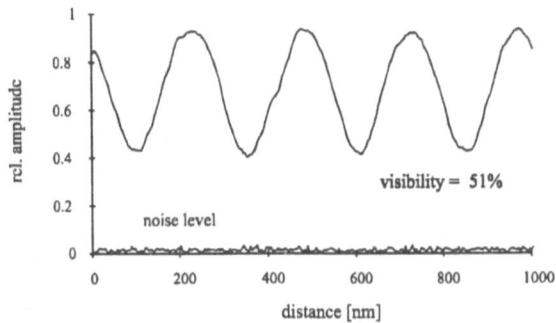

Figure 16. Image of standing evanescent wave on prism surface. Observed fringe visibility is 50 %. Measurement performed in air with a He-Ne laser (633 nm) and an evanescent field decay length of 136 nm.

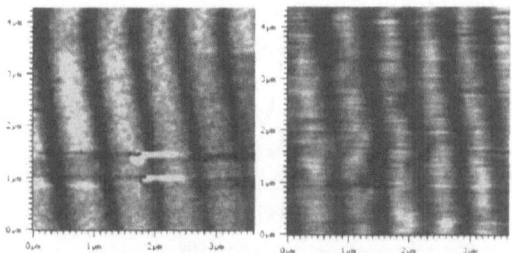

Figure 17. Optical and charge grating on BaTiO$_3$. Left: standing light wave (illumination 0.05 W/cm^2), right: charge grating (illumination 5 W/cm^2). The image size is 3.5 x 3.5 μm^2. The light intensity to measure the charge grating was 100 times that of the optical image.

The resolution depends on the tip geometry as in all scanning probe techniques. An increase of the measured fringe contrast was observed when using a shorter decay length of the evanescent field in the TIR setup. Optical superresolution in a scanning force microscope is possible. Combining the optical contrast with classical scanning force microscopy gives a powerful new near field optical microscope.

3.3 APPLICATION TO PHOTOREFRACTIVE MATERIALS

The most attractive feature of this setup is that measurement modes such as magnetic field detection or electrical charge detection can be combined with optical microscopy with one detector. We show here that charges and light intensities can be measured simultaneously. The light induced space charge gratings[48] in a photorefractive crystal[49] are ideal candidates to test the performance of our microscope. A standing wave pattern in the photorefractive crystal creates a spatial modulation of the charge distribution via photo excitation and recombination of free charge carriers. The space charge fields modulate the index of refraction by the electrooptical effect[50,51]. The charge grating is spatially shifted with respect to the generating light intensity grating with a phase depending on the transport mechanism. A phase shift of 90° is indicative of diffusion dominated transport. On the surface, these charge gratings can be imaged with the SFM[52]. A total internal reflection geometry was used to illuminate the photorefractive sample with two laser beams from the back. The sample was optically connected to the coupling prism with index matching liquid.

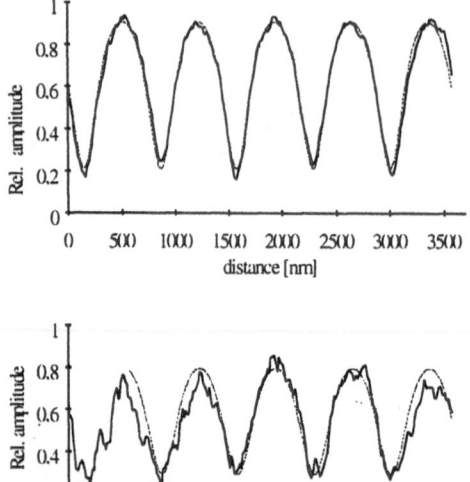

Figure 18: Cross section showing the phase relationship between the optical grating (top) and the charge grating (bottom) at the surface. The fringe visibility is 83 % for the light and 68% for the charges. The phase shift between the two gratings is 60°.

Figure 17 shows the result of a measurement of a charge grating on BaTiO$_3$. The image size is 3.5 x 3.5 µm^2. On left, the light intensity grating is displayed. While the writing beams are on, the charge buildup proceeds. Since the surface photovoltaic effect saturates at intensities of 25 pW one can switch off all the influences of the light by increasing the intensity. The sensitivity for the optical signal drops whereas that of the charge signal increases. Hence, at high optical intensities the force microscope is only sensitive to the charge (see Figure 17b). The image clearly reveals a spatial modulation with the same period as the light intensity distribution.

Finally, Figure 18 shows a first simultaneous measurement of the surface charge grating and the optical grating. A relatively small phase shift between the two gratings is obvious, in strong contrast to the theoretical expectation. A fit gives $\varphi \cong (15 \pm 5)°$. The intensity used in the experiment is the same as that of Figure 17. A cross talk between the charge and the optical signal can not be entirely excluded. However, the bulk phase shift of $\varphi \cong 90°$ was never observed in measurements with the scanning force microscope. A complicated transport mechanism with a pronounced surface might be the reason for this observation.

4 Summary

We have shown in this article that near field optical microscopes based both on the classical aperture/shear force scheme and on the scanning force microscope scheme can be applied to physical problems in materials science. A compact, rugged piezoelectrical shear force detection scheme working at ambient and with water covered samples was essential for the aperture based experiments. The topographical resolution was $5pm/\sqrt{Hz}$ for an oscillation amplitude of 5-10 nm (peak to peak) of the fiber. Spectrally resolved nearfield images of transversal modes in near-IR vertical cavity modes surface emitting semiconductor lasers allow a comparison of the topographical information to the optical one. Near-field spectral imaging of laser diodes appears to be a very promising means in the optimization of the fabrication process with respect to an optimal current injection into the active area of the devices.

Cleavage surfaces of InP wavers with overgrown structures were imaged to characterize the active layers. The poor resolution of 4 µm in illumination mode is indicative of charge carrier diffusion. Near filed optical images can be interpreted if the diffusion length is 4.4 µm. Diffusion effects can be suppressed in the collection mode, greatly improving the lateral resolution to 500 nm at a wavelength of 1115 nm.

Finally, it was demonstrated that the surface photovoltage effect in a semiconducting probe tip offers a detection mechanism to easily detect light via electrostatic forces at low power densities. A lateral resolution of 70-110 nm is routinely achieved. In a TIR geometry, the detection allows an alternative approach to optical near-field microscopy on transparent samples by force detection. The detection mechanism can be combined with other standard techniques in a SFM to simultaneously detect local light intensity and surface charge distributions. As an example the phase shift of the optical and of the charge grating in the photorefractive material $BaTiO_3$ was measured. A value of $15°±5°$ was found, in contrast to the bulk value of 90°. It is hypothesized that surface specific transport mechanisms might be the reason.

Photons and local probes are powerful tools because they combine the best of two worlds. Optics provides a high energy resolution; local probes are responsible for the spatial resolution.

5 Acknowledgments

The experiments described in this paper were carried out both at the University of Ulm and at the University of Konstanz. We gratefully acknowledge the help and the cooperation of H. Bielefeldt, J. Koenen (Ulm). G. Krausch (Konstanz), J. Mertz (Konstanz). M. Pietralla (Ulm), A. Simon (Ulm), and G. Volswinkler (Ulm). Samples were provided by the groups of H. Heinecke (Ulm) and K.-J. Ebeling (Ulm). This work was supported in part by the Deutsche Forschungsgemeinschaft (Sonderforschungsbereich 306 (Konstanz) and Sonderforschungsbereich 239 (Ulm)) and the German Bundesministerium für Bildung. Wissenschaft. Forschung und Technologie (BMBF) under Grant No. 13N6522 (Ulm).

172

6 References

[1]   W. Demtroeder, *Laser spectroscopy*, 2nd ed. (Springer, Berlin ; Heidelberg. 1996).

[2]   R. Wiesendanger, *Scanning probe microscopy and spectroscopy* (Cambridge Univ. Press, Cambridge, 1994).

[3]   E. H. Singe, "Suggested method for extending microscopic resolution into the ultra-microscopic regime," *Phil. Mag.* **6**, 356 (1928).

[4]   J. A. O'Keefe, "Resolving power of visible light," *J. Opt. Soc. Am. A* **46** (5), 359 (1956).

[5]   E. A. Ash and G. Nichols, "Super resolution aperture scanning microscope," *Nature* **237**, 510 (1972).

[6]   U. C. Fischer and H. P. Zingsheim, "Submicronic contact imaging with visible light by energy transfer," *Appl. Phys. Lett.* **40** (3), 195-197 (1982).

[7]   D. W. Pohl, W. Denk, and M. Lanz, "Optical stethoscopy: image recording with $\lambda/20$." *Appl. Phys. Lett.* **44**, 651-653 (1984).

[8]   A. Lewis, M. Isaacson, A. Harootunian, and M. Muray, "Development of a 500 Å resolution light microscope," *Ultramicroscopy* **13**, 243-312 (1984).

[9]   U. Dürig, D. W. Pohl, and F. Rohner, "Near-Field Optical-Scanning Microscopy," *J. Appl. Phys.* **59**, 3318-3327 (1986).

[10]   E. Betzig, H. Barshatzky, A. Lewis, M. Isaacson, and K. Lin, "Super-Resolution Imaging with Near Field Scanning Optical Microscopy (NSOM)," *Ultramicroscopy* **25**, 155-163 (1988).

[11]   N. F. van Hulst and M. H. P. Moers, "Biological applications of near-field optical microscopy," *IEEE Engineering in Medicine and Biology Magazine* **15** (1), 51-58 (1996).

[12]   W. M. Duncan, "Near-field scanning optical microscope for microelectronic materials and devices," *Journal of Vacuum Science & Technology A   Vacuum Surfaces and Films* **14** (3), 1914-1918 (1996).

[13]   D. A. Van den Bout, J. Kerimo, D. A. Higgins, and P. F. Barbara, "Spatially resolved spectral inhomogeneities in small molecular crystals studied by near-field scanning optical microscopy," *J. Phys. Chem.* **100** (29), 11843-11849 (1996).

[14]   J. Barenz, A. Eska, O. Hollricher, O. Marti, M. Wachter, U. Schöffel, and H. Heinecke, "Near field luminescence measurements on GaInAsP/InP doubleheterostructures at room temperature," (1997), in preparation.

[15]   I. Hörsch, R. Kusche, O. Marti, B. Weigl, and K. J. Ebeling, "Spectrally resolved near-field mode imaging of vertical cavity semiconductor lasers," J. Appl. Phys. 79 (8 Part 1), 3831-3834 (1996).

[16]   J. D. Pedarnig, M. Specht, and T. W. Hänsch, "Fluorescence Lifetime Variations and Local Spectroscopy in Scanning Near-Field Optical Microscopy," in *Photons and Local Probes*, edited by Othmar Marti and Rolf Möller (Kluwer Academic Publishers. Dordrecht. The Netherlands, 1995), Vol. E 300, pp. 151-163.

[17]   J. D. Pedarnig, M. Specht, M. Heckl, and T. W. Hänsch, "Scanning Plasmon Near-Field Microscope," in *Near Field Optics*, edited by D. W. Pohl and D. Courjon (Kluwer Academic Publishers, Dordrecht. The Netherlands, 1993), Vol. E 242, pp. 273-280.

[18] F. Zenhausern, M. P. O'Boyle, and H. K. Wickramasinghe, "Apertureless near-field optical microscope," *Appl. Phys. Lett.* **65** (13), 1623-1625 (1994).

[19] J. Mertz, M. Hipp, J. Mlynek, and O. Marti, "Optical Near Field Imaging with a Semiconductor Probe Tip." *Appl. Phys. Lett.* **64**, 2338-2340 (1994).

[20] M. Hipp, J. Mertz, J. Mlynek, and O. Marti, "Optical Near-Field Imaging by Force Microscopy," in *Photons and Local Probes*, edited by Othmar Marti and Rolf Möller (Kluwer Academic Publishers, Dordrecht, Netherlands, 1995), Vol. E 300, pp. 109-122.

[21] N. F. van Hulst, M. H. P. Moers, O. F. J. Noordman, T. Faulkner, F. B. Segerink, K. O. van der Werf, B. G. de Grooth, and B. Bölger, "Operation of a scanning near field optical microscope in reflection in combination with a scanning force microscope," in *Scanning Probe Microscopies* (SPIE, Bellingham, WA 98227, USA, 1992), Vol. 1639, pp. 36-43.

[22] R. Toledo-Crow, P. C. Yang, Y. Chen, and M. Vaez-Iravani, "Near-field differential scanning optical microscope with atomic force regulation," *Appl. Phys. Lett.* **60** (24), 2957-2959 (1992).

[23] E. Betzig, P. L. Finn, and J. S. Weiner, "Combined Shear Force and near-field scanning optical microscopy," *Appl. Phys. Lett.* **60** (20), 2484 (1992).

[24] S. Vieira, "The behaviour and calibration of some piezoelectric ceramics used in STM," *IBM J. Res. Develop.* **30**, 553 (1986).

[25] G. Binnig and D. P. E. Smith, "Single-tube three-dimensional scanner for scanning tunneling microscopy," *Rev. Sci. Instrum.* **57**, 1688 (1986).

[26] D. W. Pohl and D. Courjon, *Near Field Optics* (Kluwer Academic Publishers, Dordrecht, 1993).

[27] O. Marti and R. Möller, *Photons and Local Probes* (Kluwer Scientific Publishers, Dordrecht, 1995).

[28] E. H. K. Stelzer and S. Lindek, "Fundamental reduction of the observation volume in far-field light microscpy by detection orthogonal," *Opt. Comm.* **111**, 536-547 (1994).

[29] PI Physik Instrumente, "PZT Model P-730.20," .

[30] K. J. Ebeling, *Integrierte Optoelektronik*, 2. Aufl. ed. (Springer, Berlin ; Heidelberg, 1992).

[31] U. Fiedler and K. J. Ebeling, "Design of VCSEL's for feedback insensitive data transmission and external cavity active mode-locking," *IEEE J. on Selected Topics in Quantum Electronics* **1** (2), 442-450 (1995).

[32] P. K. Hansma, J. P. Cleveland, M. Radmacher, D. A. Walters, P. E. Hillner, M. Bezanilla, M. Fritz, D. Vie, H. G. Hansma, C. B. Prater, J. Massie, L. Fukunaga, J. Gurley, and V. Elings, "Tapping Mode Atomic Force Microscopy in Liquids," *Appl. Phys. Lett.* **64**, 1738-1740 (1994).

[33] J. P. Spatz, S. Sheiko, M. Möller, R. G. Winkler, and O. Marti, "Forces affecting a substrate in tapping mode," *Nanotechnology* **6**, 40-44 (1995).

[34] R. G. Winkler, J. P. Spatz, S. Sheiko, M. Moller, P. Reineker, and O. Marti, "Imaging material properties by resonant tapping-force microscopy: A model investigation," *Physical Review B Condensed Matter* **54** (12), 8908-8912 (1996).

[35] J. Barenz, O. Hollricher, and O. Marti, "An easy-to-use non-optical shear-force distance control for near-field optical microscopes," *Rev. Sci. Instrum.* **67** (5), 1912-1916 (1996).

174

[36]   R. Brunner, A. Bietsch, O. Hollricher, and O. Marti, "Distance Control in Near-Field Optical Microscopy with electrical shear force detection suitable for imaging in liquids," *Rev. Sci. Instrum.* **68** (4), 1769-1772 (1996).

[37]   C. Girard and M. Spajer, "Model for reflection near field optical microscopy," *Appl. Opt.* **29**, 3726 (1990).

[38]   G. Krausch, S. Wegscheider, A. Kirsch, H. Bielefeldt, J. C. Meiners, and J. Mlynek, "Near field microscopy and lithography with uncoated fiber tips: A comparison," *Opt Commun* **119** (3-4), 283-288 (1995).

[39]   V. Sandoghdar, S. Wegscheider, G. Krausch, and J. Mlynek, "Reflection scanning near-field optical microscopy with uncoated fiber tips: How good is the resolution really?," *J. Appl. Phys.* **81** (6), 2499-2503 (1997).

[40]   S. Alexander, L. Hellemans, O. Marti, J. Schneir, V. Elings, P. K. Hansma, M. Longmire, and J. Gurley, "An atomic-resolution atomic-force microscope implemented using an optical lever," *J. Appl. Phys.* **65**, 164 (1989).

[41]   G. Meyer and N. M. Amer, "Novel optical approach to atomic force microscopy," *Appl. Phys. Lett.* **53** (12), 1045-1047 (1988).

[42]   R. J. Hamers and K. Markert, "Atomically Resolved Carrier Recombination at Si(111)-(7x7) Surfaces," *Phys. Rev. Lett.* **64**, 1051 (1990).

[43]   J. M. R. Weaver and H. K. Wickramasinghe, "Semiconductor characterization by scanning force microscope surface photovoltage microscopy," *J. Vac. Sci. Technol. B* **B9** (3), 1562-1565 (1991).

[44]   M. Nonnenmacher, M. P. O'Boyle, and H. K. Wickramasinghe, "Kelvin Probe Force Microscopy," *Appl. Phys. Lett.* **58**, 2921 (1991).

[45]   W. H. Brattain and J. Bardeen, "Surface properties of germanium," *Bell Syst. Tech. J.* **32**, 1 (1953).

[46]   M. H. Hecht, "Photovoltaic effects in photoemssion studies of Schottky barrier formation," *J. Vac. Sci. Technol B* **8**, 1018 (1990).

[47]   J. Mertz, O. Marti, and J. Mlynek, "Regulation of a Microcantilever Response by Active Control," *Appl. Phys. Lett.* **62**, 2344-2346 (1993).

[48]   H.-J. Eichler, P. Guenter, and D. W. Pohl, *Laser-induced dynamic gratings* (Springer, Berlin ; Heidelberg ; New York ; Tokyo, 1986).

[49]   P. Guenter, *Electro-optic and photorefractive materials* (Springer, Berlin ; Heidelberg, 1987).

[50]   P. Guenter, *Fundamental phenomena* (Berlin ; Heidelberg, 1988).

[51]   P. Guenter, Survey of applications (Berlin ; Heidelberg, 1989).

[52]   B. D. Terris, J. E. Stern, D. Rugar, and H. J. Mamin, "Contact electrification using force microscopy.," *Physical Review Letters* **63** (24), 2669-72 (1989).

QUO VADIS, NEAR-FIELD OPTICS?

D. W. POHL
IBM Research Division, Zurich Research Laboratory,
CH-8803 Rüschlikon, Switzerland

B. HECHT
Laboratory for Physical Chemistry, ETH Zürich,
CH-8092 Zurich, Switzerland

AND

H. HEINZELMANN
Institute of Physics, University of Basel,
CH-4056 Basel, Switzerland

Abstract. Scanning near-field optical microscopy (SNOM) has pushed today's optical resolution limit to about 20 nm, with a potential for improvement by another order of magnitude still remaining. SNOM provides chemical specificity based on spectral (amplitude, phase), polarization, and/or fluorescence contrast; it also allows dynamic studies with femtosecond time resolution and photochemistry on the nanometer scale. In spite of these prospects and considerable ongoing research efforts, progress in near-field optical microscopy has been rather slow in the past few years.

1. Introduction

The first scanning near-field optical microscope (SNOM) became operational as early as 1983/84 [1], but the technique attracted little attention at the time, remaining a Sleeping Beauty for almost ten years. In the early nineties, however, interest in combining high-resolution topographic imaging and local chemical characterization began to grow. One of the instruments offering this capability was obviously SNOM, which allows the entire tool box of optical characterization methods to be applied to objects of nanoscale size. The NATO ARW on "Near Field Optics" at Arc-et-Senans, France, in October of 1992 [2] may have been the event at which Sleeping

175

N. García et al. (eds.), Nanoscale Science and Technology, 175–183.

Beauty was kissed awake (with perhaps E. Betzig in the role of the prince [3]).

2. The three concepts

In the meantime, Sleeping Beauty has become a reasonably attractive young adult, though she might have had a little too much fast food in her youth. Like a teenager, she likes to confuse the minds of her admirers, for instance concerning which form of electromagnetic field confinement would be best suited for *near-field optical* (NFO) microscopy: Is it the *proximity field* that can build up next to an illuminated small aperture? Or is the evanescent wave of the type generated by total internal reflection (TIR) sufficient? The resolving capability would scale with the dimensions of the aperture in the first case, and with the TIR decay length, which is a sizable fraction of the wavelength, in the second. Furthermore, with respect to proximity fields, why not use a radiating molecular dipole, a Mie scatterer, or another highly curved surface structure as NFO probe?

2.1. APERTURE SNOM

The first concept led to the development of *aperture* SNOM [1, 3, 4]. This is the most commonly and best understood SNOM technique today. The next section of the present paper will therefore be devoted to this technique only.

The main advantage of aperture SNOM is a perfect confinement of the probe light field to the aperture area. Major disadvantages are the fragility of the aperture probe, its low throughput, and the difficult fabrication process. An inherent limit with respect to resolution is imposed by the finite transmissivity of the metal coating around the aperture; it restricts the minimum achievable *optical* diameter to about 10 nm.

2.2. "TUNNEL" SNOM

The second concept led to the development of a technique known by the acronyms *STOM* or *PSTM* [5-7] which stand for scanning tunneling optical microscopy (or microscope), and photon STM, respectively. This type of microscopy has the advantage of requiring merely a transparent probe tip without coating. However, the resolving power of this technique is not yet well understood. Some experimental results seem to indicate a resolution capability of as little as 20 nm [8]. In the majority of published results, however, the presented resolution does not fall below 100...200 nm, which corresponds to $\lambda/6 - \lambda/3$, where λ is the wavelength of the probe light.

Such a resolution can also be achieved with classical optical microscopes and therefore is not very attractive for practical applications.

From a theoretical point of view, one may argue that the process of *photon tunneling*, i.e. frustrated TIR, from the evanescent wave will extend over a distance of the decay length, say $\lambda/2\pi$, cf. Fig. 2 of [9], resulting in a lateral resolution capability of similar size, i.e. in the 100-nm range. If, however, light was coupled very effectively into the transparent probe tip at its highly curved apex, then a resolution limit equal to the radius of curvature, say 10 to 20 nm, may be expected. In this case, however, TIR-type evanescence would play only an auxiliary role, and the proximity fields would again determine the imaging process. The available experiments and numerical simulations do not yet allow an assessment of the significance of these two effects.

2.3. SCATTERING PROBE SNOM

Scattering probes hold promise for achieving higher resolution than aperture probes because the apex radius of curvature can be made arbitrarily small. This third concept resulted in a number of interesting exploratory efforts [10-14] but has not yet led to an established technique.

It has long been recognized and was demonstrated as early as 1989 [10, 11] that the highly curved surfaces of small particles or the apex of a pointed tip efficiently confine light in their immediate vicinity when illuminated by a light source in an appropriate way [15, 16]. In order to build up a strong, confined field in a tip/sample system, it is necessary that the incident light be polarized in the direction of the tip axis. Particularly strong field enhancement occurs in such a geometry if the conditions for generation of a *tip plasmon* can be met [15]. Light polarized perpendicular to the axis of the tip, on the other hand, is attenuated next to the tip apex [9] as a consequence of the boundary conditions for the electric field. The imaging properties of the recently introduced "apertureless" SNOM [14] are quite surprising in view of this result.

3. The topographic artifact

Another source of confusion in the short history of the Sleeping Beauty stems from its flirt with other scanning probe microscopies. As a result of simultaneous NFO and force or tunnel microscope operation, a significant artifact is frequently generated in the NFO image [17-21]. A look at SNOM literature shows that at least half of all published SNOM images are suspected of being influenced or even dominated by this effect.

The artifact generates topographic features in SNOM images with the resolution of the nonoptical distance control mechanism (frictional "shear" force or tunneling, in general), independent of the NFO resolving power of the probe. The features generated in this way represent the path of the probe rather than the optical properties of the sample. They therefore may feign a high resolution capability even if an instrument is completely inappropriate for NFO imaging. Some examples of such behavior are discussed in Ref. [18].

It is to be feared that Sleeping Beauty might fall into disgrace in the eyes of many admirers after this prank..., which would be a pity indeed because she still is full of promise: SNOM can be used in a variety of operating modes providing a wealth of information with the help of spectral, polarization, and/or fluorescence contrast. Furthermore it is the only scanning probe microscopy suitable for the study of ultrafast phenomena when combined with short-pulse laser techniques. Finally, SNOM may make it possible to extend the techniques developed for CD ROM storage into the near-field optical (NFO) regime.

4. Imaging with the aperture SNOM

Exploitation of *proximity fields* is the basis of *aperture* SNOM, the most popular branch of NFO microscopy. Analytic calculations of ideal apertures [22-25] as well as numerical computations of real apertures [26] indicate that the electric energy density $|\vec{E}^2|$ of the transmitted light decays with approximately the sixth power of the distance from the exit plane. The high-intensity "proximity" region is confined to a roughly hemispherical volume next to the probe tip apex whose diameter equals that of the aperture. One would hence expect that the resolution limit would be given by the aperture diameter. It seems, however, that nature is gracious in this case because we consistently observe a resolution of 20 to 50 nm (see below) with apertures that appear to have diameters of 50 to 100 nm [27], see Fig. 1. The reason for this difference is not obvious; it cannot be ruled out that the contours seen with the electron microscope do not correspond exactly to the physical rim of the aperture.

The aperture SNOM images depicted in Fig. 2 more or less illustrate the state of the art. The transmission images obtained with (a) amplitude [18], (b) phase [18], (c) fluorescence [18], and (d) polarization [28] contrast show details that are completely invisible even under the best classical optical microscopes; and the fluorescent and polarization-active structures could not be detected with other scanning probe microscopes nor with electron microscopes. The objects pictured here are: (a) a *latex sphere projection pattern* [29], consisting of nearly triangular aluminum patches of 15 nm thickness, lateral extension $\approx$ 50 nm, nearest-neighbor distance of 220 nm; (b) a glass grating consisting of nearly rectangular elevations 8 nm high and

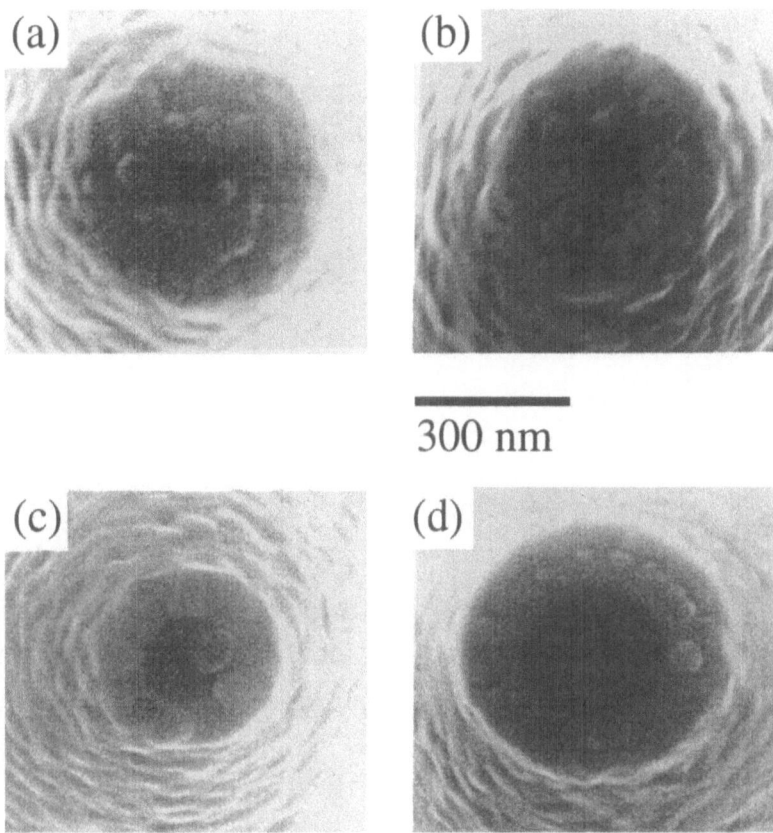

(a)

(b)

300 nm

(c)

(d)

Figure 1. Aperture SNOM probes with aluminum coating, end-on view, scanning electron microscopy, illustrating the problems of grain formation and reproducibility [27].

190 nm wide, interleaved with flat valleys of the same width (prepared by B. Curtis, PSI Zurich); (c) the surface of a T3 fibroplast cell (image made in collaboration with P. Descouts and M. Jobin, University of Geneva), and (d) a Langmuir–Blodgett film whose molecules were "brushed" into a preferential direction within the rectangular area by means of an atomic force microscope (AFM) [28].

Although we show the above results with some pride, we must admit that images of this quality are still rare because very good aperture SNOM probes cannot yet be produced with sufficient reliability. Even if the aperture appears small in the scanning electron microscope image the resolution capability may be low — it takes only one grain of the aluminum coating protruding from the probe tip apex by a few tens of nanometers to prevent the effective opening of the aperture from getting sufficiently close to the

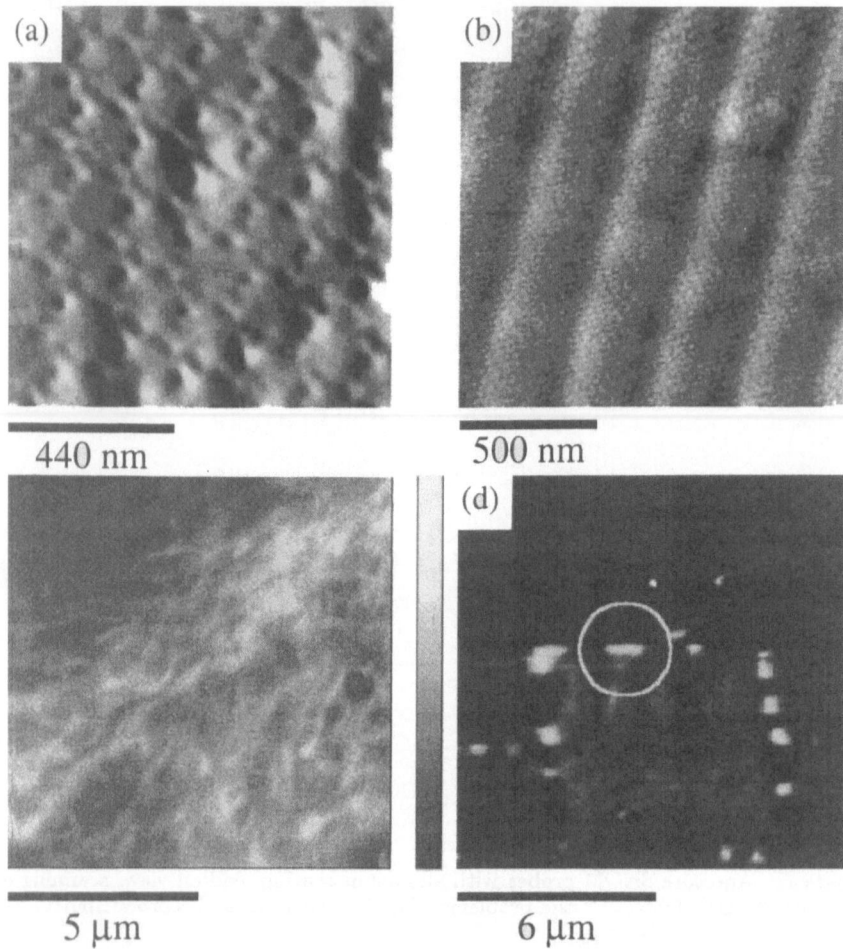

Figure 2. Illustration of SNOM imaging capability: (a) latex sphere projection pattern, aluminum patch diameter ≈50 nm, thickness 15 nm, $\lambda = 633$ nm, amplitude contrast, forbidden light (FL), constant height mode (CHM) [18]; (b) glass grating, 8 nm height, 380 nm period [27], $\lambda = 633$ nm, phase contrast, FL, CHM; (c) T3 fibroplast cell, cytoskeleton fluorescently labeled, fluorescence contrast, excitation at $\lambda = 488$ nm, constant gap width mode (CGM) [18]; (d) Langmuir–Blodgett film, molecular orientation inside rectangle different from environment, birefringence change along borders made visible by polarization contrast, $\lambda = 633$ nm, allowed light, CGM [28].

sample surface. In such a situation, the resolving power sometimes increases abruptly during operation, indicating that the protruding grain was torn off the NFO probe. In very rare cases, the resolution was found to improve to a surprisingly great extent.

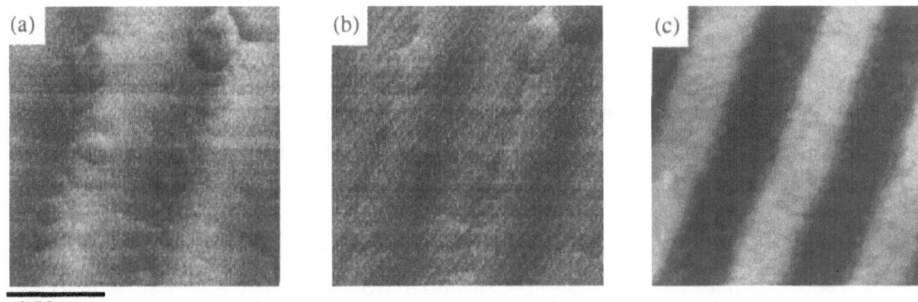

250 nm

Figure 3. Sign of hope: (a) allowed and (b) forbidden-light images of the grating shown in Fig. 2b with unusually high resolution ($\approx$10 nm) recorded simultaneously, $\lambda = 633$ nm, phase contrast [18], (c) AFM image of the same grating (not the same area, tapping mode) [27].

Figures 3a and b show an example of such unexpectedly high resolution of the simultaneously recorded "allowed" and "forbidden"-light NFO images of the same glass grating as in Fig. 2b. Details 10 nm in size can be clearly recognized in these constant-height-mode images. For comparison, Fig. 3c depicts a tapping-mode AFM image of the same grating (though not of the same area). It is interesting to see that the grainy features are visible both in SNOM and AFM images although their appearance differs distinctly in the three pictures. The interpretation of the obviously different informational content of allowed and forbidden-light images is not yet completely clear but it seems that small phase objects appear with reversed contrast in the two types of images, whereas amplitude objects have the same contrast.

The unusually high resolving power might be generated by a small metal grain torn off the coating during scanning and/or dithering, and settling in the middle of the aperture. It would act as a Mie scatterer and therefore enhance the light confinement at the center. The eventual generation of such high-resolution images nourishes hope that increasing mastery of nanoscale technology may provide NFO probes with $\leq$10 nm resolving power on a routine basis. For apertures with unstructured inner parts, however, the finite penetration depth of light into a metal coating imposes a natural limit on the minimum diameter and maximum resolving power. For aluminum, one of the optically best metals, this limit is $\approx$8 nm for green light.

5. Summary

1. "Sleeping Beauty" SNOM is definitely awake and is being besieged by an increasing number of courters. As usual in such a situation, many

efforts are being replicated, and the pace of progress is not yet quite in balance with the efforts invested.

2. A good deal of these efforts — startups in particular — may have been in vain because of the topographic artifact.

3. In spite of this shortcoming, the considerable number of true SNOM results firmly proves the potential of SNOM as a high-resolution scanning probe microscopy with chemical and structural specificity.

4. The available experimental data as well as theoretical insight confirm a resolution capability of 20 nm with potential for extension into the 3-nm range. The ultimate limit will not be determined by field confinement but by discrimination against stray light.

5. Progress towards this goal requires better mastering of nanofabrication techniques on the one hand, and a better understanding of optical near fields on the other. With this in mind, the next chapter in the development of SNOM might have the title *The Taming of the Shrew...*

References

1. Pohl, D.W., Denk, W. and Lanz, M. (1984) Optical stethoscopy: Image recording with resolution $\lambda/20$, *Appl. Phys. Lett.* **44**, 651–653.
2. *Near Field Optics* (1993) D.W. Pohl and D. Courjon (eds.), Proceedings of the NATO ARW on Near Field Optics (NFO I), Arc-et-Senans, France, October 1992, *NATO ASI Series E: Applied Sciences*, Vol. 242, Kluwer Academic Publishers, Dordrecht.
3. Betzig, E. (1993) Principles and applications of near-field scanning optical microscopy (NSOM), in [2], pp. 7–15.
4. Betzig, E., Lewis, A., Harootunian, A., Isaacson, M., and Kratschmer, E. (1986) Near-field scanning optical microscopy (NSOM), *Biophys. J.* **49**, 269–279.
5. Courjon, D., Sarayeddine, K. and Spajer, M. (1989) Scanning tunneling optical microscopy, *Optics Commun.* **71**, 23–27.
6. Reddick, R.C., Warmack, R.J. and Ferrell, T.L. (1989) New form of scanning optical microscopy, *Phys. Rev. B* **39**, 767–770.
7. de Fornel, F., Goudonnet, J.P., Salomon, L. and Lesniewska, E. (1989) An evanescent field optical microscope, in *Proc. SPIE*, Vol. 1139, SPIE, Bellingham, pp. 77–84.
8. Bainier, C., Leblanc, S., and Courjon, D. (1993) Scanning tunneling optical microscopy: Application to very low relief objects, in [2], pp. 97–104.
9. Novotny, L., Pohl, D.W., and Hecht, B. (1996) Light confinement in scanning near-field optical microscopy, *Ultramicroscopy* **61**, 1–9.
10. Fischer, U.Ch. and Pohl, D.W. (1989) Observation on single-particle plasmons by near-field optical microscopy. *Phys. Rev. Lett.* **62**, 458–461.
11. Fischer, U.Ch., Dürig, U. and Pohl, D.W. (1989) Scanning near-field optical microscopy (SNOM) in reflection or scanning optical tunneling microscopy (SOTM), *Scanning Microscopy* **3**, 1–7.
12. Fischer, U.Ch. (1990) Resolution and contrast generation in scanning near-field optical microscopy, in it Scanning Tunneling Microscopy and Related Methods, R.J. Behm, N. Garcia and H. Rohrer (eds.), *NATO ASI Series E: Applied Sciences*, Vol. 184, Kluwer Academic Publishers, Dordrecht, pp. 475–496.
13. Koglin, J., Fischer, U.Ch., Brzoska, K.D., Goehde, W. and Fuchs, H. (1995) The tetrahedral tip as a probe for scanning near field optical microscopy, in O. Marti and

R. Möller (eds.), *Photons and Local Probes, NATO ASI Series E: Applied Sciences*, Vol. 300, Kluwer Academic Publishers, Dordrecht, pp. 79–92.

14. Zenhausern, F., Martin, Y. and Wickramasinghe, H.K. (1995) Scanning interferometric apertureless microscopy: Optical imaging at 10 angstrom resolution, *Science* **269**, 1083–1085.

15. Denk, W. and Pohl, D.W. (1991) Near-field optics: microscopy with nanometer-size fields, *J. Vac. Sci. Technol. B* **9**, 510–513.

16. Martin, O.J.F. and Girard, Ch. (1997) Controlling and tuning strong optical field gradients at a local probe microscope tip apex, *Appl. Phys. Lett.* **70**, 705–707.

17. Valaskovic, G.A., Holton, M., Morrison, G.H. (1995) Image contrast of dielectric specimens in transmission mode near-field scanning optical microscopy: Imaging properties and tip artefacts, *J. Microscopy* **79**, 29–54.

18. Hecht, B., Bielefeldt, H., Novotny, L., Inouye, Y. and Pohl, D.W. (1997) Facts and artifacts in near-field optical microscopy, *J. Appl. Phys.* **81**, 2492–2498.

19. Sandoghdar, V., Wegscheider, S., Krausch, G. and Mlynek, J. (1997) Reflection scanning near-field optical microscopy with uncoated fiber tips: How good is the resolution really? *J. Appl. Phys.* **81**, 2499–2503.

20. Williamson, R.L., Bereton, L.J., Pidduck, A.J. and Miles, M.J. (1997) Are artefacts in scanning near-field optical microscopy related to the misuse of shear-force? *Ultramicroscopy* (in press).

21. Bozhevolnyi, S. (1997) Topographical artifacts and optical resolution in near-field optical microscopy, *J. Opt. Soc. Am. B* **14**(9) (in press).

22. Bethe, H.A. (1944) Theory of diffraction by small holes, *Phys. Rev.* **66**, 163–182.

23. Bouwkamp, C.J. (1950) On Bethe's theory of diffraction by small holes, *Philips Res. Rep.* **5**, 321–332.

24. Leviatan, Y. (1986) Study of near-zone fields of a small aperture, *J. Appl. Phys.* **60**, 1577–1583.

25. Klimov, V.V. and Letokhov, V.S. (1996) Atom optics in the laser near field, *Laser Phys.* **6**, 475–500.

26. Novotny, L., Pohl, D.W. and Hecht, B. (1995) Scanning near-field optical probe with ultrasmall spot size, *Opt. Lett.* **20**, 970–972.

27. Hecht, B., Heinzelmann, H.. and Pohl, D.W. Influence of detection optics on imaging properties of a scanning near-field optical microscope (in preparation).

28. Lacoste, T., Huser, T., Heinzelmann, H. and Güntherodt, H.J. (in preparation).

29. Fischer, U.Ch. and Zingsheim, H.P. (1981) Submicroscopic pattern replication with visible light, *J. Vac. Sci. Technol.* **19**, 881–885.

LIGHT EMISSION IN SILICON NANOSTRUCTURES

D.J. LOCKWOOD
Institute for Microstructural Sciences
National Research Council,
Ottawa, ON, Canada K1A 0R6

ABSTRACT. Interest in obtaining useful light emission from silicon-based materials has never been greater. This is because there is a strong demand for optoelectronic devices based on silicon and also because there has recently been significant progress in materials engineering methods. Here we review the latest developments in this work, which is aimed at overcoming the indirect band gap limitations in light emission from silicon. Subjects covered include optical band gap engineering through quantum confinement and Brillouin zone folding in silicon-based superlattices and heterostructures, light emission from isoelectronic and erbium impurity centres in silicon, and luminescence in silicon nanoparticles and porous silicon. One promising new approach, based on thin-layer Si/SiO_2 superlattices, is reviewed in detail. The incorporation of these different materials into devices is described and future device prospects are assessed.

1. Introduction

The ubiquitous silicon microelectronics 'chip' is taken for granted in modern society. There has been much research involved in producing these high technology marvels and such research continues unabated at a faster and faster pace. Despite the often stated announcement that 'the age of GaAs has arrived', it never quite has, and continued developments in Si and, more recently, $Si_{1-x}Ge_x$ alloy technology [1,2] continue to advance the frontiers of microminiaturization, complexity, and speed. This continued advance has been driven by application requirements in switching technology and high-speed electronics. Gallium arsenide and other compound semiconductors have, however, maintained a significant role in the construction of optoelectronic and purely photonic devices where the medium of switching and communication is light itself [3].

The merging of Si-based electronics with photonics has largely required the pursuit of hybrid technologies for light emitters and modulators, which are often both expensive and complicated to produce. The most satisfactory solution and still mostly a dream as far as light sources are concerned would be optoelectronic devices created entirely from Si-based materials, where the extensive experience in Si fabrication and processing could be put to best use [4]. Already, a wide range of optoelectronic integrated circuits (OEICs) incorporating Si or $Si_{1-x}Ge_x$ as a detector or waveguide have been elaborated [1,4–6]. Nevertheless, the major deficiency in Si-based optoelectronic devices remains the lack of suitable light emitters and especially lasers. The general requirements in Si-based light sources are efficient light emitting diodes, lasers, and optical amplifiers for use in optical commu-

N. García et al. (eds.), Nanoscale Science and Technology, 185–210.
© 1998 NRC.

nications technologies such as fiber optics and displays. Operating wavelengths in the range 0.45–1.6 μm are needed to cover both full colour displays and the fiber optic operating wavelengths of 1.3 and 1.55 μm.

Many quite different approaches to alleviating the miserable light emission in Si (~10^{-4} quantum efficiency at 300 K) have been proposed and are actively being explored [7–9]. Some, such as $Si_{1-x}Ge_x$ quantum well or Si/Ge superlattice structures, rely on band structure engineering, while others rely on quantum confinement effects in low dimensional structures, as typified by quantum dots or porous Si (π-Si [10]). Still another approach is impurity mediated luminescence from, for example, isoelectronic substitution or by the addition of rare earth ions. An overview of results obtained with these and other methods is given below. However, in order to understand more fully the reasons why such different approaches are necessary, it is appropriate to review first what creates the optical emission problem in crystalline Si (c-Si).

2. Physical Properties of Silicon

Silicon crystallizes in the diamond structure [11], which consists of two interpenetrating face-centered cubic lattices displaced from each other by one quarter of the body diagonal. In zinc blende semiconductors such as GaAs, the Ga and As atoms lie on separate sublattices, and thus the inversion symmetry of Si is lost in III-V binary compounds. The energy band structure in semiconductors is derived from the relationship between the energy and momentum of a carrier, which depends not only on the crystal structure but also on the bonding between atoms, the respective bond lengths, and the chemical species. The band structure is often quite complex and can only be calculated empirically. The results of such calculations [12] for Si and GaAs are shown in Fig. 1. The figure shows the dispersion relations for the energy $E(k)$ of an electron (positive energy) or hole (negative energy) for wave vectors k within the first Brillouin zone.

The valence band structure is much the same for many semiconductors and exhibits a maximum at the Brillouin zone centre or Γ point (*i.e.*, at $k = 0$). The notable difference between Si and GaAs is that the degeneracy in the $Γ_{25'}$ band maximum at $k = 0$ is removed in the case of GaAs, because of the spin-orbit interaction, into $Γ_6$ and $Γ_7$ subbands. In general, $E(k)$ has maxima or minima at zone centre and zone boundary symmetry points, but additional extrema may occur at other points in the Brillouin zone (see Fig. 1). In the case of Si, the lowest point in the conduction band occurs away from high symmetry points near the X point at the Brillouin zone boundary (along <001>), whereas in GaAs it occurs at the Γ point. The energy gap in a semiconductor is defined as the separation between this conduction band minimum and the valence band maximum at the Γ point. For GaAs, the energy gap is classified as direct, because a transition can occur directly at $k = 0$ between initial and final states having the same wave vector. Correspondingly, Si is termed an indirect gap semiconductor, because the initial and final states have different wave vectors.

In direct gap GaAs, an excited electron at the bottom of the conduction band can relax spontaneously back into the valence band by emitting a photon at the band gap energy. This electron-hole radiative recombination process can only occur in Si if momen-

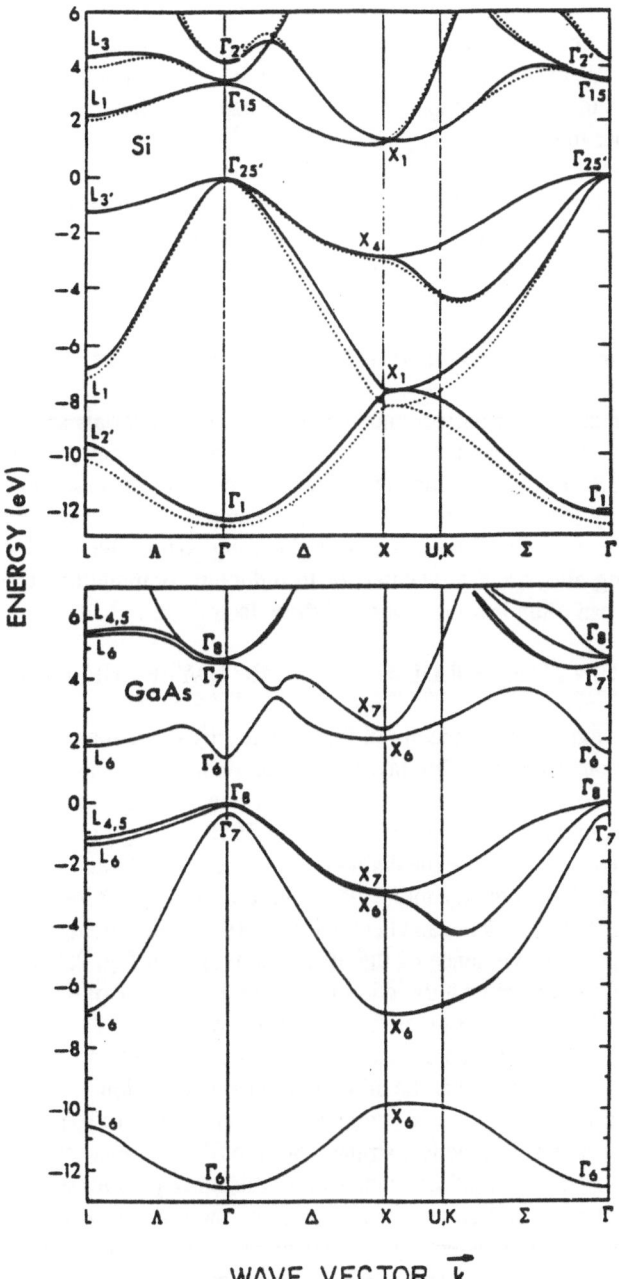

Figure 1. Theoretical band structures of Si and GaAs. In the case of Si, results are shown for nonlocal (solid line) and local (dashed line) pseudopotential calculations. [After Chelikowsky, J.R., and Cohen, M.L. (1976) Nonlocal pseudopotential calculations for the electronic structure of eleven diamond and zinc-blende semiconductors, *Phys. Rev. B* **14**, 556–582.]

tum is conserved, *i.e.*, the excited electron wave vector must be reduced to zero. This, in pure Si, occurs via the transfer of momentum to a phonon that is created with equal and opposite wave vector to that of the initial state in the conduction band. Such a three-body process is quite inefficient compared with direct gap recombination [8,13], which is why Si is such a poor light emitter.

Electron-hole pairs may bind to each other to form excitons, which can be either free or tied to impurities or defects [8,13]. The decay of such excitons can lead to light emission that may be tuneable by, for example, quantum confinement. Such excitonic emission is thus under active investigation in quantum well, wire, and dot structures [14].

3. Overcoming the Indirect Band Gap Limitations in Silicon

Materials engineering, a relatively new phenomenon in materials science, is now being actively applied to Si in an attempt to overcome the indirect band gap limitations in light emission from Si. In these various attempts, the aim is to increase the efficiency of the luminescence by increasing the overlap of the electron and hole wavefunctions via, for example, confinement and band structure engineering, to tune the wavelength of the emission by forming alloys and molecules, or to induce recombination at impurity centres. Such attempts can often involve several of these factors.

3.1. BRILLOUIN ZONE FOLDING IN ATOMIC LAYER SUPERLATTICES

Some 20 years ago it was conjectured theoretically by Gnutzmann and Clausecker [15] that Brillouin zone folding in thin layer superlattices where the layer thicknesses were of the order of the unit cell dimensions could result in a direct (or quasi-direct, as it is now termed) band gap structure. The growth in the 1980s of high-quality $(Si_mGe_n)_p$ atomic layer superlattices (*m* and *n* are the number of monolayers of Si and Ge in each period and *p* is the number of periods) by molecular beam epitaxy [1] led impetus to this concept, which was revisited by Jackson and People [16] in 1986 and, subsequently, by a number of other theoreticians. The essence of the idea is conveyed in Fig. 2. The new superlattice periodicity *d* along the growth direction results in a smaller Brillouin zone of size $\pm\pi/d$ compared with that of the original lattice ($\pm 2\pi/a$, where *a* is the lattice constant). The electronic band structure is then folded back into this new reduced Brillouin zone. For this simple model, it is apparent that the minimum in the conduction band in bulk Si is folded into the Brillouin zone centre for $d \approx 5a/2$, which corresponds to 10 monolayers of Si, and a direct gap is evident. In practice, strains within a Si_mGe_n superlattice together with the band offsets at the heterointerfaces compromise this naïve picture (see, for example, Refs. 1 and 17–20). Theory has shown that for certain superlattice periods and when the Si layers are strained a direct energy gap is expected in Si_mGe_n superlattices, but the transition probability is still several orders of magnitude below that of GaAs.

The first experimental evidence of modifications to the Si and Ge band structures in such superlattices was obtained from electroreflectance measurements of Si_4Ge_4 superlattices grown on (001) Si [21]. However, it was not until later on when strain-symmetrized Si_mGe_n superlattices were grown on strain-relaxed thick $Si_{1-x}Ge_x$ alloy buffer

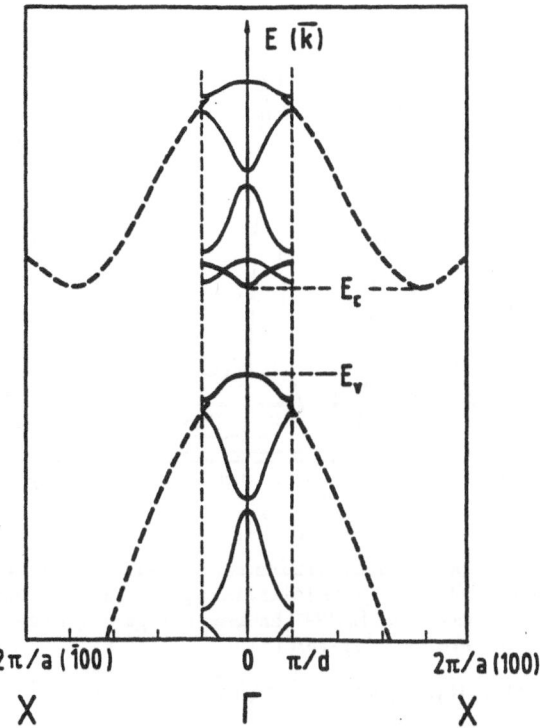

Figure 2. Schematic representation of the Brillouin zone folding concept in Si resulting from the new superlattice periodicity in the growth direction. Here, the conduction band minimum along the X direction is folded back into the Γ point when the superlattice period is about 10 monolayers of Si. [Reprinted from Kasper, E. and Schäffler, F. (1991) Group-IV compounds, in T.P. Pearsall (ed.), *Strained Layer Superlattices: Materials Science and Technology*, Academic Press, Boston, pp. 223–309.]

layers on Si that first indications of the predicted photoluminescence (PL) intensity enhancement and reduced energy gap were obtained [1,22]. Improvements in crystal growth conditions subsequently led to a positive identification of these new features of Si_mGe_n superlattices [23]. As shown in Fig. 3, the PL no-phonon (NP) peak clearly shifts to lower energy and increases in intensity with increasing superlattice periodicity, as compared with a $Si_{0.6}Ge_{0.4}$ alloy layer of the same average composition as the $Si_{3i}Ge_{2i}$ ($i = 1,2,3$) superlattices.

From a device point of view, although infrared emission can readily be obtained at low temperature from such Si_mGe_n structures at energies useful for fibre-optic transmission work, the PL and electroluminescence (EL) are essentially quenched at room temperature [24,25]. Unless there are further major improvements in material quality, it is more likely that these atomic layer superlattices will find eventual use as infrared detectors rather than as emitters [26,27]. However, some promising steps towards room temperature EL structures have been reported recently [28,29].

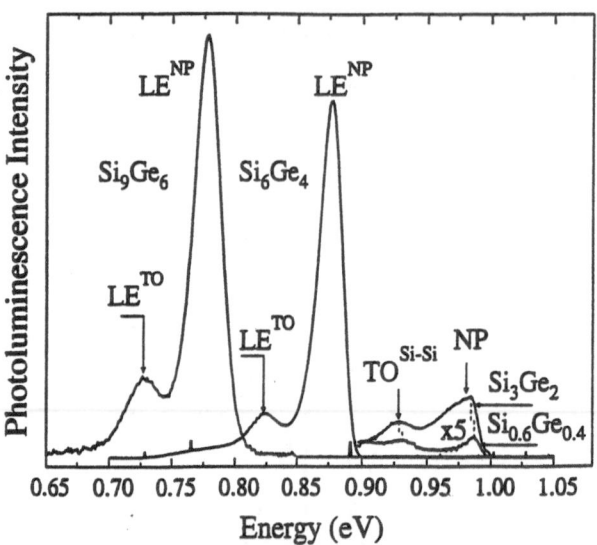

Figure 3. Low temperature PL spectra of strain-symmetrized Si_mGe_n superlattices and of a $Si_{0.6}Ge_{0.4}$ alloy layer grown on a step-graded SiGe buffer layer on Si. [After Menczigar, U., Abstreiter, G., Olajos, J., Grimmeiss, H.G., Kibbel, H., Presting, H., and Kasper, E. (1993) Enhanced band-gap luminescence in strain-symmetrized $(Si)_m/(Ge)_n$ superlattices, *Phys. Rev. B* **47**, 4099–4102.]

3.2. BAND STRUCTURE ENGINEERING VIA ALLOYING

Alloying of Ge or C with Si allows engineering of the electronic band structure, where the energy gap may be varied with alloy composition and strain [27,30]. This is shown, for example, in Fig. 4 for strained $Si_{1-x}Ge_x$ on Si, where the tuneability range is appropriate for fibre-optic communications. Unfortunately, because of heterostructure stability limitations, the $Si_{1-x}Ge_x$ layer thickness must be kept below the critical thickness, which decreases rapidly with increasing x (see Fig. 4). Thus the absorbing/emitting regions in infrared detectors/emitters are necessarily small. Also, the band gap remains indirect. Despite these severe limitations much research has been carried out on the optical properties of $Si/Si_{1-x}Ge_x$ heterostructures [27], which exhibit type I band alignment [30,31], and, to a lesser extent, on $Si/Si_{1-x}C_x$ or even $Si/Si_{1-x-y}Ge_xC_y$ [32–34], and the properties of infrared emitting devices are being explored.

Electroluminescence and PL have been observed from $Si_{1-x}Ge_x$ in both single layer and superlattice form with increased intensity compared with Si, as shown, for example, in Fig. 5. The strong broad peak seen in PL and EL (~0.5% internal quantum efficiency) at 0.89 eV in Fig. 5 is typical of results obtained from related studies (see, for example, Refs. 36–38): The luminescence energy tracks the alloy composition dependence shown in Fig. 4, but is at a lower energy. The recombination mechanism varies depending on the alloy layer thickness and perfection resulting in near band edge and/or excitonic luminescence [39–41].

In the earliest work, the EL from $Si_{1-x}Ge_x/Si$ p-n diodes was quenched by increasing the temperature above 80 K [35], but EL was soon reported at temperatures up to 220 K in

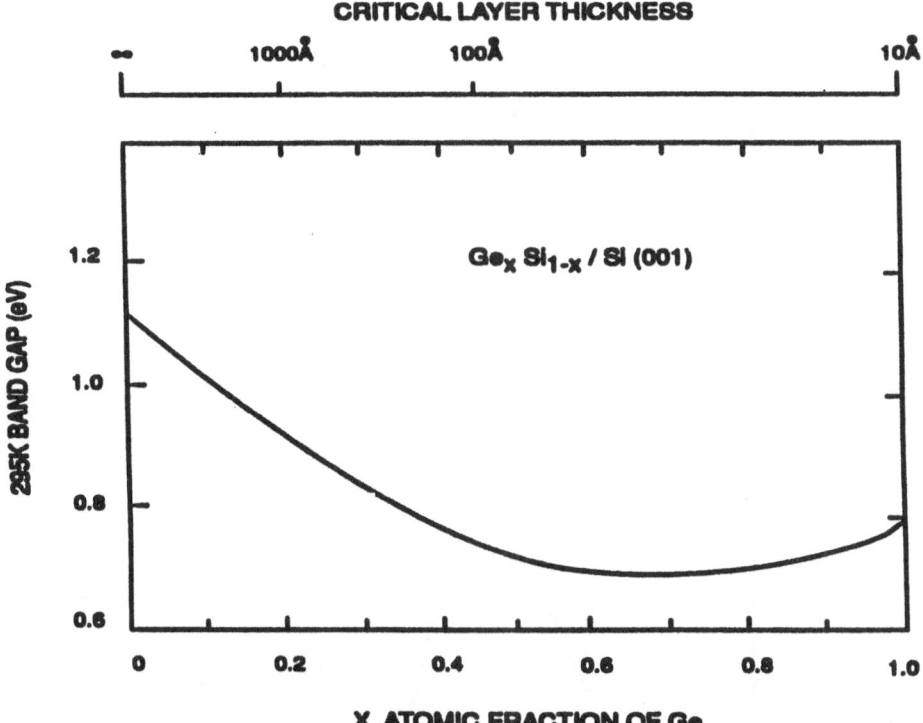

Figure 4. The band gap at room temperature of strained Si$_{1-x}$Ge$_x$ on Si. Also shown is the critical layer thickness as a function of x. [Reprinted from Pearsall, T.P. (1994). Electronic and optical properties of Ge-Si superlattices, *Prog. Quantum Optics* **18**, 97–152 with kind permission from Elsevier Science Ltd., The Boulevard, Langford Lane, Kidlington OX51GB, UK.]

p-i-n diode structures [42]. Progress in materials quality and device design has continued to improve EL device performance (see, for example, Refs. 43–47) such that room temperature EL has now been reported at wavelengths near 1.3 μm [43,47]. The major problem with such devices for practical purposes at present is their low efficiency at room temperature [43,47].

3.3. LUMINESCENCE AT IMPURITY CENTRES

Another approach to increasing the EL efficiency of an indirect band gap semiconductor is to introduce an impurity that localizes the electron and hole, as pioneered in GaP [48]. This has been done in Si EL diodes by using, for example, rare earth impurities [49], carbon complexes [50], and sulphur/oxygen complexes [51] as localization centres for electron-hole recombination. Extrinsic luminescence in Si can arise from a variety of sources [8,52]. Here, we concentrate on isoelectronic and rare-earth extrinsic centres, as these are presently the most promising for device applications.

Isoelectronic centres are created by doping Si with electrically neutral impurities such as the isovalent elements C, Ge, and Sn or a multiple-atom complex with no dangling

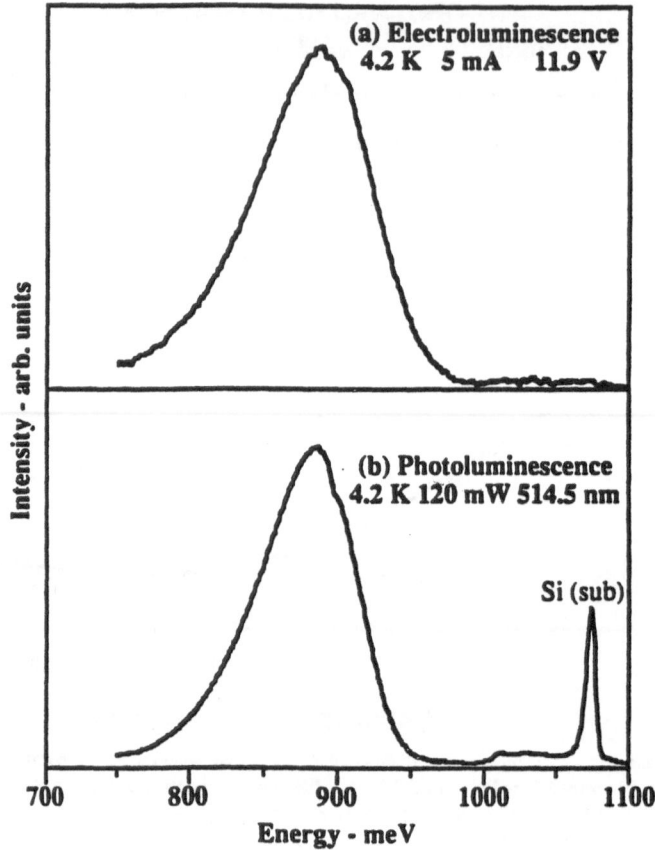

Figure 5. Broad EL and PL from a $Si_{0.82}Ge_{0.18}$ p-n heterostructure at 4.2 K. A sharper emission line from the Si substrate is also evident. [Reprinted with permission from Rowell, N.L., Noël, J.-P., Houghton, D.C., and Buchanan, M. (1990) Electroluminescence and photoluminescence from $Si_{1-x}Ge_x$ alloys, *Appl. Phys. Lett.* **58**, 957–958. Copyright 1991 American Institute of Physics.]

bonds. Isoelectronic impurities bind free excitons in Si, which can increase the probability of electron-hole recombination due to spatial confinement of the particles. The resultant recombination energy may appear as light or disappear through phonon generation and other nonradiative decay channels [8]. An example of isoelectronic bound exciton emission is shown in Fig. 6 for Si implanted with In. The characteristic sharp NP excitonic emission in PL and EL occurs at 1.11 μm (1.12 eV), which is just below the Si indirect band gap of 1.17 eV at 14 K. The optical emission intensity decreases with increasing temperature [53].

The optical properties of a variety of such isoelectronic impurity centres including In, Al–N, Be, S, and Se have been studied both in Si and $Si_{1-x}Ge_x$ alloys [8,52]. A luminescence external quantum efficiency of 5% and a lifetime greater than 1 ms have been reported for the S complex emission at 1.32 μm in Si at low temperatures [54], but the PL intensity and lifetime decrease sharply with increasing temperature. This variation with

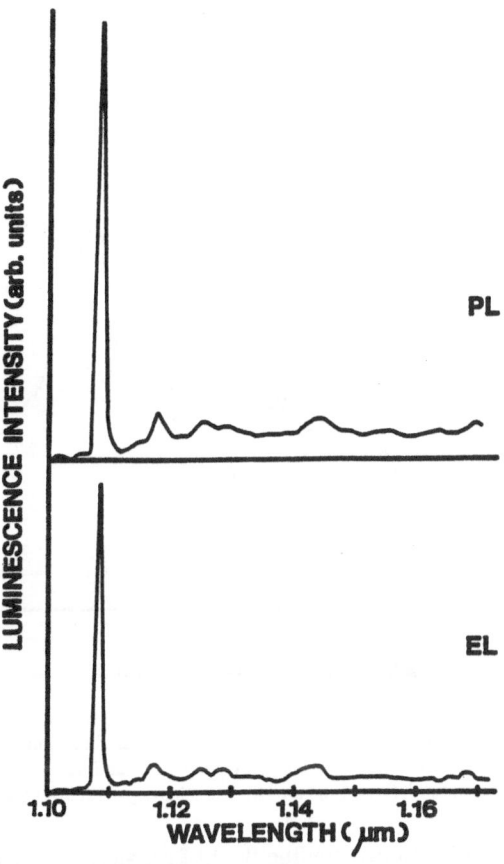

Figure 6. Sharp PL and EL at 1.11 μm from a quenched Si:In sample at 14 K. [Reprinted with permission from Brown, T.G., and Hall, D.G. (1986) Observation of electroluminescence from excitons bound to isoelectronic impurities in crystalline silicon, *J. Appl. Phys.* **59**, 1399–1401. Copyright 1986 American Institute of Physics.]

increasing temperature is due to exciton dissociation and competing nonradiative recombination processes. The low bound-exciton emission intensity at room temperature militates against isoelectronic-impurity based EL devices at present.

The optical properties of rare earth ions in solids have been investigated in great detail and are generally well understood [55]. The optical emission of the Er^{3+} ion is of particular interest for semiconductor device applications, because it occurs near 1.5 μm. The Er^{3+} ion emits photons at 1.54 μm in Si (see Fig. 7) by intracentre transitions between Er^{3+}-ion discrete states ($I_{13/2} \rightarrow I_{15/2}$ transition within the 4f electron shell). The excitation of the Er^{3+} ions is a complicated process [8,57] involving first electron-hole carrier generation in Si, then exciton formation, and finally Er excitation by an intracentre Auger process, with a number of competing pathways in the excitation process. The excited state relaxation then occurs via photon emission or, with increasing temperature, via nonradiative backtransfer processes [57,58]. This results in a low quantum efficiency and a marked quenching of the luminescence for temperatures above approximately 150 K [57,59]. Nev-

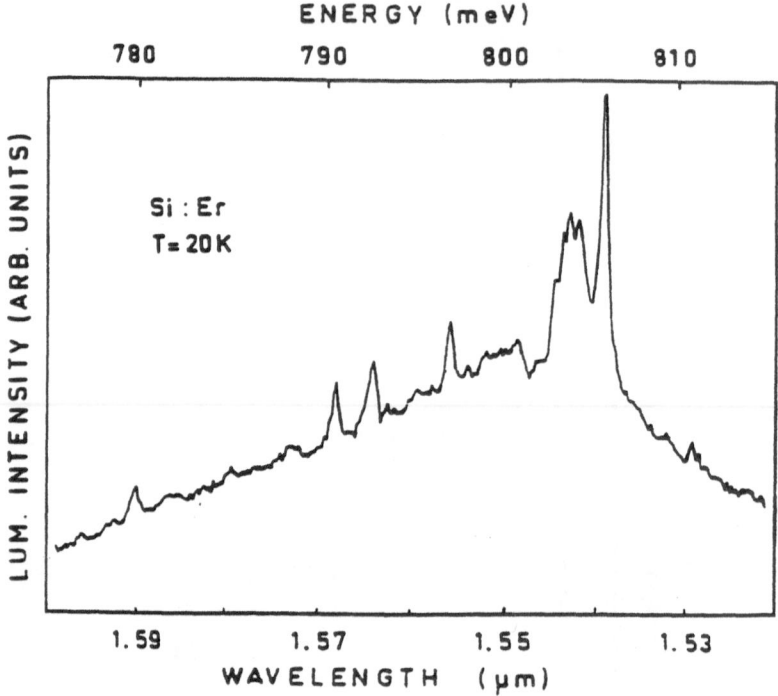

Figure 7. The low-temperature PL spectrum of Er implanted and annealed Si. [Reprinted with permission from Ennen, H., Schneider, J., Pomrenke, G., and Axmann, A. (1983) 1.54-μm luminescence of erbium-implanted III-V semiconductors and silicon, *Appl. Phys. Lett.* **43**, 943–945. Copyright 1983 American Institute of Physics.]

ertheless, research continues on overcoming the Si:Er materials system constraints such as the low solid solubility of Er in Si and the low optical efficiency at room temperature, and room temperature EL devices with improved performance through the use of an oxygen codopant are now emerging (see, for example, Refs. 58, 60, and 61).

3.4. SILICON NANOSTRUCTURES

Research on the quantum confinement of carriers in silicon-based nanostructures including π-Si, nanoclusters, and quantum wells, wires, and dots forms a large part of the work on light emission in silicon. Much of this work was stimulated by the discovery of bright visible light emission at room temperature in π-Si reported in 1990 [62]. The number of papers published per year on π-Si alone has been approximately 500 for the last several years. The interest in nanostructures of Si stems from the effects of confinement on carrier wave functions when the crystallite diameter is less than the size of the free exciton Bohr radius of 4.3 nm [14] in bulk c-Si. The quantum confinement increases the electron-hole wave function overlap, resulting in increased light emission efficiency, and shifts the emission peak to higher energy [8,62].

3.4.1. Porous Silicon

Porous silicon was discovered over 35 years ago by Uhlir [63]. The porous material is created by electrochemical dissolution in HF-based electrolytes. Hydrofluoric acid, on its own, etches single-crystal Si extremely slowly, at a rate of only nanometers per hour. However, passing an electric current between the acid electrolyte and the Si sample speeds up the process considerably, leaving an array of deep narrow pores that generally run perpendicular to the Si surface. Pores measuring only nanometers across, but micrometers deep, have been achieved under specific etching conditions.

In July 1989, Canham conceived the idea of fabricating Si quantum wires in π-Si by reverting to the much slower chemical HF etch after electrochemically etching c-Si. In this way Canham proposed to join up the pores leaving behind an irregular array of undulating free standing pillars of c-Si only nanometers wide. In 1990, Canham [62] observed intense visible PL at room temperature (see Fig. 8) from π-Si that had been etched under carefully controlled conditions. Visible luminescence ranging from green to red in colour was soon reported by Canham *et al.* for other π-Si samples and ascribed to quantum size effects in wires of width ~3 nm [62,64]. Independently, Lehmann and Gösele [65] re-

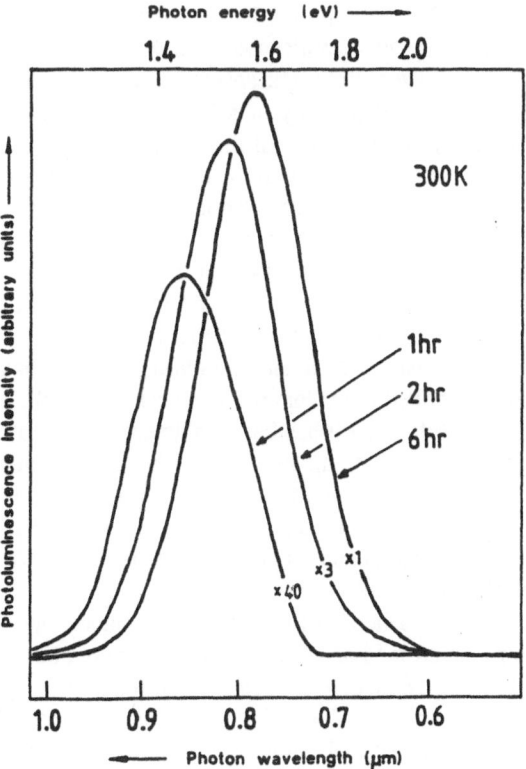

Figure 8. Room-temperature PL from anodized Si after immersion in 40% aqueous HF for the times indicated. [Reprinted with permission from Canham, L.T. (1990) Silicon quantum wire array fabrication by electrochemical and chemical dissolution of wafers, *Appl. Phys. Lett.* **57**, 1046–1048. Copyright 1990 American Institute of Physics.]

ported on the optical absorption properties of π-Si. They observed a shift in the bulk Si absorption edge to values as high as 1.76 eV that they also attributed to quantum wire formation. Visible PL in π-Si at room temperature was soon also reported by Bsiesy et al. [66], Koshida and Koyama [67], and Gardelis et al. [68], while visible electroluminescence was observed by Halimaoui et al. [69] during the anodic oxidation of π-Si and, later, by Koshida and Koyama [70] with a diode cell. Tremendous activity on research into the physical and associated chemical characteristics of π-Si has ensued from these early reports with, unfortunately, considerable duplication of effort. It is impossible to mention all of this work here and interested readers are directed to recent reviews and books [10,71–77] for further aspects of this work.

A strong PL signal has been observed from π-Si at wavelengths from the near infrared through the visible to the blue depending on the sample porosity and the surface chemical treatment. It has even been possible using specialized preparation techniques to produce "white" light emitting π-Si [71]. For discussion purposes, it is convenient to divide these wavelength regions into three: near infrared, red-yellow, and blue.

The most widely studied PL is in the far-red to orange-yellow region, which we shall denote simply as the "red" PL. As evident in Fig. 8, this PL shifts to shorter wavelength with increasing chemical dissolution time. It was soon found that much smaller immersion times were required to produce noticeable blue shifts when the chemical dissolution was carried out in the presence of light. The spectra also show a blue shift with increasing anodization current density. The porosity of π-Si increases with increasing anodization current density. Therefore, the behaviour of the red PL spectra qualitatively reflects the differences in sample porosity and hence in the dimensions of Si nanocrystallites within π-Si. The blue shift of the PL and optical absorption with increasing porosity provided the first important evidence that quantum confinement effects could be playing a role. Nevertheless, after much research, the controversy over the origin(s) of the red PL in π-Si persists. This is because the PL peak wavelength and intensity are sensitive to the surface chemistry of π-Si, particularly with regard to the relative amounts of hydrogen and oxygen on the surface. Thus, besides the quantum confinement mechanism, various surface state models have been invoked to explain the various results [10]. Although evidence of quantum confinement effects in π-Si has been obtained via optical absorption measurements [78], the problems in explaining the PL in such a way are amply demonstrated by the data of Fig. 9. The π-Si samples in this case had a sphere-like morphology (spherites) and the optical gap is seen to be in good agreement with theoretical predictions for quantum dots, but there is a substantial and, as yet, unexplained energy difference between the absorption and emission data.

Oxidation of the π-Si surface has been shown to produce blue PL [10]. The blue PL is quite weak in as-prepared π-Si and becomes intense only after strong oxidation. The blue PL has a much faster decay than the red PL and requires surface oxidized π-Si. Its origin is of some debate at present. Models currently under consideration include band-to-band recombination in Si nanocrystals, emission from oxide, and emission due to surface states. Present indications are that while the red PL possibly originates from the near-surface region of the Si crystallites, the blue PL may emanate from the small c-Si core region.

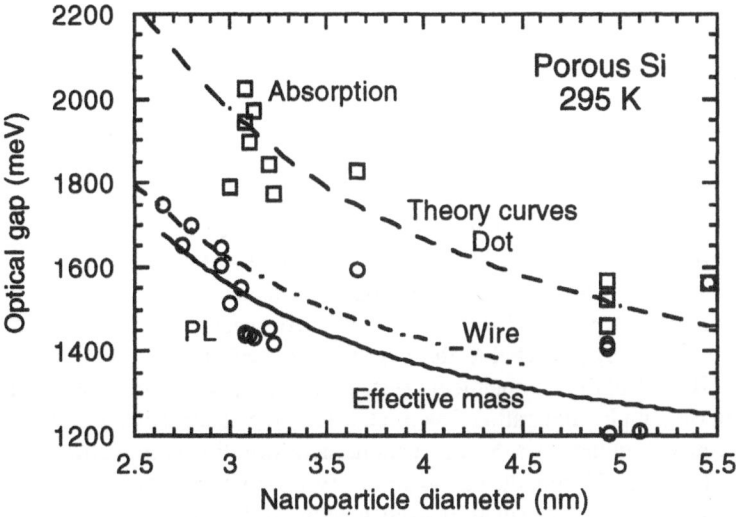

Figure 9. Dependence of the optical absorption energy gap and PL peak energy on spherite diameter in π-Si samples at room temperature. The solid line is the effective mass model prediction for the optical gap in c-Si spheres, while the broken and short dash-dot lines are theoretical predictions based on a linear combination of atomic orbitals framework for quantum dots and wires, respectively. [After Lockwood, D.J. and Wang, A.G. (1996) Photoluminescence in porous silicon due to quantum confinement, in D.J. Lockwood, P.M. Fauchet, N. Koshida, and S.R.J. Brueck (eds.), *Advanced Luminescent Materials*, The Eelectrochemical Scoeity, Pennington, pp. 166–172.]

The near-infrared PL [10] at ~0.8 eV (below the bulk Si band gap) exhibits complex nonexponential dynamics, with a wide distribution of decay times, and has been assigned to deep level transitions associated with dangling bonds on the surface of the Si nanocrystallites.

From these considerations it is apparent that PL in π-Si is very sensitive to the chemistry of π-Si production and treatment. Crystalline-Si wires, c-Si spherites, and amorphous Si (a-Si) material, or any combination of them, may be formed in a given sample. The π-Si layers thus formed may be far from uniform, which adds to the difficulties in analyzing their optical properties. Other light emitting species may also be formed on the surfaces of the anodized and otherwise chemically treated Si.

Despite all these disadvantages, the ease of production of π-Si and the facts that the room temperature PL is very efficient (1–10% quantum efficiency) and that it is tunable through blue to near infrared wavelengths have led to impressive efforts to produce practical room-temperature devices. The latest generation of red light emitting diodes (LEDs) have external quantum efficiencies of 0.1% and lifetimes of the order of months [76]. Most recently, π-Si LEDs have been integrated into Si microelectronic circuits to provide an addressable LED display [80]. However, improvements in efficiency and power dissipation are necessary for display applications, while an increased modulation frequency (presently ~1 kHz) is required for optical interconnects. One way to improve the EL efficiency, to narrow the band width, to improve the directionality, and to increase the long-term stability is to insert the LED into a π-Si resonant cavity [81]. The long switching times (up to milliseconds) observed in present π-Si LEDs may yet prove to be an Achilles' heel in optoelectronic applications.

3.4.2. Silicon Nanoclusters

Rather than produce nanometer-size Si crystallites by etching, as in π-Si, there have been numerous attempts at growing them either directly from a gas phase or indirectly by recrystallization within a matrix [8,14,82]. In fact, the observation of a nanoparticle size dependence of the PL energy in very small Si crystallites passivated with hydrogen [83] predates the similar finding in π-Si [61]. Takagi *et al.* [83] found that the PL peak energy varied as $1/d^2$ ($3 < d < 5$ nm), where d is the Si nanoparticle diameter, in accordance with quantum confinement effects predicted by a simple effective mass model. As for π-Si, however, the emitted light energy falls below that expected from calculations of the energy gap for Si spheres [10]. Also, the confinement effect is seen [83,84] or not seen [85] in emission depending on sample preparation. Interpretation of the nanoparticle PL spectra suffers from the same ambiguities as π-Si, i.e., nanoparticle size distribution effects and surface chemistry effects. In addition, the nanoparticle crystal structure may deviate from the cubic diamond structure for very small Si nanoclusters [8]. Recent calculations [86] have shown that luminescence in Si nanocrystallites can be due to excitons trapped at the surface, which is passivated by hydrogen or silicon oxide, while the optical absorption is characteristic of quantum confinement effects. In recent definitive experiments [87], the indirect nature of the Si band gap has been seen from PL and absorption spectra for small (1–2 nm in diameter) surface-oxidized nanocrystals. The red PL quantum efficiency and lifetime is similar to that found for π-Si [87], indicating a similar light emission mechanism involving quantum-confined nanocrystal states.

The controversy surrounding the interpretation of the PL in Si nanoparticles and in π-Si is displayed throughout the literature. The vagaries and complexities of the nanocrystal-interface-surface system are proving difficult to unravel in the short term.

Nanocrystals of Si trapped in some matrix form an attractive system for device fabrication when compared with π-Si, because of the increased surface stability and material rigidity. Recently, visible EL has been observed, for example, from Si nanocrystallites embedded in films of a-Si:H [88] and from an electrochemically-formed nanocrystalline Si thin film deposited on SnO_2 [89]. In the latter case the p-i-n LED at room temperature emitted orange-red light (1.8 eV) that was readily visible to the eye. The light emission is ascribed variously to near surface states [88] and the quantum size effect [89]. Substantial progress in the development of such EL structures can be expected over the next few years.

3.4.3. Quantum Wells, Wires, and Dots

One of the major problems involved in π-Si and Si-nanocluster research and development work is the inhomogeneity of the material. Such inhomogenous broadening effects in the PL and EL can be minimized by preparing *uniform* Si structures in the form of quantum wells, wires, or dots. Such structures can readily be produced directly by modern epitaxial growth techniques such as planar epitaxy, quantum wire formation along wafer steps, and dot self-assembly, or indirectly by etching appropriate planar structures in the case of wires and dots. The predicted Si transition energies [90] due to the different degrees of quantum confinement are shown in Fig. 10 (more sophisticated pseudopotential calculations [91] give qualitatively similar results), where it can be seen that appreciable confinement effects are seen only for diameters less than 3 nm. Etched structures of this size have been difficult to produce in Si until very recently.

Wells. The simplest approach is to grow thin quantum wells of Si separated by wide band gap barriers. Suitable barrier candidates are SiO_2, CaF_2, and Al_2O_3 [92], and although a number of Si/barrier superlattices have been produced in the past [93] none has produced convincing evidence for quantum confinement induced emission until recently. In 1995, Lu *et al.* [94] reported visible light emission at room temperature from ultrathin-layer Si/ SiO_2 superlattices grown by molecular beam epitaxy that exhibited a clear quantum confinement shift with Si layer thickness, as shown in Fig. 11. According to effective mass theory and assuming infinite potential barriers, which is a reasonable approximation since wide gap (9 eV) SiO_2 barriers are used, the energy gap E for one-dimensionally confined Si should vary as

$$E = E_g + \frac{\pi \hbar^2}{2d^2}\left(\frac{1}{m_e^*} + \frac{1}{m_h^*}\right),\qquad(1)$$

where E_g is the bulk material band gap and m_e^* and m_h^* are the electron and hole effective masses [95]. This simple model is a reasonable first approximation to compare with ex-

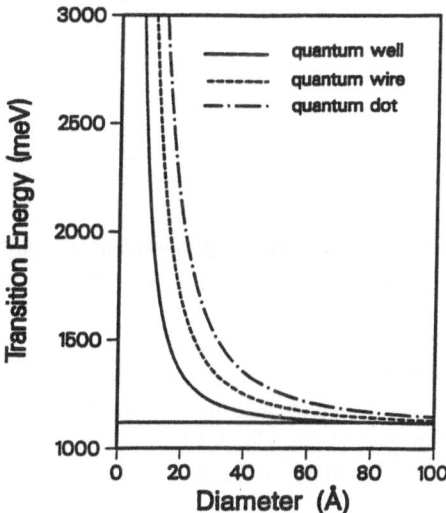

Figure 10. Optical gap in Si quantum wells, wires, and dots versus system diameter. The transition energy is calculated for the lowest electron and heavy hole eigenenergies for infinite confining potentials. The horizontal line is the bulk Si band gap at room temperature. [Reprinted from Lockwood, D.J., Aers, G.C., Allard, L.B., Bryskiewicz, B., Charbonneau, S., Houghton, D.C., McCaffrey, J.P., and Wang, A. (1992) Optical properties of porous silicon, *Can. J. Phys.* **70**, 1184–1193.]

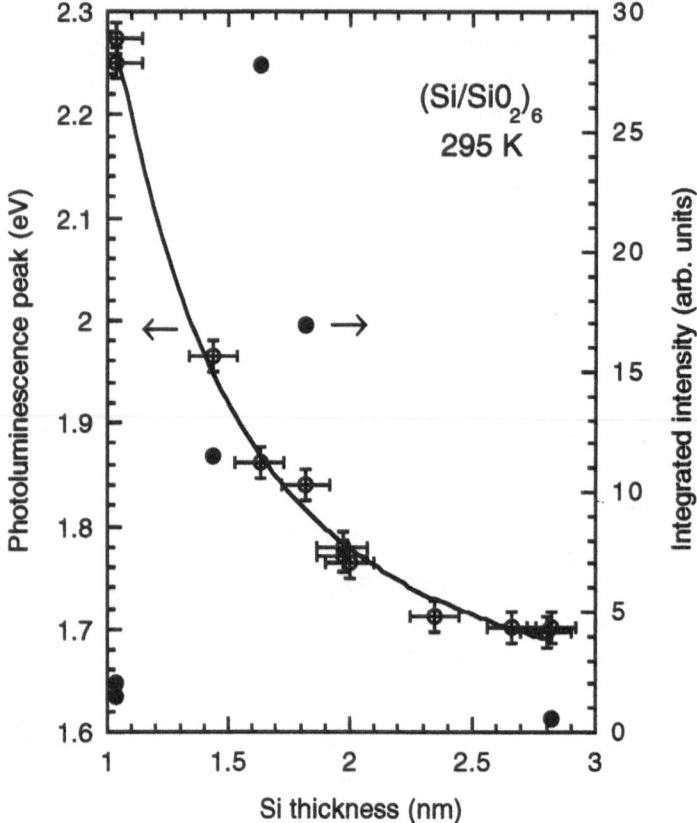

Figure 11. The PL peak energy (o) and integrated intensity (•) at room temperature in (Si/SiO₂)₆ superlattices as a function of Si layer thickness. The solid line is the fit with effective mass theory. [Reprinted from Lockwood, D.J., Lu, Z.H., and Baribeau, J.-M. (1996) Quantum confined luminescence in Si/SiO₂ superlattices, *Phys. Rev. Lett.* **76**, 539–541.]

periment for quantum wells [91]. The shift in PL peak energy with Si well thickness d is well represented by Eq. (1), as can be seen in Fig. 11, with

$$E(\text{eV}) = 1.60 + 0.72d^{-2}. \tag{2}$$

The very thin layers of Si ($1 < d < 3$ nm) are amorphous, but nearly crystalline, owing to the growth conditions and the huge strain at the Si–SiO₂ interfaces [96]. The fitted E_g of 1.60 eV is larger than that expected for c-Si (1.12 eV at 295 K), but is in excellent agreement with that of bulk a-Si (1.5–1.6 eV at 295 K). The indications of direct band-to-band recombination were confirmed by measurements via x-ray techniques of the conduction and valence band shifts with layer thickness [94,97]. The fitted confinement parameter of 0.72 eV/nm² indicates $m_e^* \approx m_h^* \approx 1$, comparable to the effective masses of c-Si at room temperature. The integrated intensity at first rises sharply with decreasing Si thickness until $d \approx 1.5$ nm and then decreases again, which is consistent with quantum well exciton emission [98]. The PL intensity is enhanced by factors of up to 100 on annealing and is also selectively enhanced and band-width narrowed by incorporation into a planar optical microcavity [99], as shown in Fig. 12. The bright PL obtained from as-grown and annealed a-Si/SiO$_2$ superlattices offers interesting prospects for the fabrication of a Si-based light emitter that can be tuned from 500 to beyond 800 nm by varying the a-Si layer thickness and/or the annealing conditions, all using available vacuum deposition technology and standard Si wafer processing techniques. The next important step is to develop LEDs based on such superlattices.

Wires. Quantum wires obtained by etching Si/Si$_{1-x}$Ge$_x$ heterostructures have been investigated by several groups (see, for example, Refs. 100 and 101). In PL measurements, wires defined by electron beam lithography and reactive ion etching have shown small blue shifts of up to 30 meV in the Si$_{1-x}$Ge$_x$ alloy peak at ~1.1 eV due to a combination of strain and confinement [100,101]. Alternatively, Si$_{1-x}$Ge$_x$ wires have been grown on V-groove patterned Si substrates [102]: The infrared emission (PL and EL) in this case exhibits a large optical anisotropy [103]. No significant intensity enhancements compared with PL from quantum well transitions have been realized in these wire structures.

It has not yet been possible to produce thin enough freestanding wires of c-Si by etching techniques to observe quantum confinement effects, although room temperature PL at wavelengths from 400–850 nm is found for pillars with diameters ~10 nm (see, for example, Refs. 104 and 105 and references therein). Recently, an EL device based on Si nanopillars has been produced: The device emitted red light that was visible to the naked eye [106].

Dots. Attention has now turned to the production of Si$_{1-x}$Ge$_x$ quantum dots, as these produce the strongest confinement effects for a given diameter or can achieve desired

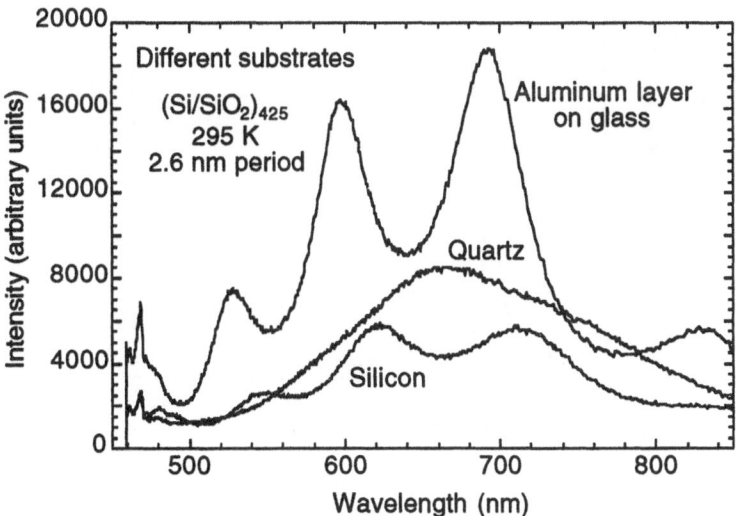

Figure 12. Room temperature PL of a (Si/SiO$_2$)$_{425}$ superlattice with a 2.6 nm periodicity deposited on Si, quartz, and Al-coated glass. [After Sullivan, B.T., Lockwood, D.J., Labbé, H.J., and Lu, Z.-H. (1996) Photoluminescence in amorphous Si/SiO$_2$ superlattices fabricated by magnetron sputtering, *Appl. Phys. Lett.* **69**, 3149–3151.]

confinements with smaller diameters than for wires (see Fig. 10). Quantum dots fabricated by etching $Si/Si_{1-x}Ge_x$ superlattices have produced 4 K PL at 0.97 eV that is 200 times brighter in 60 nm dots compared with the unetched superlattice PL [107]. Similar studies of $Si_{1-x}Ge_x$ dots fabricated by self-assembling island growth on Si have shown an increased luminescence efficiency due to the localization of excitons in the dots [108]. In the latter case, the dots were buried in Si, which has the advantage of minimizing surface defect recombination. In both cases, EL has been observed from diode structures at low temperatures [108,109] and at room temperature [109], as shown in Fig. 13. The infrared EL at 4.2 K in the dot is two orders of magnitude higher in intensity than in the as-grown superlattice (see Fig. 13). At room temperature, the dot EL at 1.3 μm is only 50% less efficient, with a threshold injection current of ~0.1 pA/dot and an electrical-input to optical-output power conversion efficiency of 0.14% [109]. Thus it is conceivable that this work could lead to a new generation of $Si/Si_{1-x}Ge_x$ optoelectronic devices at the optical fiber communication wavelength of 1.3 μm.

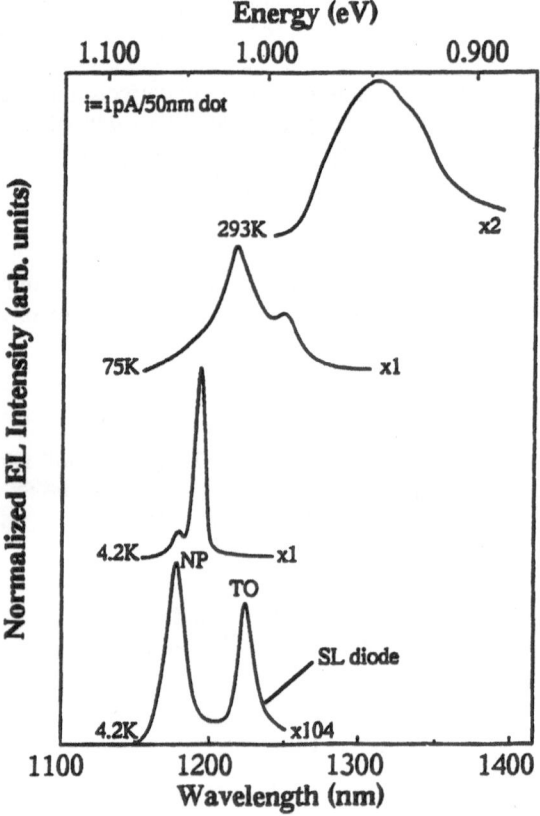

Figure 13. Temperature dependence of EL spectra of a 50 nm $Si/Si_{0.7}Ge_{0.3}$ quantum dot diode under reverse bias of 0.5 V and an injection current of 1 pA/dot. A reference spectrum from a superlattice (SL) diode is also shown. [After Tang, Y.S., Ni, W.-X., Sotomayor Torres, C.M., and Hansson, G.V. (1995) Fabrication and characterisation of $Si-Si_{0.7}Ge_{0.3}$ quantum dot light emitting diodes, *Electron. Lett.* 31, 1385–1386. Reproduced by permission of The Institution of Electrical Engineers.]

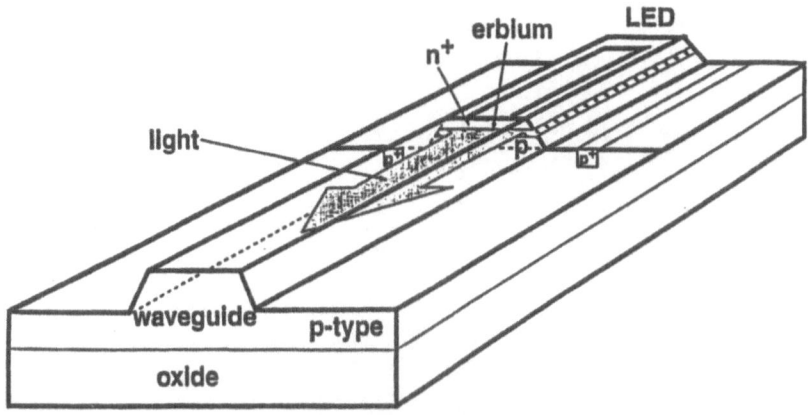

Figure 14. Schematic representation of a Si:Er edge emitting LED integrated with a Si waveguide. [Reprinted from Michel, J., Zheng, B., Palm, J., Ouellette, E., Gan, F., and Kimerling, L.C. (1996) Erbium doped silicon for light emitting devices, *Mat. Res. Soc. Symp. Proc.* **422**, 317–324.]

4. Prospects for Silicon-Based Optoelectronic Devices

Although a considerable number of optical detectors and waveguide structures have been created from Si-based materials [4,110,111], there is still a paucity of LEDs constructed from Si and, most importantly for many all-Si optoelectronic applications, no lasers. Of the materials systems reviewed here, LEDs made from Si:Er show the most immediate promise for device applications at 1.54 µm. A schematic picture of an optoelectronic device [58] comprising an edge emitting Si:Er LED integrated with a Si waveguide on a Si-on-insulator substrate is shown in Fig. 14. The EL linewidth of such LEDs at room temperature is approximately 10 nm [112]. This narrow linewidth and the fixed emission wavelength augers well for optical fiber communication systems with high bandwidth capacity. Optical gain at 1.54 µm should be obtainable in suitable Si waveguide structures and even laser emission, if the room temperature quantum efficiency can be improved.

Porous Si LEDs emitting at orange-red wavelengths are no longer just a curiosity with the announcement of LEDs having reasonable external quantum efficiencies (0.1%), lifetimes of the order of months, and low driving voltages (2–5 V) in forward bias [113] and also of devices with integrated Si transistor drivers, as shown schematically in Fig. 15. Apart from display applications [80], however, the long lifetime and broad linewidth of the optical emission will limit optical communications applications of π-Si LEDs, and it is not clear whether current injection lasers will ever be made from π-Si. The need to be compatible with existing large-scale Si processing also leads to difficulties with electrochemically created π-Si. It may be that oxidized Si nanoparticles will eventually prove to be superior to π-Si in this regard and also in device stability, but the long lifetime and wide band width of the emitted light are still going to limit device performance. Nevertheless, π-Si is a versatile material and offers extremely diverse optoelectronic functionality to Si in the areas of infrared and visible waveguiding, photodetection, and photomodulation [114].

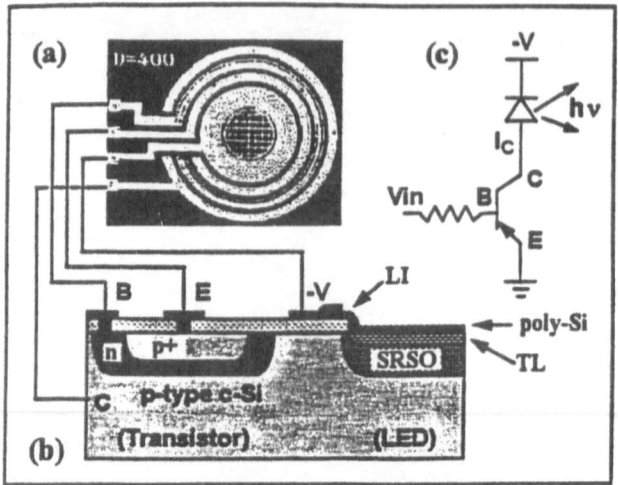

Figure 15. Integrated π-Si LED/bipolar transistor device operational at room temperature: (a) plan view, (b) cross section, and (c) equivalent circuit. The LED is in the centre of the structure and has a 400 μm diameter light-emitting area. Partial oxidation of π-Si in a dilute oxygen ambient has produced Si nanoclusters within an oxide matrix—Si-rich Si oxide (SRSO). Reprinted with permission from Hirschman, K.D., Tsybeskov, L., Duttagupta, S.P., and Fauchet, P.M. (1996) Silicon-based visible light emitting devices integrated into microelectronic circuits, *Nature* **384**, 338–341. Copyright 1996 Macmillan Magazines Limited.]

Light emission from quantum well and dot structures may yet hold the most promise for producing lasers at wavelengths across the visible into the infrared. The Si/SiO$_2$ multiple quantum well structures [93] are well suited for visible wavelength lasers at room temperature. Their optical absorption characteristics are ideal for optical pumping in a quantum microcavity, but it is not yet certain if their electrical characteristics are amenable to injection laser design. Quantum dot LEDs made from Si/Si$_{1-x}$Ge$_x$ [109] show considerable potential for laser applications at 1.3 μm. However, much more research and development work on these structures is required before this potential can be realized.

In conclusion, considerable progress has been made over the last decade on obtaining efficient light emission from a wide variety of Si-based materials. This work has led to the development of light emitting devices that are just now reaching useful performance levels. The intensity of research and development on light emission in Si is increasing as a result of these stimulating advances in materials engineering and technology. It is likely that a Si-based laser will emerge from this research in the near future, although the actual active laser material could be none of those discussed here, because of the burgeoning diversification [6] in Si-based materials.

References

1. Kasper, E. and Schäffler, F. (1991) Group-IV compounds, T.P. Pearsall (ed.), in *Strained-Layer Superlattices: Materials Science and Technology*, Academic Press, Boston, pp. 223–309.
2. Abstreiter, G. (1992) Engineering the future of electronics, *Physics World* **5** (3), 36–39.
3. Saleh, B.A. and Teich, M.C. (1991) *Fundamentals of Photonics*, Wiley, New York.
4. Soref, R.A. (1993) Silicon-based optoelectronics, *Proc. IEEE* **81**, 1687–1706.
5. Kasper, E. and Presting, H. (1990) Device concepts for SiGe optoelectronics, *SPIE Proc.* **1361**, 302–312.

6. Soref, R.A. (1996) Silicon-based group IV heterostructures for optoelectronic applications, *J. Vac. Sci. Technol. A* **14**, 913–918.
7. Iyer, S.S. and Xie, Y.-H. (1993) Light emission from silicon, *Science* **260**, 40–46.
8. Kimerling, L.C., Kolenbrander, K.D., Michel, J., and Palm, J. (1997) Light emission from silicon, *Solid State Phys.* **50**, 333–381.
9. Lockwood, D.J. (1997) *Light Emission in Silicon*, Academic, Boston.
10. Lockwood, D.J. (1994) Optical properties of porous silicon, *Solid State Commn.* **92**, 101–112.
11. *Properties of Silicon* (1988) INSPEC, London.
12. Chelikowsky, J.R. and Cohen, M.L. (1976). Nonlocal pseudopotential calculations for the electronic structure of eleven diamond and zinc-blende semiconductors, *Phys. Rev. B* **14**, 556–582.
13. Pankove, J.I. (1971) *Optical Processes in Semiconductors*, Dover, New York.
14. Yoffe, A.D. (1993) Low-dimensional systems: Quantum size effects and electronic properties of semiconductor microcrystallites (zero-dimensional systems) and some quasi-two-dimensional systems, *Advan. Phys.* **42**, 173–262.
15. Gnutzman, U. and Clausecker, K. (1974) Theory of direct optical transitions in an optical indirect semiconductor with a superlattice structure, *Appl. Phys.* **3**, 9–14.
16. Jackson, S.A. and People, R. (1986) Optical absorption probability for the zone-folding induced quasi-direct gap in Ge(x)Si(1-x)/Si strained layer superlattices, *Mat. Res. Soc. Symp. Proc.* **56**, 365–370.
17. People, R. and Jackson, S.A. (1987) Indirect, quasidirect, and optical transitions in the pseudomorphic (4x4)-monolayer Si-Ge strained-layer superlattice on Si (001), *Phys. Rev. B* **36**, 1310–1313.
18. Brey, L. and Tejedor, C. (1987) New optical transitions in Si-Ge strained superlattices, *Phys. Rev. Lett.* **59**, 1022–1025.
19. Froyen, S., Wood, D.M., and Zunger, A. (1987) New optical transitions in strained Si-Ge superlattices, *Phys. Rev. B* **36**, 4547–4550.
20. Hybertsen, M.S. and Schlüter, M. (1987) Theory of optical transitions in Si/Ge (001) strained-layer superlattices. *Phys. Rev. B* **36**, 9683–9693.
21. Pearsall, T.P., Bevk, J., Feldman, L.C., Bonar, J.M., Mannaerts, J.P., and Ourmarzd, A. (1987) Structurally induced optical transitions in Ge-Si superlattices, *Phys. Rev. Lett.* **58**, 729–732.
22. Zachai, R., Eberl, K., Abstreiter, G., Kasper, H., and Kibbel, H. (1990) Photoluminescence in short-period Si/Ge strained-layer superlattices, *Phys. Rev. Lett.* **64**, 1055–1058.
23. Menczigar, U., Abstreiter, G. Olajos, J., Grimmeiss, H.G., Kibbel, H., Presting, H., and Kasper, E. (1993) Enhanced band-gap luminescence in strain-symmetrized $(Si)_m/(Ge)_n$ superlattices, *Phys. Rev. B* **47**, 4099–4102.
24. Presting, H., Menczigar, U., Abstreiter, G., Kibbel, H., and Kasper, E. (1992) Electro- and photoluminescence from ultrathin SimGen superlattices, *Mat. Res. Soc. Symp. Proc.* **256**, 83–88.
25. Menczigar, U., Brunner, J., Freiss, E., Gail, M., Abstreiter, G., Kibbel, H., Presting, H., and Kasper, E. (1992) Photoluminescence studies of $Si/Si_{1-x}Ge_x$ quantum wells and Si_mGe_n superlattices, *Thin Solid Films* **222**, 227–233.
26. Presting, H., Kibbel, H., Jaros, M., Turton, R.M., Menczigar, U., Abstreiter, G., and Grimmeiss, H.G. (1992) Ultrathin Si_mGe_n strained layer superlattices — a step towards Si optoelectronics, *Semicond. Sci. Technol.* **7**, 1127–1148.
27. Pearsall, T.P. (1994) Electronic and optical properties of Ge-Si superlattices, *Prog. Quantum Optics* **18**, 97–152.
28. Engvall, J., Olajos, J., Grimmeiss, H.G., Presting, H., Kibbel, H., and Dasper, E. (1993) Electroluminescence at room temperature of a Si_mGe_n strained-layer superlattice, *Appl. Phys. Lett.* **63**, 491–493.
29. Engvall, J., Olajos, J., Grimmeiss, H.G., Kibbel, H.., and Presting, H. (1995) Luminescence from monolayer-thick Ge quantum wells embedded in Si, *Phys. Rev. B* **51**, 2001–2004.
30. People, R. and Jackson, S.A. (1990) Structurally induced states from strain and confinement, in T.P. Pearsall (ed.), *Strained Layer Superlattices: Physics*, Academic Press, Boston, pp. 119–174.
31. Houghton, D.C., Aers, G.C., Yang, S.-R.E., Wang, E., and Rowell, N.L. (1995) Type I band alignment in $Si_{1-x}Ge_x/Si(001)$ quantum wells: Photoluminescence under applied [110] and [100] uniaxial stress, *Phys. Rev. Lett.* **75**, 866–869.
32. St. Amour, A., Liu, C.W., Sturm, J.C., Lacroix, Y., and Thewalt, M.L.W. (1995) Defect-free band-edge photoluminescence and band gap measurement of pseudomorphic $Si_{1-x-y}Ge_xC_y$ alloy layers on Si(100), *Appl. Phys. Lett.* **67**, 3915–3917.

206

33. Orner, B.A., Olowolafe, J., Roe, K., Kolodzey, J., Laursen, T., Mayer, J.W., and Spear, J. (1996) Band gap of Ge rich $Si_{1-x-y}Ge_xC_y$ alloys, *Appl. Phys. Lett.* **69**, 2557–2559.
34. Soref, R.A., Atzman, Z., Shaapur, F., Robinson, M., and Westhoff, R. (1996) Infrared waveguiding in $Si_{1-x-y}Ge_xC_y$ upon silicon, *Optics Lett.* **21**, 345–347.
35. Rowell, N.L., Noël, J.-P., Houghton, D.C., and Buchanan, M. (1990) Electroluminescence and photoluminescence from $Si_{1-x}Ge_x$ alloys, *Appl. Phys. Lett.* **58**, 957–958.
36. Noël, J.-P., Rowell, N.L., Houghton, D.C., and Perovic, D.D. (1990) Intense photoluminescence between 1.3 and 1.8 μm from strained $Si_{1-x}Ge_x$ alloys, *Appl. Phys. Lett.* **57**, 1037–1039.
37. Sturm, J.C., Manohoran, H. Lenchyshyn, L.C., Thewalt, M.L.W., Rowell, N.L., Noël, J.-P., and Houghton, D.C. (1991) Well-resolved band edge photoluminescence of excitons confined in strained $Si_{1-x}Ge_x$ quantum wells, *Phys. Rev. Lett.* **66**, 1362–1365.
38. Lenchyshyn, L.C., Thewalt, M.L.W., Sturm, J.C., Schwartz, P.V., Prince, E.J., Rowell, N.L., Noël, J.-P., and Houghton, D.C. (1992) High quantum efficiency photoluminescence from localized excitons in $Si_{1-x}Ge_x$, *Appl. Phys. Lett.* **60**, 3174–3176.
39. Noël, J.-P., Rowell, N.L., Houghton, D.C., Wang, A., and Perovic, D.D. (1992) Luminescence origins in molecular beam epitaxial $Si_{1-x}Ge_x$, *Appl. Phys. Lett.* **61**, 690–692.
40. Lenchyshyn, L.C., Thewalt, M.L.W., Houghton, D.C., Noël, J.-P., Rowell, N.L., Sturm, J.C., and Xiao, X. (1993) Photoluminescence mechanisms in thin $Si_{1-x}Ge_x$ quantum wells, *Phys. Rev. B* **47**, 16655–16658.
41. Rowell, N.L., Noël, J.-P., Houghton, D.C., Wang, A., Lenchyshyn, L.C., Thewalt, M.L.W., and Perovic, D.D. (1993) Exciton luminescence in $Si_{1-x}Ge_x/Si$ heterostructures grown by molecular beam epitaxy, *J. Appl. Phys.* **74**, 2790–2805.
42. Robbins, D.J., Calcott, P., and Leong, W.Y. (1991) Electroluminescence from a pseudomorphic $Si_{0.8}Ge_{0.2}$ alloy, *Appl. Phys. Lett.* **59**, 1350–1352.
43. Mi, Q., Xiao, X., Sturm, J.C., Lenchyshyn, L.C., and Thewalt, M.L.W. (1992) Room-temperature 1.3 μm electroluminescence from strained $Si_{1-x}Ge_x/Si$ quantum wells, *Appl. Phys. Lett.* **60**, 3177–3179.
44. Fukatsu, S., Usami, N., Chinzei, T., Shiraki, Y., Nishida, A., and Nakagawa, K. (1992) Electroluminescence from strained SiGe/Si quantum well structures grown by solid source Si molecular beam epitaxy, *Jpn. J. Appl. Phys.* **31**, L1015–L1017.
45. Kato, Y., Fukatsu, S., and Shiraki, Y. (1995) Postgrowth of a Si contact layer on an air-exposed $Si_{1-x}Ge_x/Si$ single quantum well grown by gas-source molecular beam epitaxy, for use in an electroluminescent device, *J. Vac. Sci. Technol. B* **13**, 111–117.
46. Förster, M., Mantz, U., Ramminger, S., Thonke, K., Sauer, R., Kibbel, H., Schäffler, F., and Herzog, H.-J. (1996) Electroluminescence, photoluminescence, and photocurrent studies of Si/SiGe *p-i-n* heterstructures, *J. Appl. Phys.* **80**, 3017–3023.
47. Presting, H., Zinke, T., Splett, A., Kibbel, H., and Jaros, M. (1996). Room-temperature electroluminescence from Si/Ge/$Si_{1-x}Ge_x$ quantum-well diodes grown by molecular-beam epitaxy, *Appl. Phys. Lett.* **69**, 2376–2378.
48. Thomas, D.G., Gershenzon, M., and Hopfield, J.J. (1963) Bound excitons in GaP, *Phys. Rev.* **131**, 2397–2404.
49. Ennen, H., Pomrenke, G., Axmann, A., Eisele, K., Haydl, W., and Schneider, J. (1985) 1.54-μm electroluminescence of erbium-doped silicon grown by molecular beam epitaxy, *Appl. Phys. Lett.* **46**, 381–383.
50. Canham, L.T., Barraclough, K.G., and Robbins, D.J. (1987) 1.3-μm light-emitting diode from silicon electron irradiated at its damage threshold, *Appl. Phys. Lett.* **51**, 1509–1511.
51. Bradfield, P.L., Brown, T.G., and Hall, D.G. (1989) Electroluminescence from sulfur impurities in a *p-n* junction formed in epitaxial silicon, *Appl. Phys. Lett.* **55**, 100–102.
52. Davies, G. (1989) The optical properties of luminescence centres in silicon, *Physics Reports* **176**, 83–188.
53. Brown, T.G. and Hall, D.G. (1986) Observation of electroluminescence from excitons bound to isoelectronic impurities in crystalline silicon, *J. Appl. Phys.* **59**, 1399–1401.
54. Brown, T.G. and Hall, D.G. (1986) Optical emission at 1.32 μm from sulfur-doped crystalline silicon, *Appl. Phys. Lett.* **49**, 245–247.
55. Dieke, G.H. (1968) *Spectra and Energy Levels of Rare Earth Ions in Crystals*, Wiley, New York.
56. Ennen, H., Schneider, J., Pomrenke, G., and Axmann, A. (1983) 1.54-μm luminescence of erbium-implanted III-V semiconductors and silicon, *Appl. Phys. Lett.* **43**, 943–945.
57. Palm, J., Gan, F., Zheng, B., Michel, J., and Kimerling, L.C. (1996) Electroluminescence of erbium-doped silicon, *Phys. Rev. B* **54**, 17603–17615.

58. Michel, J., Zheng, B., Palm, J., Ouellette, E., Gan, F., and Kimerling, L.C. (1996) Erbium doped silicon for light emitting devices, *Mat. Res. Soc. Symp. Proc.* **422**, 317–324.
59. Michel, J., Benton, J.L., Ferrante, R.F., Jacobson, D.C., Eaglesham, D.J., Fitzgerald, E.A., Xie, Y.-H., Poate, J.M., and Kimerling, L.C. (1991) Impurity enhancement of the 1.54-μm Er^{3+} luminescence in silicon, *J. Appl. Phys.* **70**, 2672–2678.
60. Stimmer, J., Reittinger, A., Abstreiter, G., Holzbrecher, H., and Buchal, Ch. (1996) Growth conditions of erbium-oxygen-doped silicon grown by MBE, *Mat. Res. Soc. Symp. Proc.* **422**, 15–20.
61. Coffa, S., Franzò, G., and Priolo, F. (1996) High efficiency and fast modulation of Er-doped light emitting Si diodes, *Appl. Phys. Lett.* **69**, 2077–2079.
62. Canham, L.T. (1990) Silicon quantum wire array fabrication by electrochemical and chemical dissolution of wafers, *Appl. Phys. Lett.* **57**, 1046–1048.
63. Uhlir, A. Jr. (1956) Electrolytic shaping of germanium and silicon, *Bell Syst. Tech. J.* **35**, 333–347.
64. Cullis, A.G., and Canham, L.T. (1991) Visible light emission due to quantum size effects in highly porous crystalline silicon, *Nature* **353**, 335–338.
65. Lehmann, V., and Gösele, U. (1991) Porous silicon formation: A quantum wire effect, *Appl. Phys. Lett.* **58**, 856–858.
66. Bsiesy, A., Vial, J.C., Gaspard, F., Herino, R., Ligeon, M., Muller, F., Romestain, R., Wasiela, A., Halmaoui, A., and Bomchil, G. (1991) Photoluminescence of high porosity and of electrochemically oxidized porous silicon layers, *Surf. Sci.* **254**, 195–200.
67. Koshida, N. and Koyama, H. (1991) Efficient visible photoluminescence from porous silicon, *Jpn. J. Appl. Phys.* **30**, L1221–L1223.
68. Gardelis, S., Rimmer, J.S., Dawson, P., Hamilton, B., Kubiak, R.A., Whall, T.E., and Parker, E.H.C. (1991) Evidence for quantum confinement in the photoluminescence of porous Si and SiGe, *Appl. Phys. Lett.* **59**, 2118–2120.
69. Halimaoui, A., Oules, C., Bomchil, G., Bsiesy, A., Gaspard, F., Herino, R., Ligeon, M., and Muller, F. (1991) Electroluminescence in the visible range during anodic oxidation of porous silicon films, *Appl. Phys. Lett.* **59**, 304–306.
70. Koshida, N. and Koyama, H. (1992) Visible electroluminescence from porous silicon, *Appl. Phys. Lett.* **60**, 347–349.
71. Bensahel, D.C., Canham, L.T., and Osscicini, S. (1993) *Optical Properties of Low Dimensional Structures*, Kluwer, Dordrecht.
72. Feng, Z.C. and Tsu, R. (1994) *Porous Silicon*, World Scientific, Singapore.
73. Kanemitsu, Y. (1995) Light emission from porous silicon and related materials, *Phys. Reports* **263**, 1–91.
74. Vial, J.C. and Derrien, J. (1995) *Porous Silicon Science and Technology*, Springer-Verlag, Berlin.
75. Hérino, R. and Lang, W. (1995) *Porous Silicon and Related Materials*, Elsevier, Amsterdam.
76. Lockwood, D.J., Fauchet, P.M., Koshida, N., and Brueck, S.R.J. (1996) *Advanced Luminescent Materials*, The Electrochemical Society, Pennington.
77. Fauchet, P.M. (1996) Photoluminescence and electroluminescence from porous silicon, *J. Lumin.* **70**, 294–309.
78. Lockwood, D.J., Wang, A., and Bryskiewicz, B. (1994) Optical absorption evidence for quantum confinement effects in porous silicon, *Solid State Commun.* **89**, 587–589.
79. Lockwood, D.J. and Wang, A.G. (1996) in D.J. Lockwood, P.M. Fauchet, N. Koshida, and S.R.J. Brueck (eds.), *Photoluminescence in porous silicon due to quantum confinement*, The Electrochemical Society, Pennington, pp. 166–172.
80. Hirschman, K.D., Tsybeskov, L., Duttagupta, S.P., and Fauchet, P.M. (1996) Silicon-based visible light emitting devices integrated into microelectronic circuits, *Nature* **384**, 338–341.
81. Pavesi, L., Guardini, R., and Mazzoleni, C. (1996) Porous silicon resonant cavity light emitting diodes, *Solid State Commun.* **97**, 1051–1053.
82. Ogawa, T. and Kanemitsu, Y. (1995) *Optical Properties of Low-Dimensional Materials*, World Scientific, Singapore.
83. Takagi, H., Ogawa, H., Yamazaki, Y., Ishizaki, A., and Nakagiri, T. (1990) Quantum size effects on photoluminescence in ultrafine Si particles, *Appl. Phys. Lett.* **56**, 2379–2380.
84. Schuppler, S., Friedman, S.L., Marcus, M.A., Adler, D.L., Xie, Y.-H., Ross, F.M., Harris, T.D., Brown, W.L., Chabal, Y.J., Brus, L.E., and Citrin, P.H. (1994) Dimensions of luminescent oxidized and porous silicon structures, *Phys. Rev. Lett.* **72**, 2648–2651.

208

85. Kanemitsu, Y., Ogawa, T., Shiraishi, K., and Takeda, K. (1993) Visible photoluminescence from oxidized Si nanometer-sized spheres: Exciton confinement on a spherical shell, *Phys. Rev. B* **48**, 4883–4886.
86. Allan, G., Delerue, C., and Lannoo, M. (1996) Nature of luminescent surface states of semiconductor nanocrystallites, *Phys. Rev. Lett.* **76**, 2961–2964.
87. Brus, L.E., Szajowski, P.F., Wilson, W.L., Harris, T.D., Schuppler, S., and Citrin, P.H. (1995) Electronic spectroscopy and photophysics of Si nanocrystals: Relationship to bulk c-Si and porous Si, *J. Amer. Chem. Soc.* **117**, 2915–2922.
88. Tong, S., Liu, X.-N., Wang, L.-C., Yan, F., and Bao, X.-M. (1996) Visible electroluminescence from nanocrystallites of silicon films prepared by plasma enhanced chemical vapor deposition, *Appl. Phys. Lett.* **69**, 596–598.
89. Toyama, T., Matsui, T., Kurokawa, Y., Okamoto, H., and Hamakawa, Y. (1996) Visible photo- and electroluminescence from electrochemically formed nanocrystalline Si thin film, *Appl. Phys. Lett.* **69**, 1261–1263.
90. Lockwood, D.J., Aers, G.C., Allard, L.B., Bryskiewicz, B., Charbonneau, S., Houghton, D.C., McCaffrey, J.P., and Wang, A. (1992) Optical properties of porous silicon, *Can. J. Phys.* **70**, 1184–1193.
91. Zunger, A. and Wang, L.-W. (1996) Theory of silicon nanostructures, *Appl. Surf. Sci.* **102**, 350–359.
92. Tsu, R. (1993) Silicon-based quantum wells, *Nature* **364**, 19.
93. Lockwood, D.J. (1997) Quantum confined luminescence in Si/SiO$_2$ superlattices, *Phase Transitions*, to be published.
94. Lu, Z.H., Lockwood, D.J., and Baribeau, J.-M. (1995) Quantum confinement and light emission in SiO$_2$/Si superlattices, *Nature* **378**, 258–260.
95. Lockwood, D.J., Lu, Z.H., and Baribeau, J.-M. (1996) Quantum confined luminescence in Si/SiO$_2$ superlattices, *Phys. Rev. Lett.* **76**, 539–541.
96. Lu, Z.H., Lockwood, D.J., and Baribeau, J.-M. (1996) Visible light emitting Si/SiO$_2$ superlattices, *Solid-State Electron.* **40**, 197–201.
97. Lockwood, D.J., Baribeau, J.-M., and Lu, Z.H. (1996) Visible photoluminescence in SiO$_2$/Si superlattices, in D.J. Lockwood, P.M. Fauchet, N. Koshida, and S.R.J. Brueck (eds.), *Advanced Luminescent Materials*, The Electrochemical Society, Pennington, pp. 339–347.
98. Brum, J.A. and Bastard, G. (1985) Excitons formed between excited sub-bands in GaAs-Ga1-xAlxAs quantum wells, *J. Phys. C. Solid State Phys.* **18**, L789–L794.
99. Sullivan, B.T., Lockwood, D.J., Labbé, H.J., and Lu, Z.-H. (1996) Photoluminescence in amorphous Si/SiO$_2$ superlattices fabricated by magnetron sputtering, *Appl. Phys. Lett.* **69**, 3149–3151.
100. Tang, Y.S., Wilkinson, C.D.W., Sotomayor Torres, C.M., Smith, D.W., Whall, T.E., and Parker, E.H.C. (1993) Optical properties of Si/Si$_{1-x}$Ge$_x$ heterostructure based wires, *Solid State Commun.* **85**, 199–202.
101. Lee, J., Li, S.H., Singh, J., and Bhattacharaya, P.K. (1994) Low-temperature photoluminescence of SiGe/Si disordered multiple quantum wells and quantum well wires, *J. Electron. Mat.* **23**, 831–833.
102. Usami, N., Mine, T., Fukatsu, S., and Shiraki, Y. (1993) Realization of crescent-shaped SiGe quantum wire structure on a V-groove patterned Si substrate by gas-source Si molecular beam epitaxy, *Appl. Phys. Lett.* **63**, 2789–2791.
103. Usami, N., Mine, T., Fukatsu, S., and Shiraki, Y. (1994) Optical anisotropy in wire-geometry SiGe layers grown by gas-source selective epitaxial growth technique, *Appl. Phys. Lett.* **64**, 1126–1128.
104. Nassiopoulos, A.G., Grigoropoulos, S., and Papadimitriou, D. (1996) Light emitting properties of silicon nonopillars produced by lithography and etching, in D.J. Lockwood, P.M. Fauchet, N. Koshida, and S.R.J. Brueck (eds.), *Advanced Luminescent Materials*, The Electrochemical Society, Pennington, pp. 296–306.
105. Zaidi, S.H., Chu, A.-S., and Brueck, S.R.J. (1996) Room temperature photoluminescence from manufactured 1-D Si grating structures, in D.J. Lockwood, P.M. Fauchet, N. Koshida, and S.R.J. Brueck (eds.), *Advanced Luminescent Materials*, The Electrochemical Society, Pennington, pp. 307–316.
106. Nassiopoulos, A.G., Grigoropoulos, S., and Papadimitriou, D. (1996) Electroluminescent device based on silicon nonopillars, *Appl. Phys. Lett.* **69**, 2267–2269.
107. Tang, Y.S., Sotomayor Torres, C.M., Kubiak, R.A., Smith, D.A., Whall, T.E., Parker, E.H.C., Presting, H., and Kibbel, H. (1995) Optical emission from Si/Si$_{1-x}$Ge$_x$ quantum wires and dots, in D.J. Lockwood (ed.), *The Physics of Semiconductors*, Vol. 2, World Scientific, Singapore, pp. 1735–1738.
108. Apetz, R., Vescan, L., Hartmann, A., Dieker, C., and Lüth, H. (1995) Photoluminescence and electroluminescence of SiGe dots fabricated by island growth, *Appl. Phys. Lett.* **66**, 445–447.
109. Tang, Y.S., Ni, W.-X., Sotomayor Torres, C.M., and Hansson, G.V. (1995) Fabrication and characterisation of Si-Si$_{0.7}$Ge$_{0.3}$ quantum dot light emitting diodes, *Electron. Lett.* **31**, 1385–1386.

110. Hall, D.G. (1993) The role of silicon in optoelectronics, *Mat. Res. Soc. Symp. Proc.* **298**, 367–378.
111. Bozeat, R. and Loni, A. (1995) Silicon-based waveguides offer low-cost manufacturing, *Laser Focus World* **31** (4), 97–102.
112. Zheng, B., Michel, J., Ren, F.Y.G., Kimerling, L.C., Jacobson, D.C., and Poate, J.M., Room-temperature sharp line electroluminescence at λ=1.54 μm from an erbium-doped, silicon light-emitting diode, *Appl. Phys. Lett.* **64**, 2842–2844.
113. Collins, R.T., Fauchet, P.M., and Tischler, M.A. (1997) Porous silicon: From luminescence to LEDs, *Physics Today* **50** (1), 24–31.
114. Canham, L.T., Cox, T.I., Loni, A., and Simons, A.J. (1996) Progress towards silicon optoelectronics using porous silicon technology, *Appl. Surf. Sci.* **102**, 436–441.

CONFINED ELECTRONS AND PHOTONS

A Domain where New Physical Phenomena, Device Concepts And Widescale Applications Converge

C. WEISBUCH, H. BENISTY, D. LABILLOY
Laboratoire de Physique de la Matière Condensée, Ecole Polytechnique, 91120 Palaiseau, France.
R. HOUDRÉ, R.P. STANLEY AND M. ILEGEMS
Institut de Micro et Optoélectronique, Ecole Polytechnique Fédérale de Lausanne, CH-1015 Lausanne Ecublens, Switzerland.

1. Introduction

The scientific fields of confined electrons and photons have become areas of major efforts worldwide. Their appeal originates in the many facets they offer in fundamental and applied science, in technology and device development, and to high technology, large-scale industries.

As is well-known the field is immense. We will only deal with those effects of interest to optical properties, even further restricting the subject to that of the impact of confined electrons and photons to light emission from solids, as it is a major challenge, and as it demonstrates well the powers of the new physical concepts originating in confined electrons and photons systems.

Thanks to the NATO programme on Nanoscale Science and Technology a summer school and a workshop were organized, in July 1993 at Erice [1] and in August 1995 in Cargese [2], respectively. This review will outline some of the main points of these two events and will complete them with the recent progress. Both proceedings certainly constitute so far the best introductions to the field.

2. Light emission from solids : a grand challenge

Solid state materials are clearly the preferred state of matter (as opposed to gases and liquids) when it comes to active light source materials, because of their compactness, cooling capacity and robustness. They also allow direct electrical injection when dealing with conducting materials. They however present several drawbacks schematically shown on Fig. 1.

(i) The problem of delocalized electrons : perfect crystalline solids have delocalized wavefunctions over the whole crystal. While this property ensures the good electron mobility of such materials, a highly desirable feature, it is contradictory with good luminescence properties : (1) travelling electrons can encounter non-radiative centers ;

N. García et al. (eds.), Nanoscale Science and Technology, 211–234.
© 1998 *Kluwer Academic Publishers.*

212

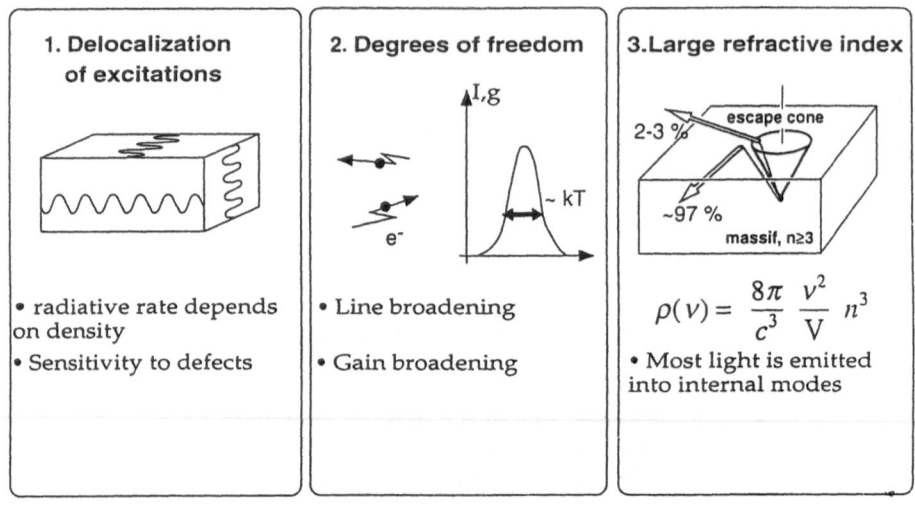

Figure 1   : The three issues in light emission from solids.

(2) electron and hole overlap is weak on average and will be represented in k-space by a product of state occupancy factors. The radiative lifetime is therefore density-dependent, and only reaches a value in the nanosecond range when saturating the occupancy factor. Such a short liftetime is necessary for device speed and efficiency (to overcome non-radiative recombination rates).

This contradiction between delocalization, beneficial for transport, and localization, good for luminescence, is seen in numerous physical systems : amorphous Silicon, crystals doped with luminescent impurities (rare-earth doped glasses, YAG, silicon, III-Vs). The contradiction even extends to polymer materials where, however, a good compromise between transport and luminescent properties is demonstrated by the excellent electroluminescence properties of small molecules such as Alq3 or PPV-based polymers [3].

(ii) The problem of kinetic energy : the translational degrees of freedom of band electrons lead to a thermal distribution of kinetic energies, through thermalization with the lattice. This translates into line broadening of the order of kT, diminishing the coherence of LEDs, leading to dispersion of signals in optical fiber systems, and most deleteriously to diminishing the optical gain for free carriers in laser materials : instead of being distributed over a linewidth given by the inverse lifetime, the gain per carrier is distributed over the thermal linewidth, translating into a much higher carrier injection to reach threshold gain at the peak of the gain curve.

(iii) The problem of high refractive index : active materials, in particular semiconductors, have high indices of refraction, ranging from 2.5 to 3.6, the higher indices corresponding to the smaller bandgaps. This leads to a small extraction efficiency for spontaneous emission : as the density of states of photon modes scales as n^3 it is clear that only a fraction n^{-3} of the emitted photons inside a high-index material will be emitted into air. A more direct calculation shows that only a fraction $(4n^2)^{-1}$ of incident light is emitted through an interface within the critical incidence angle $\theta \approx n^{-1}$, the

remainder being totally-internally reflected. Thus, a good GaAs emitter might only yield a 2 % external quantum efficiency.

Of course, physicists and engineers have offered various solutions to these problems. Internal quantum efficiencies above 80 % on an industrial basis, through sophisticated growth techniques (MBE and MOCVD) on highest-quality substrates. Geometrical shaping helps gather light emitted under various angles within the critical angle [4]. Commercial LEDs use mirrors on all sides of emitting materials, and also use polymeric hemispheric domes of index 1.5 to diminish the effective index radio to 3.6 / 1.5, thus increasing the outcoupled light by a total factor of 12 [5]. In addition, some of the light might be recycled leading to LEDs which can reach efficiencies up to 40 % [6]. However, such designs lead to good performance only at the single device level (they do not lend easily to integration) and they also diminish the source brightness.

One should remark that if photon recycling were perfect it would lead to unit efficiencies as in the end every photon would escape, however at the cost of a large speed decrease. In the real world where some loss mechanisms are always present, one requires very high internal quantum efficiencies (IQE) to achieve the many photon cycles needed to extract the light. Yablonovitch' s group demonstrated [7] a 70 % external quantum efficiency (EQE) LED with 99.7 % IQE.

All told, one can see why novel physical inputs are required in the field of light emitters : although lasers, in particular in their VCSEL form, have many attributes of good sources, they are not available for all uses and their fabrication might also be too demanding. LEDS, while rather efficient at the single device level, still suffer from a number of weaknesses.

3. Confined electrons

The optical properties of 2D quantum well (QW) structures are by now quite-well documented [8, 9] as well as their applications [10-12], with the most spectacular achievement being the various designs of quantum cascade lasers by Capasso and his collaborators [13, 14]. We will thus focus on the properties of 1D quantum wires (QWRs) and 0D quantum dots (QDs), mainly stressing the newer points compared to 3D bulk semiconductors and 2D QWs.

Since the early proposal [15] that QDs should prove useful for laser action, great efforts where devoted to reach the improvements predicted from various modelling efforts [16 , 17] : higher modulation speeds, reduction of the temperature dependance, smaller linewidth enhancement factor leading to lower noise, narrowed linewidth etc... All these improvements originate in the full energy quantization of energy levels in QDs. While this quantization has indeed been widely demonstrated in the past few years at the **single QD** level, progress towards the sharp emission lines of large **ensembles of QDs,** needed in devices, has been slow and is today the main difficulty to solve. Also, detailed analysis of QD-based devices reveal a number of critical issues that do not appear in simple-minded, first-order descriptions. On the other hand, QDs can be useful in other applications such as those based on the electron localization properties of QDs, or those based on the diminished energy relaxation in QDs (the "phonon bottleneck " effect). We will describe in the following our perception of the field today.

3.1 FABRICATION TECHNIQUES FOR QDs

Five main techniques are being favoured today to fabricate low-D electron systems, while many more have been explored. They each yield specific QD systems.

The direct growth technique by self-organization is the most widely used [18]. It relies on the fact that strained growth of lattice-unmatched systems leads to a **3D cluster-like growth** of islands.

Self-organized growth on **patterned substrates** combines the advantages of the low defect density of direct-growth methods and the ability to design the geometric pattern of the grown structures [19]. They are particularly used for quantum wires (QWRs) growth, and have led to the best QWR lasers (see e.g. [20]).

Direct nanofabrication by **lithographic techniques** is also widely used. Beyond the usual etching techniques which often lead to fabrication damage, more "gentle" fabrication schemes are being implemented, such as stressor-induced localization or localized impurity-assisted interdiffusion under light beam driving. These latter techniques have provided well-defined single QD structures [21].

One can **nucleate** and grow QDs in an ambient which provides the atoms forming the QDs. Beyond nucleation of nanoclusters in supersaturated glasses, recent techniques use vapor phase [22] or liquid phase [23] growth. These methods require some self-limitation in the growth process so as to reduce statistical size fluctuations. Various effects have been used : limited growth in micelles, size selection by spectrometry of mass, selection by centrifugation or sedimentation of liquid-phase grown QDs.

Quantum-well (QW) size fluctuations have recently emerged as a naturally-occuring system made of QDs : due to the irregular structure of QW interface, confinement energies vary along the QW plane, resulting in 3D confinement in pseudo-QDs [24, 25]. These systems have allowed very fine measurements of excited state spectroscopy, inter-box transfer, ground-state interactions...

3.2 BASIC OPTICAL PROPERTIES

3.2.1. Electron-hole pair transitions
Electron confinement in 2-D QWs, 1-D QWRs or zero-D QDs progressively restrict the number of degrees of freedom for electrons (Fig. 2). This translates into density-of-states (DOSs) with sharper and sharper features. Optical transitions involving **unbound electron-hole pairs** are characterized by an oscillator strength which is unchanged with dimensionality and therefore retains the 3D value f_{eh} **per transition**. Therefore, diminishing dimensionality sharpens the DOS, but **to retain a large light-matter interaction one requires to use a large number of QDs**. Dealing with equal number of injected carriers, the advantage of concentrating the DOS in QDs is approximately given by the ratio of kT to the QD linewidth (fig. 2).

3.2.2. Exciton transitions
Excitons dominate optical spectra of low-dimensional semiconductor structures in most instances (see e.g. [8]) : it is only when exciton relaxation mechanisms have associated rates faster than the exciton internal frequency (elevated temperatures for phonon mechanisms for instance), or when the exciton binding energy is diminished (phase-space filling at high electron or exciton density) that excitonic effects are replaced by e-h pair phenomena. Exciton effects are predicted to increase with diminishing

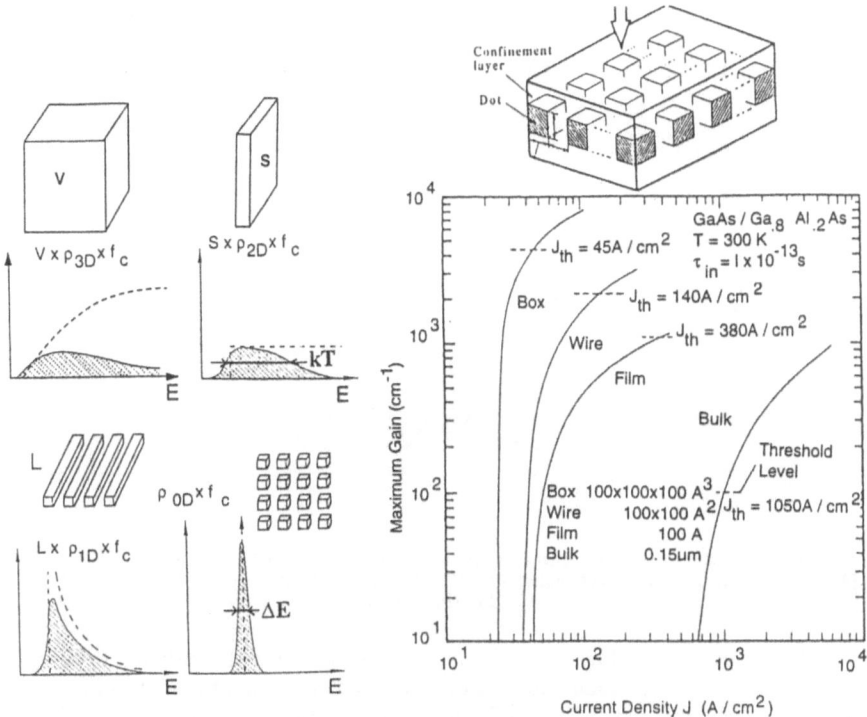

Figure 2 : (left) Schematics of various low-D systems, their associated density-of states (DOS) and gain ;
(right) the gain - current curve and threshold (from [17])

dimensionality. While this has been verified in T-shaped QWRs experiments [26], excitons are roughly unchanged in V-groove QWRs [27]. The exciton 1s oscillator strength is increased when compared to 2D QW excitons at the expense of the oscillator strength of the unbound e-h pairs (the so-called Sommerfeld factor) [28]. In addition, the electron-hole correlation **smoothes the 1D singularity** of the DOS for the unbound e-h pairs, and this smoothing also persists at high electron densities [29].

In 0D, the situation depends on the QD size relative to the exciton radius, and one defines three regimes of confinement, weak, intermediate, strong, corresponding to smaller and smaller QDs. In the weak confinement regime, the exciton motion is being quantized (for $L_{box} \geq$ the exciton thermal wavelength) and a "giant" oscillator strength can develop due to the coherence of the exciton wavefunction in the QD [30-32]. This leads to shortened lifetimes with increasing QD size, which have been observed [33].

In the strong confinement regime the confining energies for electrons and holes are larger than the exciton binding energy and therefore the Coulomb interaction only acts as a perturbation to the confining potential. It shifts the energy level without modifying the wavefunctions.

The issue of excitonic lifetime opens a very wide range of fascinating phenomena due to the interplay of the dimensionalities of electron and photon systems. It will be described below in §4.3.

3.3 QD SPECTROSCOPY

The main result of the recent years has indeed been the observation of narrow emission lines associated with single QDs, proving the freezing of the electrons translational degrees of freedom (Fig. 3). Linewidths down to 30 meV for InAs QDs in GaAs [34] or QD islands in GaAs QWs [35] have been reported.

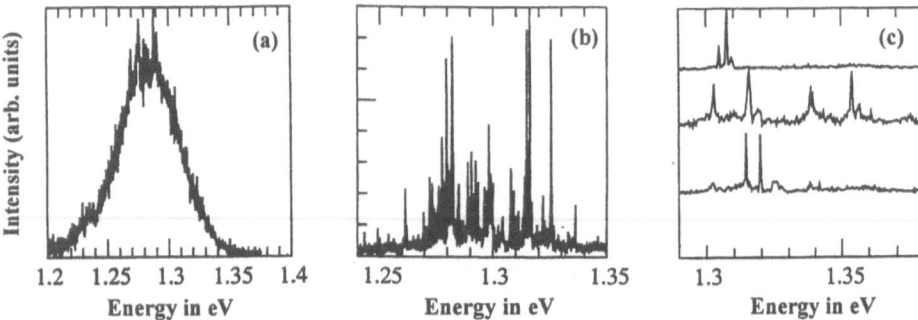

Figure 3   : Photoluminescence spectra from mesa-shaped ensembles of InAs quantum dots imbedded in GaAs (a) : 5000 nm diameter mesa, containing ≈ 8000 dots ; (b) 500 nm mesa ; (c) 200 nm mesas, containing ≈ 15 dots (From [34])

While such linewidths would be in agreement [35] with predicted lifetimes in the 30-50ps range for large QDs [30-32], they do not correspond to the usual measured times in the nanosecond range [36]. Influence of phonon interactions is not yet known. Composition fluctuations might also play a major role [37]. In II-VI QDs, linewidths down to 120 meV have been reported [38]. There are also QDs with much wider lines (≈ 2 meV) such as InP in an GaInP matrix [39].

3.4 THE ISSUE OF ENERGY RELAXATION IN QUANTUM DOTS

The full quantization of electron states in QDs raises a major issue about relaxation mechanisms of carriers to the ground state of QDs, which should depart from higher-dimensionality systems : due to the discreteness of energy levels in 0D QDs, as soon as the lateral box-size is below 1000 Å or 300 Å in the cases of electron or hole (and exciton) relaxations respectively, it is not possible to conserve both energy and momentum in LO or LA phonon-induced transitions, when using continuum phonon modes [40 , 41]. One then expects carrier accumulation in excited states, therefore broadening the emission band and gain curve, and diminishing of the overall quantum efficiency if some competing nonradiative recombination channels are present [41].

Experimentally the situation is not quite clear : steady-state PL measurements in III-V QDs often show strong excited state emission from the lowest set of states [21 , 42], compatible with a phonon relaxation bottleneck. Direct excitation of PL in the QD excited levels often shows multi-phonon sharp lines and excited-level PL of the same order as the ground -state PL [43 , 44] (Fig. 4).While this might also be proof of a relaxation bottleneck, care must be exerted [45] : this could only be due to the very weak absorption of states above the QD ground state and the resulting relative increase of Raman peaks. Some time-resolved experiments show a relaxation ladder and a slowdown of carrier relaxation [46], but most measurements [36] show a fast (sub-ns) risetime of photoluminescence (PL). Such efficient carrier relaxation as revealed by a fast onset time of PL might be due to higher-order effects (like Auger recombination [47 , 48] or multi-phonon scattering [49]). The PL efficiency is sometimes often strongly diminished when QD size is below ≈ 1000 Å, which can be indeed interpreted as proof of a relaxation bottleneck [43] or as due to increased surface defect density. However, in high-quality-growth samples, the efficiency seems very high, which is still compatible with a bottleneck effect if one assumes that non-radiative recombination is diminished when compared with higher-D systems, a distinct possibility arising from the carrier localisation effect of QDs (see below §3.6) and if QDs are defectless.

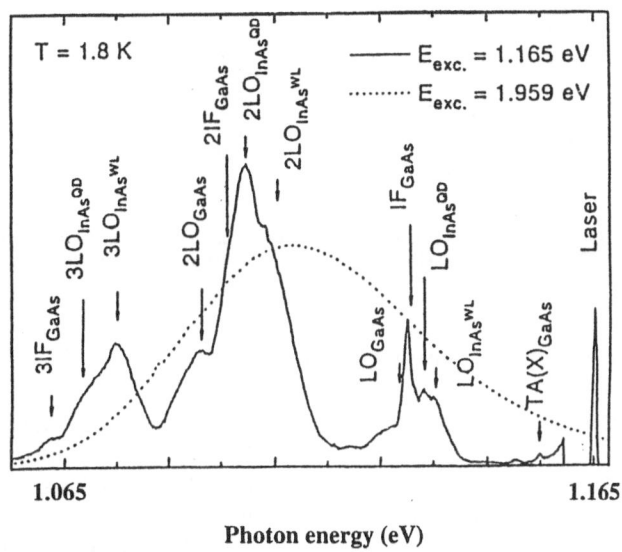

Figure 4 : Observed spectra for InAs quantum dots directly (full line) or indirectly (dashed line) excited [44]

Taking an opposite approach, one can try to take advantage of the relaxation bottleneck in conditions under which it exists. It can be put to good use for intersubband-based devices such as the quantum cascade laser or IR detectors as they are often limited by fast LO-phonon-induced decay of excited QW states (Fig. 4). For such devices, positive action of 3D quantization occurs as soon as phonon relaxation is hindered, i.e. at dot sizes in the 700 Å range. These "phonon - quantized" structures should therefore be quite easier to manufacture than electronic QDs (dot size ≈ 200 Å).

The above discussion dealt with the "usual" III-V quantum dots embedded in a well lattice-matched semiconductor matrix, hence the use of a continuum phonon model. When dealing with II-VI QDs in solution, or in a glass or polymer matrix, PL relaxation is always extremely fast whether observed c.w. or in pulsed regime [50]. Steady-state PL efficiency, even at low densities, can be extremely high [23 , 51]. Such results imply efficient, intrinsic relaxation mechanisms. Relaxation by surface states might play a role [23]. Instead of continuum phonon models, one can also consider new phonon modes such as interface, QD-localized or ripple modes. Some confined phonon models lead to relaxation rates which increase with decreasing QD radius as R^{-2} [52], at complete variance with the decreasing rate as R^8 for continuum models [41]. A recent evaluation of ripple modes in InGaAs QDs embedded in GaAs also points out the importance of such confined modes [53]. Clearly, one should develop the knowledge of electron-phonon interaction in confined systems, in addition to the study of low-D phonons [54].

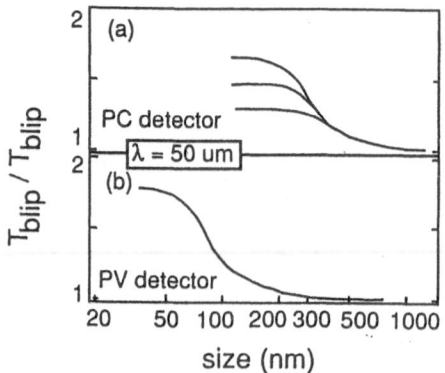

Figure 5 : Impact of the phonon relaxation bottleneck on the BLIP temperature of infrared photoconductive (PC) or photovoltaic (PV) intersubband QW detectors, (H. Benisty and C. Weisbuch).

3.5. APPLICATIONS TO QWR AND QD LASERS

The major thrust for lower dimensionality systems has been the search for better lasers. Fig. 2 shows the expected threshold current density [17] for laser structures where the active material is 3D, 2D, 1D or 0D. Asada's modelling assumes a constant level broadening of 6meV corresponding to a scattering rate of 10^{-13} s. Also, the optical confinement factors Γ are respectively 0.4, 0.037, 0.018 and 0.009 for the 3D, 2D, 1D 0D cases. This requires fractional surface coverages of 0.5 and 0.25 for QWRs and QDs respectively, i.e. a center-to-center distance is equal to their lateral size. Let us recall the threshold gain equation for a laser [10 , 12] :

$$\Gamma g = \alpha + \frac{1}{2L} \text{Log} \frac{1}{R_1 R_2} \qquad (1)$$

where g is the volume gain of the active medium, α the losses of the guided optical wave, L the laser cavity length and R_1, R_2 the end mirrors reflectivities. The increase in gain required to reach threshold seen in Fig. 2 merely reflects the diminishing of Γ with dimensionality.

As can be seen from the figure, the current required to reach the threshold gain value has two components : one is the transparency current required to reach net gain (it is given by the intersect of gain curve of Fig. 2 with the x axis). It represents the current needed to achieve population inversion between conduction and valence bands states, so that stimulated optical emission overcomes absorption. The second component above the transparency current represents the efficiency of additional carriers to create gain. It is on that component that lowering the dimensionality plays a major role (Fig. 2) by sharpening the DOS, thus enhancing the **gain per injected carrier above transparency**.

Therefore the major parameter of low D systems for lasers is the sharpness of the DOS, in particular for QDs. As already mentioned, this is limited by size fluctuations, up to now resulting in gain linewidths of the order of 50 meV, i.e. equivalent to those of QWs lasers. **The number of carriers above transparency required to reach threshold gain should then be the same as in QWs.** The transparency current might be improved somewhat, due to the roughly equivalent number of electron and holes states to be inverted [10]. Therefore, for present-day structures, only minor improvement in threshold current density is expected. To significantly improve it, one would need significantly lower widths ; the 6meV linewidth used for the simulation in Fig. 2 would require fluctuations of the order of a single monolayer in each direction, a difficult feat to achieve.

Fig. 2 also points to a major event : the main improvement to be expected in threshold reduction occured when switching from 3D bulk to 2D QW material. It was the large reduction in transparency current from 700 to 40 A.cm^{-2}, due to the adjustment of the number of quantum states in the active region to the needed electronic states for laser action [10, 55].

Beyond these intrinsic evaluations one has to consider real structures. In these, the confining layer plays a major role. It was already observed that in QW lasers it has a major influence on the threshold current density and its temperature dependence [56]. It even plays a more important role in QD lasers as the QD DOS is weaker than that of a QW, therefore increasing the relative importance of the quantum states of the confining layer. Assuming a 100 meV confinement energy for each type of carrier and a Boltzmann occupancy factor of the confining layer states, one finds that $\approx 10^{12}$ cm^{-2} levels are populated in a 1000 Å confining layer! This first means that carriers in the confining layer are numerous enough to efficiently relax QD carriers by the Auger mechanism mentioned in §3.4, but also that a "parasitic" current of 50 A cm^{-2} needs to be injected in the confining layer to ensure occupancy of the QD states, current much larger than that directly needed to populate the QDs, bringing the total threshold current back to values typical of QW lasers, even if the QDs had an ultimate-narrow linewidth.

How do experimental results compare with these first-order descriptions ? Fig. 6 represents a recent sum-up of the best results in QD lasers. They are indeed in the same range as QW lasers. The two obvious trends point to the validity of the above analysis : systems are improved when incorporating many QD layers and when using higher bandgap discontinuities between QDs and confining layer (Al-based materials). The latter point shows indeed the importance of the energy levels of the confining layer as the essentiel parameter for diminishing the thermal escape of carriers, hence avoiding large leakage currents. The former shows how an increase in QD density allows a lower occupancy factor of QD level to yield the theshold gain, hence yields a lower quasi-Fermi level and again a lower occupancy factor for confining-layer states.

This raises the question of future progress in QD lasers. Clearly, if either QD density, band discontinuities and size uniformity were improved one would still improve the performance, the latter parameter being the most important. It might prove quite a challenge to find good materials other than GaInAs/GaAlAs (for the specific case of QDs in GaInN/GaN see below §3.6), due to the required large band discontinuities.

QWR and QD lasers suffer from other limitations : the large population in confining layers sets limits to their speed of operation [57]. Gain saturation, due to both state filling and carrier heating within the small active volume also contributes to the speed limitation [58].

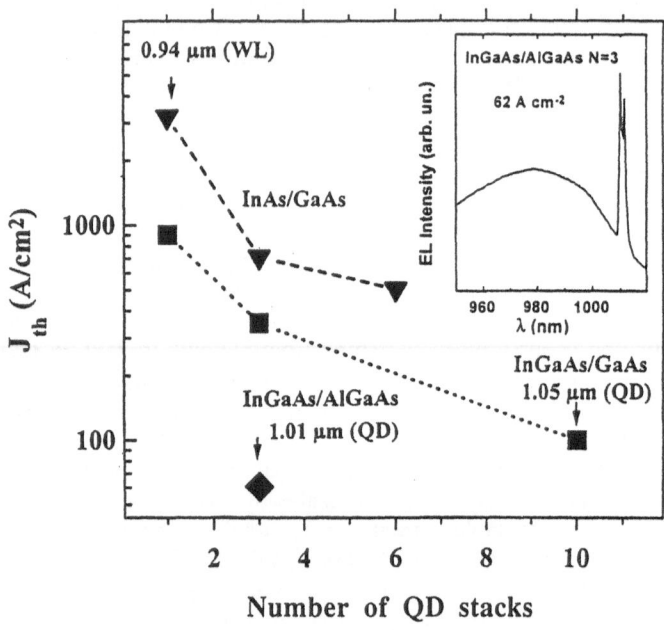

Figure 6 : Threshold current density for various QD laser structures. Inset shows an emission spectrum. Note its linewidth of ≈60 meV, the laser improvement with increasing QD confinement through deeper dots in GaAs confining layer or same dots in GaAlAs confinement (from Ledentsov [111])

3.6.   LOCALIZATION OF CARRIERS BY QDs

A frequently overlooked advantage of QDs is their ability to efficiently capture carriers in a highly radiative localized state. In that case, provided that the QD density is large compared to the non radiative (NR) defect density, carriers will be captured in the QD before experiencing an encounter with a NR defect. Radiative efficiency will increase ! This is indeed observed for InAs QDs grown on Si substrates. Whereas the luminescence is severely degraded for a test GaInAs QW grown on Si as compared to one grown on GaAs, QDs grown on either substrate have strong luminescence [59]. Recently, the same idea has been put to work for GaAs QD lasers grown on Si [60]. As the QD density is much larger than the defect density, the degradation rate of such lasers is much slower than that of QW lasers on Si.

The localization of carriers also plays a major role in the excellent properties of GaN. Although such materials exhibit huge densities of dislocations ($\approx 10^{10}$ cm^{-2}) they still have good luminescence properties [61]. While there might be several origins for such an effect in pure homogeneous material [62], including localization on donor-acceptor pairs, localization on compositionally-induced QDs states seem to play a major role for the widely-use GaInN QWs [63 , 64]. Indium-rich QDs have a 5.10^{11} cm^{-2} density, way higher than that of dislocations, and therefore act as efficient radiative centers.

4. Confined Photon Structures

Optical resonators with dimensions of an optical wavelength limit the number of optical modes that can interact with optically active material in the cavity. As in electronic systems, decreasing dimensionality of the optical system will concentrate optical intensity into a narrower spectral range (Fig. 7). While most of the activity deals with planar microcavities due to their ease of fabrication and their already good performance, ultimate systems will incorporate 0D microcavities.

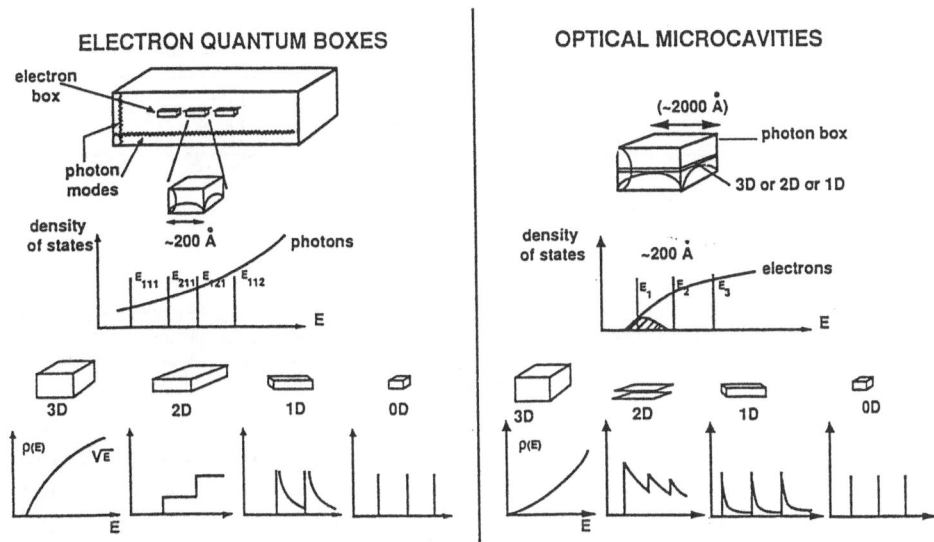

Figure 7 : Schematics of electron or photon quantization in electronic or photonic low-D systems.

4.1.   THE BASIC PHENOMENON : INTERFERENCE ENHANCEMENT AND CONTROL OF SPONTANEOUS EMISSION.

Photon confinement in wavelength-scale structures originates from the phenomenon of amplitude build-up due to multiple, in-phase reflections of optical waves (Fig. 8). These reflections either occur on **plane localized mirrors** in microcavities or on **interfaces distributed on a lattice** and acting as coherent light scatterers in photonic bandgap (PBG) materials [65-67]. As it can be shown that both types of structures lead to a similar degree of light localization [68], we will focus on microcavity structures.

Consider a **planar Fabry-Perot cavity**, with mirror transmission and reflectivity coefficients T and R, respectively. The intensity of a resonant mode is enhanced inside the cavity by a factor $4/(1-R)$ compared to an incoming wave, while that of the non-resonant modes (i.e. at other frequencies) is suppressed by a factor $(1-R)$. If one were to evaluate field amplitudes at a given frequency but at different incoming angles θ, for which the resonant mode frequency would change as $(\cos \theta)^{-1}$, the same suppression factor $(1-R)$ would be obtained for non-resonant modes, i.e. at angles different from that of the resonant mode : the unique feature of microcavities (and PBG materials) is therefore to concentrate the field intensity into the resonant mode by as much as they

222

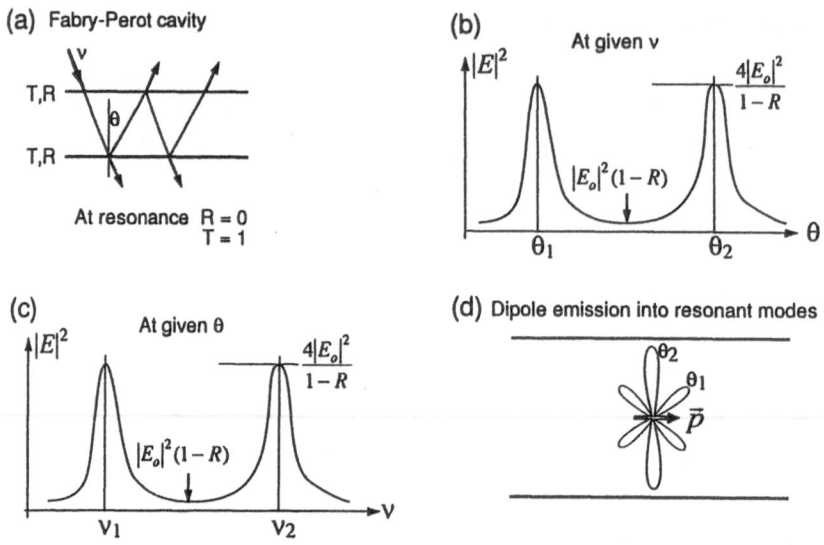

Figure 8 : Schematics of the basic phenomenom of resonant electric field increase in a planar Fabry-Perot microcavity (a) interference buildup ; (b) angular density at given frequency ; (c) spectral density at given angle ; (d) emission diagram for a third order planar cavity, for a monochromatic source, showing the angular selectivity of spontaneous emission due to the resonant enhancement of vacuum field.

suppress it in non-resonant modes. The detailed analysis [69 , 70] leads to an increased E-field intensity of an allowed mode by the quality factor $Q \approx (1-R)^{-1}$, an allowed mode width of $\Delta\omega/\omega \approx Q^{-1}$, and rejection of the antiresonant mode intensity by a factor Q^{-1}. As this is also true for the electric field of vacuum-field fluctuations, **spontaneous emission will be modified.** Most often, its changes can be accounted for by the modification of the photonic DOS entering Fermi's golden rule.This is called the **weak-coupling** regime of the light-matter interaction and it describes well how light can be preferentially emitted in one resonant mode while suppressed in other modes. It is of paramount importance that the resonant mode is enhanced as otherwise cavity discretization of optical modes would certainly select modes, but at the same time would generate light in these modes at unusefully low rates if the lifetime were just proportional to the number of active modes. This is a major difference with electron quantization where single QD systems which have few electrons (due to the Pauli exclusion principle), hence have very little action on an optical beam (see § 5 for a more detailed comparison between electron and photon mode quantization).

For **atoms in 0D cavities**, the spontaneous emission rate can be suppressed, or strongly increased by a factor $\lambda^3 Q/4\pi^2 V$ (where V is the cavity volume), depending on the overlap of the cavity resonance with the atomic transition [69 , 71 , 72]. In the solid-state, this only occurs for 0D electronic and photonic systems for which the atomic linewidth is narrower than the cavity linewidth. When dealing with 2D optical microcavities, the modes are unconfined laterally. This leads to lifetime changes of the order of 50%, diminishing when emission is broad [69, 73-75]. Even for semiconductors in 0D optical microcavities, for which the effective linewidth is kT, the modifications of spontaneous lifetime are quite small [75, 76]. It is only for sharp lines in solids, such as uniform QDs, that the lifetime change will significantly (see below § 5).

In some special cases where only one photon mode will interact with one electronic transition, Fermi's golden rule does not apply as the coupled system will oscillate indefinitely.This **strong - coupling** regime is a very important type of light - matter interaction and occurs in many instances (see below table 1). Its occurence in planar MCs will be described below in §4.3.

4.2. THE WEAK-COUPLING MICROCAVITY : LEDS AND LASERS.

The build-up of light emission in a given direction in planar 2D-MCs could have dramatic impact in LEDs and displays in that it solves difficulty #3 (Fig.1) in light extraction from solids : microcavities select an escape cone of the order of π (1-R). Another useful effect is due to the spectral width of allowed photon modes $\approx$ (1-R) , quite narrower than the usual thermally-broadened emission. This has been put to good use to increase by a factor of three the transmission capacity of LED-based optical fiber systems where chromatic dispersion is the limiting factor [77].

Mode selection does not obviously lead to an increase in **light extraction efficiency** in the resonant mode. One has to ascertain that competing spontaneous emission channels have been suppressed, or at least greatly diminished.

One would expect a 100% control if there were only one optical mode in the system. Whereas this might be achieved with 3D MCs, one has several modes in a 2D MC. Then, assuming at first order that the various modes (resonant, guided, "leaky", see below Fig. 11) have equal integrated amplitudes, the resonant outgoing mode will carry a larger fraction of spontaneous emission than the "natural" escaping fraction $1/4n^2$ (see §2) only if the number of modes in the cavity is less than $4n^2$. The number of modes being defined as the cavity order m_C, this defines the transition between macrocavities (no change in extraction efficiency) and genuine MCs as [75] $m_C=1/4n^2$.

4.2.1. *Microcavity LEDs (MC-LEDs)*

The factor of merit describing the control of spontaneous emission into a desired mode is called the spontaneous emission factor β defined by the ratio of the intensity emitted in the laser mode to that emitted in all modes.

To bring β close to unity in a vertical planar cavity geometry amounts to a fight against recombination into all other modes (guided, leaky, oblique) competing with the vertical resonant mode (Fig. 11) [78 , 79].

Progress towards high efficiency emission is well under way [78]. An extraction efficiency of 22% is achieved in the planar geometry shown in Fig. 9a. The modelling of spontaneous emission in the various modes is shown in Fig. 9c. Extraction is optimized when the cavity is slightly off-resonance with the center frequency resonant for a mode escaping at $\approx$ 45° outside the cavity [75, 78, 80]. The use of a metallic mirror on one side, also serving as an electrode, suppresses the "leaky" modes in one half-space. In addition, the high index of the metal mirror "repells" the in-plane guided mode away from the active region. Hence, the vertical emission mode has been enhanced at the expense of the guided and leaky modes. The Bragg mirror reflectivity is adjusted to 60%. This value shows that optimal outcoupling in LEDs is obtained for structure designs very far from that of VCSELs (which have reflectivities in the 99%-plus range): a bad laser does not make a good LED !

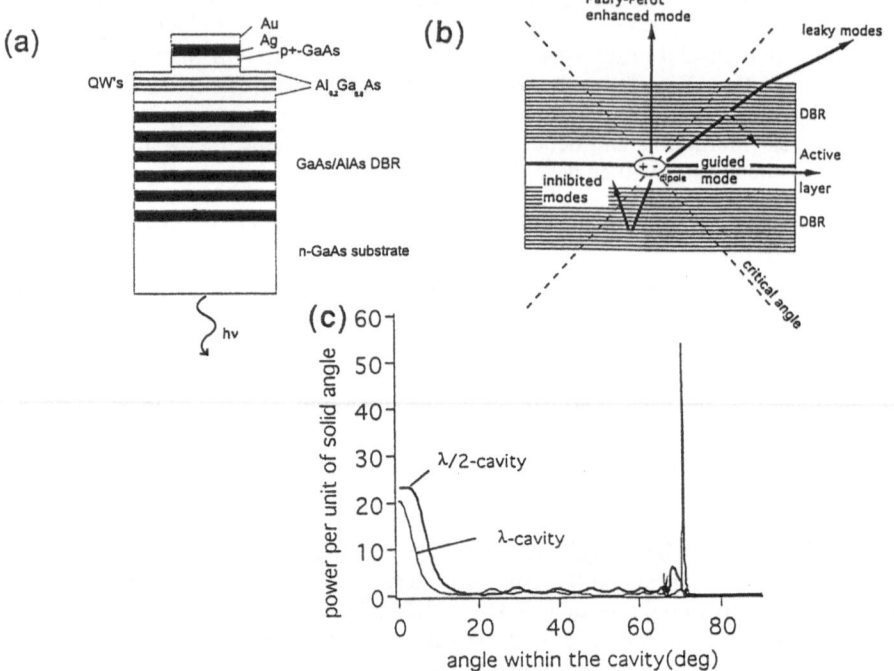

Figure 9 : Schematics of the operation of a high-efficiency LED : (a) LED structure ; (b) outline of the various competing modes ; (c) calculated emission diagram for λ/2 or λ cavities. The lobe at zero angle is the resonant vertical emission mode, modes around 70° are the waveguided modes and the weaker modes in-between are the leaky modes (From [112]).

Scaling models can be made to evaluate the extraction efficiency for such structures, as a function of the four scaling parameters : the emission linewidth, the refractive indices of the metal mirror, and the refractive indices of the DBR materials. Calculated extraction coefficients are shown in Fig. 10a as a function of DBR materials indices for a source with a 3% linewidth and a metal mirror index of 0.3+5i.

The above analysis assumed that photons which undergo total internal reflection are lost for extraction. It should be reminded that some of these can be recycled and therefore reemitted outside the crystal. We already mentioned the impact of recycling in §2 for standard emitters.What is its effect in MC LEDs ? present results [78] point to the existence of some degree of recycling. Actually, modelling predicts that recycling effects should be much more important in MC-LEDs than usually, as a photon needs much less cycles to escape than in a cavity-less LED with 2-10% extraction efficiency per cycle (Fig. 10b) [75].

4.2.2. Microcavity Lasers

In the field of lasers, the control of spontaneous emission can have dramatic effects : if β = 1, the emission quantum efficiency in the unique photon mode does not change when switching from spontaneous to stimulated emission with photon number. One reaches a "thresholdless" laser operation [81-84]. What is changing is that the spontaneous emission due to the transparency current is being split into β and (1 - β), representing the emission rates into and out of the laser mode. In such an analysis, only $(1 - \beta)$ is a loss. The rate equation modelling then leads to linear characteristic curves. In addition, photon emission in the mode becomes deterministic, leading to the generation of **photon-number squeezed states** which should allow noiseless optical communications [65, 85].

Experimentally however, the search for thresholdless lasers has been rather disappointing [69]: one needs to suppress all modes other than the laser mode. This requires 3D cavities (see the discussion in §4.4). The theoretical evaluations [70, 76] point out the severe requirements on mirror range and reflectivity, cavity size, and emission linewidth.

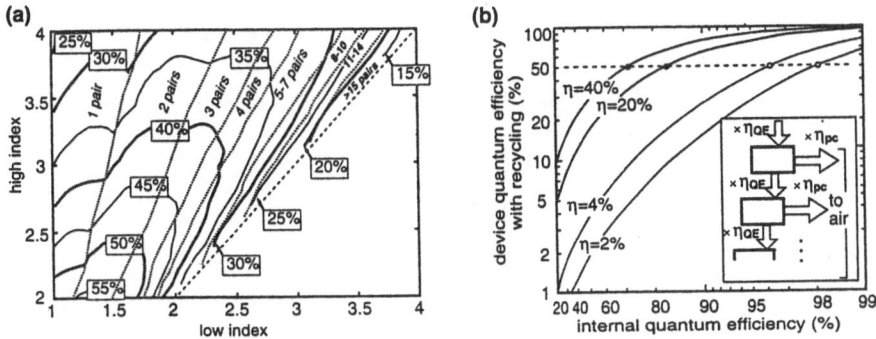

Figure 10 : (a) Calculated constant extraction efficiency curves as a function of refractive indices of the DBR materials ; (b) Overall efficiency as a function of internal quantum efficiency for various single-event extraction efficiencies, taking photon recycling into account [75].

4.3.   STRONG LIGHT-MATTER INTERACTION : CAVITY-POLARITONS.

Coupled systems whereby both electrons and photons have the **same dimensionality** with the **same translational symmetry** have a very special property : as the light-matter interaction does not change the translational symmetry, it will conserve the quantum number which describes this symmetry, i.e. the translational momentum (or translational wavevector). Then, in such systems, both **energy and wavevector are conserved** in an optical transition, which leads to the coupling of a **unique** exciton state to a photon state with the same energy and wavevector and vice-versa, although both systems can have continua of states. The system oscillates back and forth between the exciton and photon states, the so-called Rabi oscillation, instead of decaying irreversibly with the usual transition rate given by Fermi's golden rule. In 3D systems, this leads to the well-known phenomenon of **excitonic polaritons** [8]. In 2D, i.e. for QWs inserted in 2D optical microcavities (planar Fabry-Perots), one deals with **cavity-polaritons** [86].

This phenomenon of strong-coupling between light and matter occurs in planar structures very similar to MC-LEDs, but however under different conditions : one requires the coupling strength, as described by the Rabi oscillation frequency, to be faster than either of the damping rates of the electromagnetic (photon) or electronic (matter) oscillator. The coupling, given by the oscillator strength of the transition, must be large, hence the need for exciton phenomenon instead of electron-hole pairs.

Low damping rates are only obtained for photon states in high-reflectivity MCs, which allow long photon lifetimes.

The simplest way to observe the coupled exciton and photon oscillators is to measure their optical response under a weak test field, i.e. to measure the reflectivity of the system. This is shown in Fig. 11 where the anti-crossing behaviour of coupled oscillators appears as a function of cavity detuning (experimentally achieved with a variable-thickness cavity).

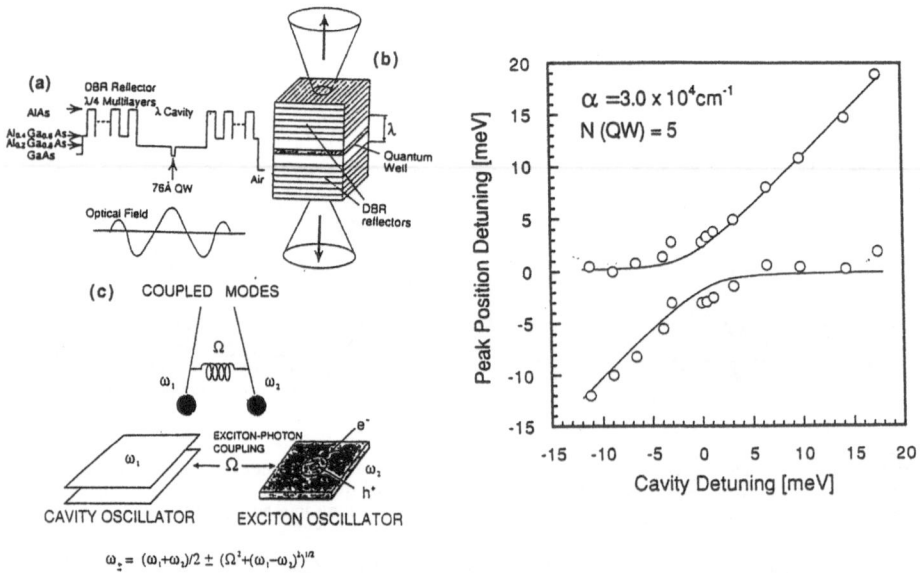

$$\omega_\pm = (\omega_1+\omega_2)/2 \pm (\Omega^2+(\omega_1-\omega_2)^2)^{1/2}$$

Figure 11 : Strong-coupling in semiconductor microcavities : (a) spatial variation of the conduction band minimum and of the optical field through the structure; (b) schematics of the structure; (c) representation of the coupled exciton and photon modes as oscillators ; (d) anti-crossing behaviour of the coupled 2D QW exciton and cavity photon modes as a function of cavity resonance energy [86].

The luminescent properties of **cavity - polaritons**, raise many questions : what are the light-emission mechanisms ? Are the dynamical processes modified ? Are such strongly-coupled systems interesting for new applications, such as thresholdless lasers or exciton-based lasers ? etc... Although a detailed answer to these questions is not possible yet, we can describe some interesting results.

A complete description of the luminescence of the resonant coupled-excitations could be complicated [87 , 88 , 89]: one must calculate both relaxation effects and emission properties of each mode. The MC polariton lineshape is however simple to describe : there is a one-to-one correspondance between a cavity polariton and an outside photon with an equal transverse-momentum. The angle-resolved emission I(E,θ) intensity is therefore proportional to the product of the absorption coefficient by an occupancy factor [90]. Conversely, observing I(E,θ) as a function of θ allows to determine the absorption peaks, directly linked to the energy levels of cavity polaritons. Doing so, one directly maps the cavity polariton dispersion curve [90].

Concerning lifetime issues, the **emission process is "quasi" forbidden in strong coupling** and is therefore of an extrinsic nature (table 1). Emission only occurs because of a breakdown of the system description, i.e. when a 3D polariton impinges at a crystal interface (the electron system is not truly 3D) or when a 2D polariton escapes the cavity because of the finite mirror reflectivity (the 2D cavity is not exactly 2D, as it is weakly coupled to a 3D photon system through mirror leakage). When the electronic (exciton) and photon systems have **different dimensionalities** there is no more momentum conservation (no polariton effect) and Fermi's golden rule does apply : for instance, for a **2D QW** in usual 3D photonic environment, wavevector conservation is no more required in the direction perpendicular to the QWs. The **recombination process is allowed**, with radiative lifetimes in the 20 ps range [8, 91]. For **1D QWRs**, the diminished phase-space leads to a decreased radiative recombination rate [92-94].

TABLE 1. Variations in exciton and photon dimensionalities

Excitons	Photons	Optical Properties
Bulk 3D	3D - Space	3D K - Conservation Mixed-Mode Exciton-Polariton Luminescence "Forbidden"
2D Quantum Well	3D - Space	No K-Conservation Luminescence "Allowed" - Fast 15 ps
2D Quantum Well	2 D - μCavity	K - Conservation Cavity-Polaritons Luminescence "Forbidden"
1D Quantum Wire	3D - Space	No K - Conservation Luminescence "Allowed" Fast
1D Quantum Wire	3D - Space	No K - Conservation Luminescence "Allowed" Fast
1D Quantum Wire	1D - μCavity	K - Conservation Luminescence "Forbidden"
0D Quantum Box	0D - μCavity	K - conservation weak or strong coupling - τ modified

It was hoped that strong-coupling would lead to a new fast and efficient recombinaison channel in semiconductors at room temperature : whereas in weak-coupling excitons are readily ionized after formation, before recombining, the strongly-coupled exciton can decay radiatively very fast before reionizing. However, the detailed study of strong coupling luminescence leads to the understanding that the **critical coupling** regime should be more efficient, as the cavity radiative loss is adjusted to the photon-exciton coupling as expressed by roundtrip absorption, a condition equivalent to an impedance matching between exciton and photon states [89]. The study of this regime, while showing improved light emission characteristics, suffers from another difficulty, like strong-coupling, namely the slow exciton energy relaxation towards fast emitting resonant states [87].

In order to solve this relaxation bottleneck problem, it has been proposed to mix semiconductor and organic materials properties [95]: organic molecules are well-known to have excellent energy relaxation and optical properties, however at the expense of transport properties. By using a mixed semiconductor-organic cavity, it should be possible to electrically-inject excitons in the semiconductor, to transfer the broad semiconductor exciton distribution to organic excitons by cavity-photon exchange. The organic exciton should relax down quickly and emit in a strongly-coupled manner. In such a system, properties of semiconductors and organics are combined, thanks to the very fast energy transfer mechanism provided by resonant cavity photons in the strong coupling regime.

Beyond such"simple","first-order" effects a unique strongly-coupled system like cavity- polaritons opens new physical possibilities : non- linear effects should bring new vistas on particle interactions. While simpler phenomena like phase- space-filling (at high temperature) [96] or exciton-exciton collisions (at low temperature) [97] have been observed, more elaborate theories or experiments have involved non-linear Boson interactions, Boson- based lasers (Bosers) etc.. A review of such novel physics has been given by Yamamoto [98]. The field of strong coupling is presently flourishing as it offers a new testground for numerous quantum optics phenomena [98]

4.4.   0D PHOTONIC STRUCTURES.

Although planar photonic systems yield numerous useful phenomena, both fundamental and applied, they are still limited by the amount of control of the light-matter interaction which they provide. For instance, the $\beta = 1$ limit is difficult to achieve in **planar microcavities** due to both the existence of "leaky" and guided modes in such 2D structures, and to the broad emission spectrum [70, 76]. One requires **0D microcavities or PBG materials** in order to suppress these modes and reach $\beta = 1$. This is presently the major present challenge in the field.

Several implementations of low-D photonic systems have already been demonstrated : 1D photonic wires [99], 0D microdisks [100], solid [101] or liquid [102] microspheres, pillars [103], etc... (Fig. 12). A detailed discussion of these various fascinating systems would take us too far and we will only point out some major results, refering the reader to an excellent introduction to the field [102].

Photonic wire lasers have been demonstrated [99]. They should have very small threshold currents due to their small volumes.They should also lead to high-monochromaticity and to some control of spontaneous emission.

Microdisks have been widely studied [100], in particular as laser structures. They allow a very good control of spontaneous emission, with $\beta= 0.23$, the highest value reported so far.They also have very low threshold currents, again due to their small volumes. They however show an increased noise level due to the strong carrier heating effects originating in the large power densities at which these devices operate.

Liquid and solid microspheres also have excellent dielectric properties due to their whispering-gallery modes. Q-factors up to 2.10^8 have been reported. Very sharp spontaneous and laser modes have been reported [104, 105]. Large changes in

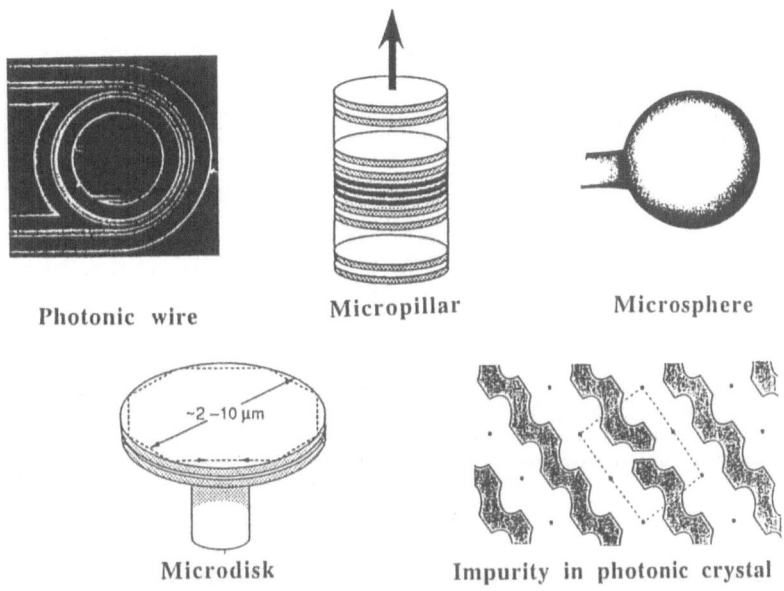

Photonic wire   Micropillar   Microsphere

Microdisk   Impurity in photonic crystal

Figure 12 : Schematics of five 0D photonic structures .

spontaneous emission lifetimes are produced by the simultaneous sharp transitions of dyes and by the high Q-values of microspheres.

Pillars provide 3D optical confinement for most of the optical modes.They are often made by etching the pillar through a planar cavity, hence retaining two end mirrors (Fig. 12). Various experiments have already been done : using QDs as active material in the pillar, Gerard was able to observe the cavity mode enhancement of the luminescence and thus map the mode energies as a function of pillar diameter [103]. Several attempts were made to obtain thresholdless lasers effects with pillars or other 3D structures [81]. So far, β's only up to 0.05 have been obtained because of the remaining modes and of the broad luminescence spectrum.

One of the most promising avenue for 3D photon confinement is that of photonic crystals [65-67]. They should provide low-loss, omnidirectional optical confinement as light with frequency within the gap will not propagate, being either reflected or diffracted [106 , 107]. While many of the predictions of photonic crystals have been verified in the microwave range, it is only recently that these effects have been observed in the visible or near infrared. The best results have been obtained on 2D crystals, as they are easier to manufacture by standard lithographic techniques than 3D ones. They should still prove useful as many of the unwanted modes in planar MCs are in or near the horizontal plane. Krauss [108] measured full bandgaps by transmission measurements of waveguided light. More recently, complete measurements of transmission, reflexion and diffraction were made [109] (Fig. 13). 3D structures for the visible have also been fabricated by lithographic techniques [110]. Such measurements confirm the basic properties of photonic crystals. They should soon be put into applications to yield ultimate light sources.

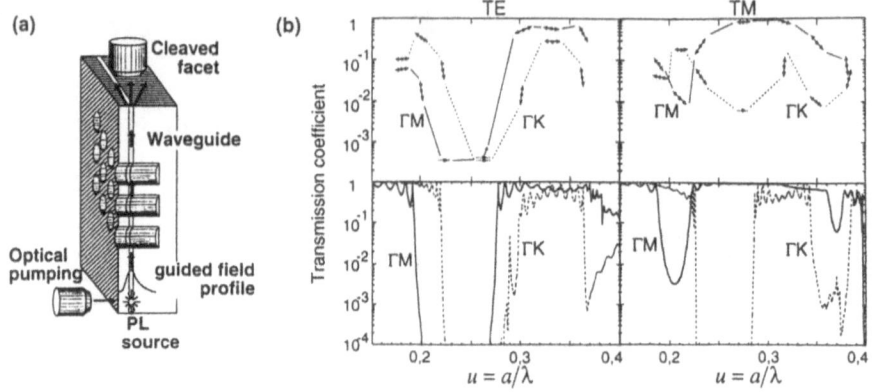

Figure 13 : (a) Schematics of the experiment to measure 2D PBG action on waveguided light ; (b) Measured (top) and calculated (bottom) transmission coefficients for a 15 period-thick triangular PBG structure (GaAs/air etched by RIE) [109].

5. A comparison between the electron and photon confinement schemes

It is useful to compare in a basic manner the two approaches of confinement. Let us consider on the one hand a single-electron quantum-box, interacting with a continuum of blackbody photon modes or with a standard 1mW light beam, and on the other hand an optical microcavity, of size $\approx(\lambda/2n)^3$ interacting with optically active medium located within the microcavity. In the first case, the electron-photon interaction of the single electron states are unmodified when compared to bulk material, at least to first-order : the lifetime and oscillator strength are unchanged. However, the optical beam has not enough interaction with the electronic system : it can only generate $\approx 10^9$ (spontaneous rate) $-10^{10,11}$ (stimulated rate) transitions per second. Moreover, the confinement factor, due to the overlap of the optical beam with the quantum box, is small, fundamentally due to the difference in wavelengths between the electron and photon. In the second case, the lifetime of the electronic excitations is almost unchanged, as well as its coupling to the optical fields : the resonance effect of the cavity increased as much the resonant electric field as it decreased all other fields. However, in this case, the active material volume ($\approx(\lambda/2n)^3$) is such that it can contain enough quantum states ($> 10^6$-10^8) in bulk, or multi-QWs, QWRs, QDs, so that they can control or generate a sizeable optical beam (10^{15}- 10^{18} transitions per second). In addition, photon spontaneous emission occurs preferentially in the singled cavity mode.

It therefore can be said that the situations of electrons or photon confinements are not equivalent : whereas both bring sharper optical features, the photon confinement scheme adds mode selectivity and a single microcavity handles enough power to achieve a useable device, whereas the electron confinement scheme requires a large number of quantum boxes to achieve sizeable effects. This difference can be traced to the Boson nature of photons, which allows many photons in a single optical mode, whereas the Fermion nature of the electron allows only one electron in a given quantum state (a single electron mode) in a quantum box.

It should be remarked that the association of QDs and 3D microcavities, i. e. electron dots in photonic dots, represents a very interesting and potentially useful system. As mentioned previously, systems which have continua of states are only partially controlled from the point of view of the light- matter interaction. For QDs in microcavities, the 10-50 μeV linewidth should lead to very large lifetime changes, even in the weak coupling regime, like for dyes or impurities in microspheres. Of course, the strong -coupling regime could also be implemented. It should be of particular interest as it would allow single quantum phenomena, therefore permitting to transfer to solid- state systems the fascinating physics already developed on atomic beams or on cooled atomic vapors, eventually using coupled or multiple cavities, generating entangled states, Schrödinger cats, and unitary transformations on quantum states for quantum computation.

6. Conclusion

There is still room for major improvements in light-emitting devices. The two fields of electron and photon confinement are alive and well, bringing us continuously new results, challenges and even surprises. The former is making spectacular progress fabrication-wise. While one can wonder about the real impact (and need) of QD lasers, QDs appear to have other very promising applications such as the possibility of intersubband, relaxation bottleneck-based devices, MC-improved QD light emitters or QD-localized emitters. Photon confinement schemes are still very young but appear to give a very powerful leverage on the photon-matter interaction. They could have a major impact on LEDs and displays. The strong coupling case is still to be evaluated for device action. Physics and fabrication of 0D microcavities or PBG materials are to be developed as they should lead to ultimate, novel physics and performance.

7. References

1. Burstein, E. and Weisbuch, C. (1995) *Confined electrons and photons*, Plenum, Boston.
2. Rarity, J. and Weisbuch, C. (1996) *Microcavities and Photonic bandgaps : physics and applications*, Kluwer, Dordrecht.
3. Dodabalapur, A. (1997) Organic light emitting diodes, *Solid State Commun.*, **102**, 259-267.
4. Carr, W. N. (1991) Photometric Figures of Merit for Semiconductor Luminescent Sources Operating in Spontaneous Mode, in S. M. Sze (ed.), *Semiconductor Devices Pioneering Papers*, World Scientific, London, p. 919.
5. Craford, M. G. (1996) Commercial Light Emitting Diode Technology : status, trends and possible future performances in [2], pp. 323-331.
6. Moon, R. L. (1997) MOVPE : is there any technology for optoelectronics ?, *J. Cryst. Growth*, **170**, 1-10.
7. Schnitzer, I. *et al.* (1993) Ultrahigh spontaneous emission quantum efficiency, 99.7 % internally and 72 % externally, from AlGaAs/GaAs/AlGaAs double heterostructures, *Appl. Phys. Lett.*, **62**, 131.
8. Andreani, L. C. (1995) Optical transitions, excitons and polaritons in bulk and low-dimensional semiconductor structures, in [1], pp. 57-112.
9. Schmitt-Rink, S., Chemla, D. S. and Miller, D. A. B. (1989) Linear and nonlinear optical properties of semiconductor quantum wells, *Adv. Phys.*, **38**, 89-188.
10. Weisbuch, C. and Vinter, B. (1991) *Quantum Semiconductor Structures: Fundamentals and applications*, Academic Press, Boston.
11. Miller, D. A. B. (1995) Quantum wells optical switching devices, in [1] pp. 675-701.
12. Coldren, L. A. and Corzine, S. W. (1995) *Diode lasers and photonic integrated circuits*, Wiley, New-York.
13. Faist, J. *et al.* (1997) Laser action by tuning the oscillator strength, *Nature*, **387**, 777-782.

232

14. Faist, J. *et al.* (1994) Quantum Cascade Laser, *Science*, **264**, 553-556.
15. Arakawa, Y. and Sakaki, H. (1982) Multidimensional quantum well laser and its temperature dependence of the current threshold, *Appl. Phys. Lett.*, **40**, 939-941.
16. Arakawa, Y. (1995) Semiconductor nanostructure lasers : fundamentals and fabrication, in [1] pp. 647-673.
17. Asada, M., Miyamoto, Y. and Suematsu, Y. (1986) Gain and the threshold of three-dimensional quantum-box lasers, *IEEE J. of Quantum Electron.*, **22**, 1915-1921.
18. Eberl, K., Petroff, P. M. and Demeester, P. (1995) *Low dimensional structures prepared by epitaxial growth or regrowth on patterned substrates*, Kluwer, Dordrecht.
19. Arakawa, Y. (1994) Fabrication of quantum wires and dots by MOCVD selective growth, *Solid-state electronics*, **37**, 523-528.
20. Tiwari, S. *et al.* (1994) High efficiency and low threshold current strained V-groove quantum-wire lasers, *Appl. Phys. Lett.*, **64**, 3536-3538.
21. Brunner, K. *et al.* (1992) Photoluminescence from a single GaAs/AlGaAs quantum dot, *Phys. Rev. Lett.*, **69**, 3316-3219.
22. Vahala, K. J. *et al.* (1993) Lower-dimensional quantum structures by selective growth and gas-phase nucleation, *J. Vac. Sc. Technol. B*, **11**, 1660-1666.
23. Bawendi, M. G. (1995) Synthesis and spectroscopy of II-VI quantum dots : an overview, in [1], pp. 339-356.
24. Brunner, K. *et al.* (1994) Sharp line photoluminescence of excitons localized at GaAs/AlGaAs quantum well inhomogeneities, *Appl. Phys. Lett.*, **64**, 3320-3322.
25. Gammon, D., Snow, E. and Katzer, D. S. (1995) Excited state spectroscopy of excitons in single quantum dots, *Appl. Phys. Lett.*, **67**, 2391-2393.
26. Wegscheider, W. *et al.* (1993) Lasing from Excitons in Quantum Wires, *Phys. Rev. Lett.*, **71**, 4071-4074.
27. Rinaldi, R. *et al.* (1997) Exciton binding energy in GaAs V-shaped quantum wires, *Phys. Rev. Lett.*, **73**, 2899-2902.
28. Ogawa, T. and Takagahara, T. (1991) Optical absorption and Sommerfeld factor of one-dimensional semiconductors : an exact treatment of excitonic effects, *Phys. Rev. B*, **44**, 8138-8156.
29. Rossi, F. and Molinari, E. (1996) Coulomb-induced suppression of band-edge singularities in the optical spectra of realistic quantum-wire structures, *Phys. Rev. Lett.*, **76**, 3642-3645.
30. Hanamura, E. (1988) Very large optical nonlinearity of semiconductor microcrystallites, *Phys. Rev. B*, **37**, 1273-1279.
31. Takagahara, T. (1987) Excitonic optical nonlinearity and exciton dynamics in semiconductor quantum dots, *Phys. Rev. B*, **36**, 9293-9296.
32. Bockelmann, U. (1993) Exciton relaxation and radiative recombination in semiconductor quantum dots, *Phys. Rev. B*, **48**, 17637-17640.
33. Nakamura, A., Yamada, H. and Tokizaki, T. (1989) Size-dependent radiative decay in CuCl semiconducting quantum spheres embedded in glasses, *Phys. Rev. B*, **40**, 8585-8588.
34. Marzin, J. Y. *et al.* (1994) Photoluminescence of single InAs quantum dots obtained by self-organized growth on GaAs, *Phys. Rev. Lett.*, **73**, 716-719.
35. Gammon, D. *et al.* (1996) Homogeneous linewidths in the optical spectrum of a single GaAs quantum dot, *Science*, **273**, 87-90.
36. Gérard, J.-M. (1995) Prospects of high-efficiency quantum boxes obtained by direct epitaxial growth, in [1], pp. 357-381.
37. Steffen, R. *et al.* (1996) Single quantum dots as local probes of electronic properties of semiconductors, *Phys. Rev.B*, **54**, 1510-1513.
38. Empedocles, S. A., Norris, D. J. and Bawendi, M. G. (1996) Photoluminescence spectroscopy of single CdSe nanocrystallite quantum dots, *Phys. Rev. Lett.*, **77**, 3873-3876.
39. Anand, S. *et al.* (1996) Sharp line injection luminescence from InP quantum dots buried in GaInP, *J. Appl. Phys.*, **80**, 1251-1253.
40. Bockelmann, U. and Bastard, G. (1990) Phonon scattering and energy relaxation in two-, one-, and zero- dimensional electron gases, *Phys. Rev. B*, **42**, 8947.
41. Benisty, H., Sotomayor-Tores, C. M. and Weisbuch, C. (1991) Intrinsic mechanism for the poor luminescence properties of quantum-box systems, *Phys. Rev. B*, **44**, 10945-10948.
42. Hessman, D. *et al.* (1996) Excited states of individual quantum dots studied by photoluminescence spectroscopy, *Appl. Phys. Lett.*, **69**, 749-751.
43. Wang, P. D. and Sotomayor Torres, C. M. (1993) Multiple-phonon relaxation in GaAs-AlGaAs quantum well dots, *J. Appl. Phys.*, **74**, 5047-5052.
44. Heitz, R. *et al.* (1996) Multiphonon-relaxation processes in self-organized InAs/GaAs quantum dots, *Appl. Phys. Lett.*, **68**, 361-363.
45. Gérard, J. M., Private communication, .
46. Adler, F. *et al.* (1996) Optical transitions and carrier relaxation in self assembled InAs/GaAs quantum dots, *J. Appl. Phys.*, **80**, 4019-4026.
47. Bockelmann, U. and Egeler, T. (1992) Electron relaxation in quantum dots by means of Auger processes, *Phys. Rev. B*, **46**, 15574-15577.

48. Efros, A. L., Kharchenko, V. A. and Rosen, M. (1995) Breaking the phonon bottleneck in nanometer quantum dots : role of Auger-like processes, *Solide state communications*, **93**, 281-284.
49. Inoshita, T. and Sakaki, H. (1992) Electron relaxation in a quantum dot : significance of multiphonon processes, *Phys. Rev. B*, **46**, 7260-7263.
50. Woggon, U. *et al.* (1996) Ultrafast energy relaxation in quantum dots, *Phys. Rev. B*, **54**, 17681-17690.
51. Murray, C. B., Norris, D. J. and Bawendi, M. G. (1993) Synthesis and characterization of nearly monodisperse CdE (E=S, Se, Te) semiconductor nanocrystallites, *J. of Am. Chem. Soc.*, **115**, 8706.
52. Nomura, S. and Kobayashi, T. (1992) Exciton-LA and TA phonon couplings in a spherical semiconductor microcrystallite, *Solid State Commun.*, **82**, 335-340.
53. Knipp, P. A. and Reinecke, T. L. (1995) Coupling between electrons and acoustic phonons in semiconductor nanostructures, *Phys. Rev. B*, **52**, 5923-5928.
54. Molinari, E. (1995) Phonons and electron-phonon interaction in low-dimensional structures, in [1], pp. 161-203.
55. Derry, P. L. *et al.* (1987) Ultralow-threshold graded-index separate confinement single quantum well buried heterostructure (Al,Ga)As lasers with high reflectivity coatings, *Appl. Phys. Lett.*, **50**, 1773-1775.
56. Nagle, J. *et al.* (1986) Threshold current of single quantum well lasers : The role of the confining layers, *Appl. Phys. Lett.*, **49**, 1325-1327.
57. Wang, J., Griesinger, U. A. and Schweizer, H. (1996) Direct determination of carrier capture times in low-dimensional semiconductor lasers : the role of quantum capture in high speed modulation, *Appl. Phys. Lett.*, **69**, 1585-1587.
58. Tiwari, S. and Woodall, J. M. (1994) Experimental comparison of strained quantum-wire and quantum-well laser characteristics, *Appl. Phys. Lett.*, **64**, 2211-2213.
59. Gérard, J. M., Cabrol, O. and Sermage, B. (1996) InAs quantum boxes : highly efficient radiative traps for light emitting devices on Si, *Appl. Phys. Lett.*, **68**, 3123-3125.
60. Egawa, T. *et al.*, (1996) First fabrication of AlGaAs/GaAs laser diodes with GaAs islands active regions on Si grown by droplet epitaxy, *IEDM*, pp. 413-416.
61. Lester, S. D. *et al.* (1996) High dislocation densities in high efficiency GaN-based light-emitting diodes, *Appl. Phys. Lett.*, **66**, 1249-1251.
62. Nakamura, S. (1997) III-V nitride based light-emitting devices, *Solid State Commun.*, **102**, 237-248.
63. Chichibu, S. *et al.* (1996) Spontaneous emission of localized excitons in InGaN single and multiquantum well structures, *Appl. Phys. Lett.*, **69**, 4188-4190.
64. Narukawa, Y. *et al.* (1997) Role of self-formed InGaN quantum dots for exciton localization in the purple laser diode emitting at 420 nm, *Appl. Phys. Lett.*, **70**, 981-983.
65. Yablonovitch, E. (1993) Photonic band-gap structures, *J. Opt. Soc. Am. B*, **10**, 23.
66. Joannopoulos, J. D., Meade, R. D. and Winn, J. N. (1995) *Photonic Crystals, Molding the Flow of Light*, Princeton University Press, Princeton, NJ.
67. Soukoulis, C. M. (1996) *Photonic bandgap materials*, NATO ASI Series, vol. 315, Kluwer, Dordrecht.
68. Stanley, R. P. *et al.* (1993) Impurity Modes in One-Dimensional Periodic-Systems: The Transition from Photonic Band-Gaps to Microcavities, *Phys. Rev. A*, **48**, 2246-2250.
69. Yokoyama, H., Nambu, Y. and Kawakami, T. (1995) Controlling spontaneous emission and microcavities, in [1], pp. 427-466.
70. Björk, G., Yamamoto, Y. and Heitmann, H. (1995) Spontaneous emission control in semiconductor microcavities, in [1] pp. 467-501.
71. Haroche, S. and Kleppner, D. (1989) Cavity Q. E. D., *Phys. Today*, **42**, 24-30.
72. Berman, P. R. (1994) *Cavity quantum electrodynamics*, Academic Press, Boston.
73. Björk, G. (1994) On the Spontaneous Lifetime Change in an Ideal Planar Microcavity - Transition from a Mode Continuum to Quantized Modes, *IEEE J. Quantum Electron.*, **QE 30**, 2314-2318.
74. Brorson, S. and Skoovgard, P. M. W. (1996) Optical mode density and spontaneous emission in microcavities in R. K. Chang and A. J. Campillo (eds.), *Optical processes in microcavities*, World Scientific, Singapore, pp. 77-99.
75. Benisty, H., De Neve, H. and Weisbuch, C. (1997) Impact of planar microcavities effects on light extraction, *to be published*,.
76. Baba, T. *et al.* (1991) Spontaneous emission factor of a microcavity DBR surrface emitting laser, *IEEE J. Quantum Electron.*, **QE 27**, 1347-1358.
77. Hunt, N. E. J. *et al.* (1992) Enhanced spectral power density and reduced linewidth at 1.3 µm in an InGaAsP quantum well resonant cavity light-emitting diode., *Appl. Phys. Lett.*, **61**, 2287-2289.
78. De Neve, H. *et al.* (1997) Recycling of guided mode light emission in planar microcavity light emitting diodes, *Appl. Phys. Lett.*, **70**, 799.
79. Björk, G., Heitmann, H. and Yamamoto, Y. (1993) Spontaneous-Emission Coupling Factor and Mode Characteristics of Planar Dielectric Microcavity Lasers, *Phys. Rev., A* **47**, 4451-4463.
80. Björk, G. (1994) On the spontaneous lifetime change in an ideal planar microcavity - transition from a mode continuum to quantized modes, *IEEE J. Quantum Electron.*, **30**, 2314.
81. Yokoyama, H. and Ujihara, K. (1995) *Spontaneous emission and laser oscillation in microcavities*, CRC Press, Boca Raton.
82. Yokoyama, H. (1992) Physics and device applications of optical microcavities, *Science*, **256**, 66-70.

234

83. Björk, G., Karlsson, A. and Yamamoto, Y. (1994) Definition of a laser threshold, *Phys. Rev. A*, **50**, 1675-1680.
84. Yamamoto, Y. and Slusher, R. E. (1993) Optical Processes in Microcavities, *Phys. Today*, **46**, 66-73.
85. Yamamoto, Y., Machida, S. and Richardson, W. H. (1992) Photon Number Squeezed States in Semiconductor Lasers, *Science*, **255**, 1219-1224.
86. Weisbuch, C. *et al.* (1992) Observation of the coupled exciton-photon mode splitting in a semiconductor quantum microcavity, *Phys. Rev. Lett.*, **69**, 3314-3317.
87. Tassone, F.,et al. (1996) Time resolved photoluminescence from a semiconductor microcavity, in [2], pp. 87-94.
88. Savona, V. and Weisbuch, C. (1996) Time-resolved light emission from polaritons in a semiconductor microcavity under resonant excitation, *Phys. Rev. B*, **54**, 10835-10840.
89. Stanley, R. P. *et al.* (1996) Cavity polariton photoluminescence in semiconductor microcavities : experimental evidence, *Phys. Rev. B*, **53**, 10995-11007.
90. Houdré, R. *et al.* (1994) Measurement of cavity-polaritons dispersion curve from angle resolved photoluminescence experiments, *Phys. Rev. Lett.*, **73**, 2043-2046.
91. Deveaud, B. *et al.* (1991) Enhance radiative recombination of free excitons in GaAs quantum wells, *Phys. Rev. Lett.*, **67**, 2355-2358.
92. Citrin, D. S. (1992) Long intrinsic radiative lifetimes of excitons in quantum wires, *Phys. Rev. Lett.*, **69**, 3393-3395.
93. Akiyama, H. *et al.* (1994) Thermalization effect on radiative decay of excitons in quantum wires, *Phys. Rev. Lett.*, **72**, 2123.
94. Gershoni, D. and Katz, M. (1994) Radiative lifetimes of excitons in quantum wires, *Phys. Rev. B*, **50**, 8930-8933.
95. Agranovitch, V. M., Benisty, H. and Weisbuch, C. (1997) Organic and inorganic quantum wells in a microcavity, *Solid State Commun.*, **102**, 631-636.
96. Houdré, R. *et al.* (1995) Saturation of the strong coupling régime in semiconductor microcavity: free carrier bleaching of cavity-polaritons, *Phys. Rev. B*, **92**, 7810.
97. Jahnke, F. *et al.* (1996) Excitonic nonlinearities of semiconductor microcavities in the nonperturbative regime, *Phys. Rev. Lett.*, **77**, 5257-5260.
98. Yamamoto, Y. (1996) Squeezing and cavity QED in semiconductors in M. Ducloy and D. Bloch (eds.), *Quantum optics of confined systems*, Kluwer, Dordrecht, pp. 201-281.
99. Zhang, J. P. *et al.* (1995) Photonic-wire laser, *Phys. Rev. Lett.*, **75**, 2678-2681.
100. Slusher, R. E. and Mohideen, U. (1996) Dynamic optical processes in microdisk lasers in R. K. Chang and A. J. Campillo (eds.), *Optical processes in microcavities*, World Scientific, Singapore, pp. 315-337.
101. Lefèvre- Seguin, V. *et al.* (1996) Very high whispering-gallery mode in Silica microspheres for cavity- QED experiments, ibid [100], pp. 101-133.
102. Chang, R. K. and Campillo, A. J. (1996) *Optical processes in microcavities*, World Scientific, Singapore.
103. Gérard, J. M. *et al.* (1996) Quantum boxes as active probes for photonic microstructures : the pillar microcavity case, *Appl. Phys. Lett.*, **69**, 449-451.
104. Campillo, A. J., Eversole, J. D. and Lin, H.-B. (1996) Cavity QED modified stimulated and spontaneous processes, in microdroplets in [102], pp. 167-207.
105. Sandhogar, V. *et al.* (1996) Very low threshold whispering-gallery-mode microsphere laser, *Phys. Rev. A*, **54**, 1777-1784.
106. Sakoda, K. (1995) Transmittance and Bragg reflectivity of two-dimensional photonic lattices, *Phys. Rev. B*, **52**, 8992.
107. Labilloy, D. *et al.* (1997) Use of guided spontaneous emission of a semiconductor to probe the optical properties of two-dimensional photonic crystals, *Appl. Phys. Lett.*, **in press**.
108. Krauss, T. F., De La Rue, R. M. and Brand, S. (1996) Two-dimensional photonic-bandgap structures operating at near-infrared wavelengths, *Nature*, **383**, 699-702.
109. Labilloy, D. *et al.* Quantitative measurement of transmission, reflection and diffraction of two-dimensional photonic bandgap structures at near-infrared wavelengths, **to be published**, .
110. Cheng, C. C. *et al.* (1996) Lithographic band gap tuning in photonic band gap crystals, *J. Vac. Sci. Technol. B*, **14**, 4110.
111. Ledentsov, N. N. *et al.* (1996) Direct formation of vertically coupled quantum dots in Stranski-Krastanow growth, *Phys. Rev. B*, **54**, 8743-8750.
112. Blondelle, J. *et al.* (1995) 16% External Quantum Efficiency from Planar Microcavity LEDs at 940 nm by precise matching of cavity wavelength, *Electron. Lett.*, **31**, 1286-1288.

QUANTUM OPTICS OF ATOMIC SYSTEMS CONFINED

IN A DIELECTRIC ENVIRONMENT

Martial DUCLOY
Laboratoire de Physique des Lasers, UMR CNRS 7538
Institut Galilée, Université Paris-Nord
F-93430 VILLETANEUSE

1. Introduction

With the advent of modern technology, new domains of optics and spectroscopy open up, in which metallic or dielectric structures (microspheres, optical fibers, semiconductor microstructures, sharp-pointed needle tips) with a characteristic size of the order of micrometer, or even tens of nanometers, are currently used. The optics of microstructures or nanostructures is a specific domain of optics, since it deals with objects whose size is commensurable or even smaller than the optical wavelength (near-field optics) [1]. The optical properties of atomic systems (or, more generally, bound electronic systems) confined in such an environment are strongly altered with respect to the free-space properties [2,3] ; this forms the core of the research field now known as cavity Quantum Electrodynamics (cavity QED [4,5]), which have strong implications in a number of emerging technologies in both Atomic Physics and Solid State Physics like scanning near-field optical microscopy and spectroscopy, near-field fluorescence microscopy, semiconductor microtechnology, microlasers, atom channeling and atom lithography [2].

In cavity QED, the interaction of atoms with the nearby solid-state surfaces is mediated by the electromagnetic (e.m.) field, via boundary-induced modifications of the e.m. vacuum modes, and the related quantum fluctuations. In the vicinity of surfaces, atoms are submitted to highly inhomogeneous surface potentials, their energy levels are shifted and their transition probabilities are altered up to the point where free-space forbidden transitions can get allowed.

One generally distinguishes two regimes of cavity QED :
(i) The *weak-coupling* regime, in which the influence of the environnement is small enough to be treated as a perturbation of the atomic system. This regime, which describes the modifications of the transition frequencies and spontaneous emission rates of an atom in front of a reflecting surface - or inside a off-resonant microscopic cavity - is characterized by a *dispersive*, and generally *irreversible*, coupling between the atom and its environment.
(ii) The *strong-coupling* regime takes over when the atom frequency approaches the oscillation frequency of a microcavity mode. In this regime, *reversible* and *quantized* energy exchange takes place between atom and microcavity field. The corresponding

N. García et al. (eds.), Nanoscale Science and Technology, 235–253.
© 1998 *Kluwer Academic Publishers.*

oscillation frequency gives rise to a line splitting in the frequency domain, usually called vacuum Rabi splitting, or normal mode splitting.

Pioneering experiments were performed in the 70's by Drexhage on the modification of spontaneous rate and radiation diagram of dye molecules in the weak coupling regime [6]. The first experimental studies of strong-coupling cavity QED have been performed in the 80's in the microwave domain, with Rydberg atom beams crossing a high-finesse superconducting cavity [7]. The last development in this field includes the generation of non-local atom - e.m. field entangled states, with their application in basic quantum mechanics (« Schrodinger cat » states, experimental evidence for « decoherence » effect, due to coupling of the microsystem to external macroscopic reservoir, and fundamental process in quantum measurement theory [8]).

Extension of strong-coupling cavity QED to the optical domain offers a number of advantages : photons are easier to detect (higher frequencies), systems can be miniaturized (wavelength scaling) leading to integrated devices, and finally to the development of thresholdless microlasers, basic components for optical technologies (optical communications, optical computing). On the other hand, new problems arise related to the increase of atom relaxation rates and e.m. field losses, and the decrease in the dipole size. Various experimental approaches to strong-regime optical cavity QED include

(i) atomic beams or dye cells in « microscopic » Fabry-Perot resonators [9]
(ii) semiconductor microcavities and photonic bandgaps [10]
(iii) development of dielectric microstructures like silica microspheres [11]

All these works show the rapid development and the emerging importance of Quantum Optics performed with confined systems (atomic or electronic systems), which has been the subject of several schools and workshops, the proceedings of which describe this recent surge [2,3].

In this article, we limit our presentation to quantum optics and cavity QED with dielectric structures, e.g. behaviour of atomic systems in the vicinity of dielectric bodies, coupling between atomic oscillators and resonant dielectric cavities. We will analyse the influence of the *intrinsic* dielectric response (dielectric dispersion, absorption bands and the related resonantly-enhanced atom-dielectric *attraction* or *repulsion* ; section 2), the various experimental approaches to long-range atom-dielectric interactions, particularly selective reflection spectroscopy, (section 3) ; the symmetry properties and surface-induced symmetry breaking (section 4) ; and finally the characteristics of the transition between the atom-photon weak coupling regime and the strong coupling in resonant dielectric microstructures (vacuum Rabi splitting induced by Mie resonances in dielectric microspheres [12], etc.)

2. Atom-dielectric interactions : The weak coupling regime

2.1. The electric image model

The interactions between isolated particles and solid-state surfaces cover a large domain of modern physics, and have important implications in many fields : collision and ion

physics, nuclear physics, surface physics, chemistry (catalysis, chemiadsorption) etc...
[13]. At short distances, the collisional character of the interaction and the surface
adsorption mechanisms are predominant. For complex particles like atomic systems, their
internal properties are deeply altered, in a way which strongly depends on the microscopic
structure of the surface. On the other hand, at larger distances, this detailed structure of
the surface gets ineffective, and the atoms retain most of their free-space characteristics.

For a perfectly conducting plane surface, the long range attraction of a neutral atom can be
viewed as the interaction between the fluctuating atomic dipole and its electric image
induced in the surface. In the *non-retarded* regime (short-range limit), the perfect
correlation between the atomic dipole and its instantaneous image leads to the well-known
van der Waals- London attraction potential scaling like z^{-3} with the atom-surface distance,
z [14]. The non-retarded dipole-image dipole interaction hamiltonian is

$$V = -\left(\vec{D}^2 + D_z^2\right)/16z^3 \qquad (1)$$

where $\vec{D}$ is the dipole operator, and D_z the dipole component normal to the surface. If the
atom is in a quantum mechanical state $|j\rangle$, the average dipole $\langle D \rangle_j$ is zero, but this is
not the case of the dipole square, leading to a surface-induced energy shift equal
to $\langle j|V|j \rangle$, which is given by a sum over all the virtual dipole couplings with other atomic
levels $|j'\rangle$

$$V_{jj} = -\sum_{j'} \left|D_{jj'}\right|^2 / 12z^3 \qquad (2)$$

with $D_{jj'} = \langle j|D|j'\rangle$. In Eq.2, one has only considered the scalar shift i.e. $\langle D_z^2 \rangle = \langle D^2 \rangle / 3$.

For a non-dispersive dielectric, with permittivity ε, there is a dielectric
«reflection coefficient» $(\varepsilon - 1)/(\varepsilon + 1)$, appearing as a multiplicative factor in Eqs. (1-2),
which describes the dielectric response of the surface [15]. The surface-induced energy
shift is responsible for (i) a state-dependent surface attraction, $-dV_{jj}/dz$, which can be
monitored by mechanical techniques : atomic beam deflection near surfaces [16, 17], atom
mirrors [18] (ii) a surface-induced frequency shift, $V_{ee} - V_{gg}$ (g, ground state ; e, excited
state) which can be probed by spectroscopic approaches - atomic spectroscopy near
surfaces, like reflection spectroscopy [19], optical spectroscopy inside a microscopic cavity
[9,20].

Note shat eqs (1)-(2) are valid in the absence of propagation effects, i.e.
$\left|\omega_{jj'} z\right| << c \left(\omega_{jj'} : j - j' \text{transition frequency}\right)$. In the long range limit, the dipole-image
dipole correlations tend to vanish. Casimir and Polder have shown that, for a ground state
atom, this retarded regime is characterized by a z^{-4} distance law [21].

2.2 van der Waals interactions with dispersive dielectrics.

Some of the atomic-transition frequencies, $\omega_{jj'}$, may lie inside the dielectric absorption range. In that case, the frequency-dependence of $\varepsilon(\omega)$, i.e. the dispersion of the dielectric response must be taken into account.

(i) Ground-state atoms

The analysis of ground state atoms interacting with a homogeneous dispersive dielectric surface has been performed by several authors [22]. One can show [23] that the dielectric response can be accounted for by introducing in Eq. 2 a *frequency-dependent* dielectric reflectivity, which is simply given by an integral over the imaginary frequency axis :

$$r(\omega > o) = \frac{2}{\pi} \int_o^{+\infty} \frac{\varepsilon(iu)-1}{\varepsilon(iu)+1} \frac{\omega \, du}{\omega^2 + u^2} \tag{3}$$

The ground state energy shift is thus given by

$$V_{gg} = -\sum_j |D_{gj}|^2 r(\omega_{gj})/12z^3 \tag{4}$$

in which a different dielectric image coefficient appears for each virtual coupling $|g> - |j>$. Due to the well-known properties of $\varepsilon(iu)$, $r(\omega)$ is a positive function, smaller than one, and monotonically decreasing with ω. As an example, Fig. 1a shows the sapphire reflection coefficient (for the ordinary index).

(ii) Excited-state atoms

In most cases, excited-state atoms interacting with dielectrics may still be described by Eqs. 3-4. This is in particular the case when the virtual dipole couplings of Eq. 6 appear in *absorption* $(\omega_{gj} > o)$, e.g. for a metastable state. This is no longer true for excited state atoms exhibiting dipole coupling in *emission* (i.e. to lower lying level). This case has been

studied in detail by Wylie and Sipe [24]. It can be shown that Eq.4 still applies, but with a dielectric reflectivity defined for negative frequencies by [23]

$$r(\omega < o) = \frac{2}{\pi} \int_o^{+\infty} \frac{\varepsilon(iu)-1}{\varepsilon(iu)+1} \frac{\omega \, du}{\omega^2 + u^2} + 2 \, \mathrm{Re} \frac{\varepsilon(|\omega|)-1}{\varepsilon(|\omega|)+1} \tag{5}$$

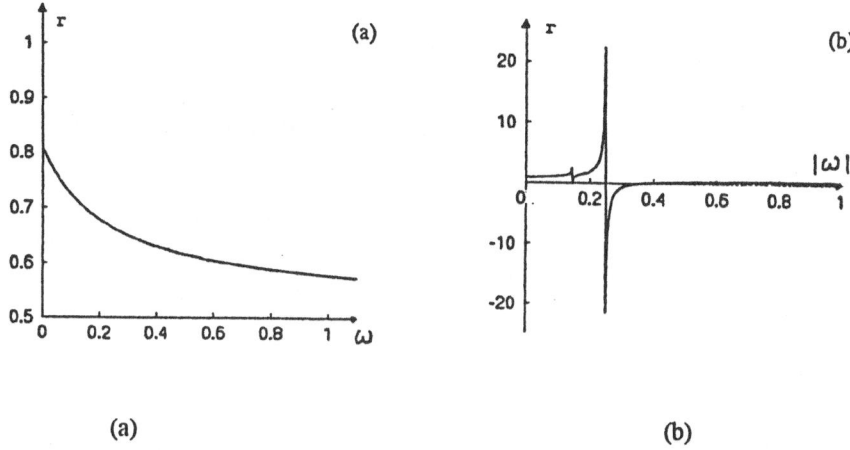

(a)   (b)

Fig. 1 : *Dielectric image reflection coefficient* $r(\omega)$ *versus frequency* ω *(in 10^{14} Hz) for the ordinary mode of sapphire. (a) absorption, (b) emission. The sapphire vibration bands induce image resonances at* $\lambda = 12\mu m$ *and* $21\mu m$ *(from Fichet et al, Ref [23]).*

Contrary to the first term of the right hand side, the second term, thanks to its dependence on real frequencies, closely reflects the dielectric dispersion at the atomic emission frequency. This implies that the dielectric resonances deeply alter the surface image coefficient, and, in particular, can enhance the dielectric response by more than one order of magnitude, as compared with a perfect reflector (r=1). This behaviour can be viewed as a resonant virtual exchange of excitation between the atom and the dielectric surface : the excited atom loses its energy which is resonantly transfered into the surface excitation. In the case of sapphire (shown on Fig. 1b for the ordinary index), *giant* van der Waals *attraction* or *repulsion* should be observable for excited atoms whose main emission coupling would be in the 11.5-12.5 μm region.

Simultaneously to this surface-enhanced shift, a resonant level broadening is also predicted, which is induced by the dissipative part of the atom-surface coupling [24, 25]. The energy coupling between excited atom and dielectric surface adds one deexcitation channel : the atom can emit either a photon into boundary-limited free space, or excites a surface phonon into the dielectric. This dissipative coupling is related to the *out-of-phase* component of the image dipole, and diverges like $1/z^3$:

$$\Gamma_e = \frac{1}{3z^3} \sum_{j,\omega_{ej}<o} \left|D_{ej}\right|^2 \operatorname{Im}\left(\frac{\varepsilon(|\omega_{ej}|) - 1}{\varepsilon(|\omega_{ej}|) + 1} \right) \tag{6}$$

Eq. 6 shows explicitly how the free-space branching ratio of a radiating state may be altered by the dielectric environment. One should note that this resonantly-enhanced atom-surface coupling yet corresponds to the weak coupling regime because the linewidth of the surface excitation channel is much larger than the atomic linewidth and the coupling strength. The energy coupling is *irreversible*, contrary to the case of the cavity QED strong coupling regime, considered in section 5 below. Finally, one should note that recently there has been an increasing interest in the behaviour of atoms embedded *inside* dielectric medium presenting absorption and dispersion [26]. All these works have strong implications in the recently developped field of single atom or molecule fluorescence detection in near field optical microscopy [27].

3. Selective reflection approach to surface interactions

3.1. Introduction

Atom-surface interactions can be monitored by either mechanical or spectroscopic approaches. Mechanical approaches rely on the deflection of atomic beams induced by metallic or dielectric cylinders [16], or atomic beam transmission inside a variable-width metallic cavity [17] : the atomic transmission decreases with smaller plate separations, due to van der Waals (*vW*) forces attracting the atoms towards the surfaces. The z^{-4} force law has been verified for Rydberg atoms, and retardation effects have been demonstrated on ground state atoms, according to the Casimir-Polder predictions [17,21].

Spectroscopic approaches rely on the measurement of transition frequencies of atoms close to surfaces. The first experimental observations of surface-induced level shift were made using selective reflection approaches [19]. Since then, laser-induced fluorescence of well-collimated atomic beams passing through a metallic cavity has been used to measure the *vW* shift of Rydberg atom, and check the z^{-3} distance law [28]. One should note that this approach is limited to long-lived atomic states (ground or metastable states, Rydberg states), contrary to selective reflection which is well adapted to short-lived radiating states.

3.2. Frequency-modulated selective reflection spectroscopy

The spectroscopy of gas-phase atoms in the vicinity of a dielectric surface can be performed through selective reflection (SR) techniques as first demonstrated in pioneering experiments by R.W. Wood in 1909 [29]. SR spectroscopy covers the field in which one monitors the spectral dependence of the optical reflection coefficient $R(\omega)$ at an interface between a dielectric window and a vapor cell. This reflection coefficient exhibits resonant variations whenever the frequency ω of the incident light beam gets into resonance with the atoms of the vapor (see fig. 2). Indeed, the optical dipoles resonantly induced in the vapor by the incident irradiation coherently reradiate an e.m. field $\Delta E_r(\omega)$ which adds to the non-resonant reflected field E_r. Under normal incidence irradiation, one gets [30]

$$\Delta E_r(\omega) = \frac{ik}{(n+1)\varepsilon_o} \int_o^{+\infty} p(z) e^{2ikz} \, dz \qquad (7)$$

where p(z) is the complex amplitude of the induced oscillating dipole and n the dielectric refractive index. In eq.7, e^{2ikz} is the atom-surface round-trip phase factor ($k = \omega / c$) appearing when evaluating the interference between the field back-radiated by the induced dipoles. This phase delay is responsible for the fact that the dominant contribution in the SR signal originates in a *typical vapor layer* $z < \lambdabar = \lambda / 2\pi$.

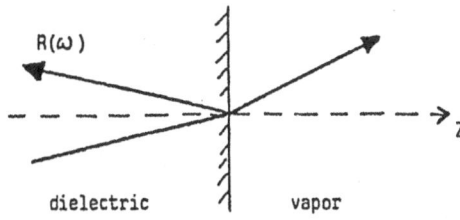

Fig. 2 :*Principle of SR Spectroscopy.*

In numerous cases, the SR signal can be evaluated through a local index model (i. e. p(z) = $\varepsilon_0 \chi E(z)$) so that SR spectroscopy becomes a way of monitoring the refractive index variations. However, at the interface with a low-pressure gas medium, the thermal motion generally invalidates the prediction of a local index theory, and vapor *spatial dispersion* must be taken into account. Indeed, when analyzing the SR lineshape, one notices a symmetric sub-Doppler contribution superimposed to the expected Doppler-broadened dispersive profile [31-32]. This narrow structure is a signature of the asymmetric behavior between arriving atoms ($v_z < 0$) resonantly driven by the incident field and already in a steady state, and the departing atoms ($v_z > 0$) in a transient evolution just following desorption.

To single out this narrow contribution, a simple frequency modulation (FM) technique can be used [33]. Demodulation of the SR signal, which yields the frequency derivative of the original SR lineshape, provides a purely Doppler-free dispersion signal (in the large Doppler approximation) [34]. This signal originates in atoms *moving parallel to the surface* ($v_z = 0$), and which remain inside the $\lambdabar$ vapor layer during the dipole lifetime. Atoms moving quickly with respect to the interface, do not stay long enough in this vapor layer to bring a sizeable contribution. Thanks to this selection of $v_z = 0$ atoms in a $\lambdabar$ layer, FM selective reflection at normal incidence allows one to perform *Doppler-free, linear*, dispersion spectroscopy of a kind of "two-dimensional atomic beam", confined inside a wavelength-deep plane.

This spectroscopic approach gives direct access to Doppler-free diagnostics of optically dense vapors and has been applied to various alkalis (Cs, Rb, K). In addition to giving access to such studies as collisional effects in dense vapor [19,35], this has allowed one to monitor long-range vW atom-dielectric interactions, by taking advantage of the great sensitivity of low-velocity atoms to frequency shifts induced by the surface potential. A general theory of FM-SR spectroscopy for two-level atoms moving into an inhomogeneous potential has been developed [34]. Specific applications of this theory to a z^{-3} scaling law for the atomic resonance frequency

$$\omega_o(z) = \omega_o - C / z^3 \tag{8}$$

has allowed one to fully interpret all the experimentally observed reflection spectra, the lineshape of which depends only on a single dimensionless parameter $A = 2Ck^3 / \gamma$ (γ:optical linewidth). A gives the vW transition shift in $\hbar\gamma$ units at a

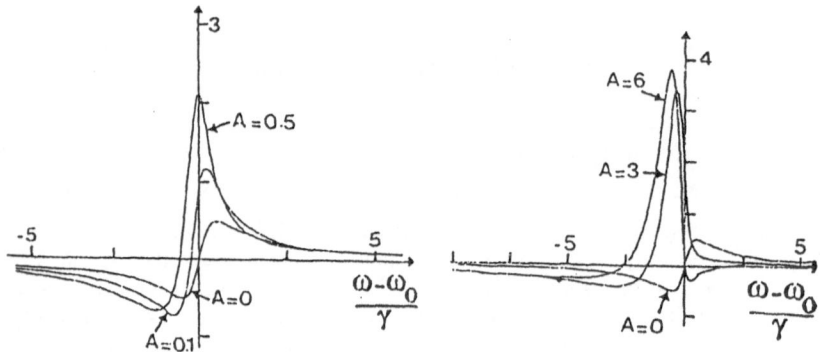

Fig 3 : *Theoretical FM-SR lineshapes for various strengths of the vW attraction. Vertical scales are in arbitrary units, normalized to A=0 (from Ducloy and Fichet [34]).*

distance λ. Two distinct regimes are observable (fig.3) [34] : (i) the weak vW regime (A<<1), which is characterized by a relatively weak red shift, as expected for an attraction process, and a small line assymetry, has been observed on the first resonance lines of Cs (D_2 [19], D_1 [36]) and Rb (D_1 [37]). (ii) In the strong vW regime (A>>1), quite anomalous lineshapes are observed, including absorption-like or even inverted dispersion-like red-shifted lineshapes, which reflect the strong spatial inhomogeneous broadening of the line. This regime appears for highly excited atomic levels e. g., on the second Cs resonance line, $6S_{1/2}$ - $7P_{3/2}$ [19], for which a sapphire surface frequency shift of 20 MHz at $z \dot= 100$nm has been measured, and theoretically interpreted [23]. All these investigations have demonstrated that FM-SR spectroscopy allows one to explore surface-induced shifts of atomic resonance lines, and is well adapted to the monitoring of the surface interaction of excited P atomic states.

Up to now, all the experimental studies have yielded a red frequency shift, corresponding to vW surface attraction. Dramatic changes are expected when an atomic emission frequency comes into resonance with dielectric surface excitation (section 2.2). This is predicted for some D levels of alkalis, and necessitates the extension of reflection spectroscopy to S or D levels, which can be performed via multiphoton reflection spectroscopy.

3.3. Nonlinear reflection spectroscopy

Nonlinear extensions of reflection spectroscopy rely on pump-probe techniques, in which the reflection of a signal beam at vapor-dielectric interface is used to monitored the refraction index changes induced by a saturating beam [38]. In this way, S and D energy levels can be probed by cascade-up three-level reflection spectroscopy [39]. Two types of signals can be monitored :
(i) stepwise resonant three-level coherent signals, in which the probe reflection is selectively enhanced for velocity groups resonant with the pump beam. This allows the monitoring of velocity distribution close to the dielectric window [40].

(ii) Off-resonant two-photon reflection signals are not much sensitive to the intermediate state (but for an overall amplitude factor), and, like in linear reflection spectroscopy, reflect the response of atoms of small relative velocity wich respect to the surface. Two-photon reflection signals have been observed on the Cs 6S-8S transition [41] : a pump beam near-resonant with the 6P-8S transition is frequency-modulated, and one monitors the modulation induced on probe reflection at 852 nm (6S-6P). Since they are mainly sensitive to $v_z = o$ atoms, such signals should allow us to monitor vW interactions of S or D levels. However there is one drawback : due to partial cancellation of the Doppler effect in two-photon spectroscopy, the active velocity group has a relatively large width. This hinders the effect of vW forces on the resonance, because they are not efficient if the atoms are moving too quickly with respect to the surface.

Another approach to vW interaction relies in excited-state reflection spectroscopy, in which the intermediate energy level is populated by broad-band excitation, and FM reflection spectroscopy is performed on the excited transition. Such an approach is presently explored on the $6P_{1/2}$ - $6D_{3/2}$ (λ = 876 nm) transition of Cesium [42], where one expects a resonantly-enhanced vW attraction or repulsion exerted by a sapphire surface on the $6D_{3/2}$ level, because of the coincidence between the $6D_{3/2} \to 7P_{1/2}$ emission resonance at 12.15μm and sapphire surface excitations (see Fig.1). The sign of the vW interactions depends dramatically on the exact position of sapphire resonances, and should be a stringent test of its dielectric function. In particular sapphire birefringence is expected to play a major role in atom-sapphire long-range interactions [23].

4. Off-diagonal van der Waals interactions; surface-induced symmetry breaking

An atomic system located in a dielectric environment loses its free-space symmetry properties. For instance, the interaction of a semi-infinite dielectric medium limited by a plane surface is in general characterized by a cylindrical symmetry along the normal to the surface, introducing new anisotropic couplings. In the previous sections, we have elaborated on surface-induced energy shifts, corresponding to the diagonal terms of the vW operator (Eqs 1-5), and mainly considered the scalar contribution, proportional to $D^2/12z^3$. However, in addition to this scalar component, there is a quadrupolar component, $D_z^2 - D^2/3$, which should be responsable for a magnetic-sublevel dependence of the energy shift (difficult to observe in selective reflection spectroscopy [19]). Its effect can be assimilated to the influence of a fictitious inhomogeneous electric field orthogonal to the surface, i. e. a wall-induced Kerr effect [43,44]. In presence of hyperfine structure, the quadrupolar interaction breaks the internal atomic symmetry and introduces a dependence of the vW shift on the hyperfine level, (F) as well as coupling between different hyperfine levels $\langle F|V|F'\rangle$. This coupling should be responsible for the appearance of additional, free-space-forbidden, hyperfine lines, like $\Delta F = \pm 2$ transitions, which are now actively looked for in selective reflection spectra.

The off-diagonal component of the vW operator in the atomic basis is also expected to have important symmetry-breaking effects on *metastable* states : vW induced couplings with radiative levels should produce a dramatic shortening of the metastable lifetime. For instance, let one consider the first excited states of rare gases, in the $np^5(n+1)s$ electronic configuration. There are two *metastable* states $^3P_2$ and $^3P_0$, and two *resonance* states $^3P_1$ and $^1P_1$, with close internal energies. Mixing between the $^3P_2$ and $^3P_1$ wavefunctions is induced by the D_z^2 component of the vW operator, via dipolar couplings to higher excited levels in the $np^5(n+1)p$ configuration. Thus second-order perturbation theory yields a relaxation rate $\gamma\left(^3P_2\right)$ proportional to z^{-6}. Because of cylindrical symmetry, there is no lifetime shortening of this kind for the $^3P_0$ state (no surface-induced $^3P_0$-$^3P_1$ mixing).

A second consequence of vW couplings is the mixing of $^3P_0$ and $^3P_2$ wavefunctions, which should be responsible for surface-induced population transfer between the two metastable states. Metastable beams « colliding » with dielectric bodies are well suited for the investigation of these peculiar predictions of atom-surface interactions. Time-of-flight measurements on a collimated supersonic beam should allow one to single out $^3P_0 \rightarrow \,^3P_2$ inelastic scattering processes produced by long-range surface interactions [45]. A calculation has been recently performed by M. Boustimi [46] for a 55 meV supersonic beam of metastable Argon scattering off a silica fiber of diameter 4 μm, after collimation and shaping by a 6 μm-wide screening slit. It is predicted that a measurable fraction (> 10^{-4}) of $^3P_0$ atoms should be transfered to the $^3P_2$ state, the excess energy (174 meV) being taken by the atom translation velocity. These investigations, which should give access to the fine details of the atom-surface interactions, is of interest in the field of metastable atom optics (mirror for metastable atoms, etc...).

5. The strong-coupling regime : atom-dielectric microsphere interactions

Up to now, we have only discussed *irreversible* atom-surface couplings which can be described in the framework of perturbative linear-response theory. The considered resonant behaviour of the dielectric response was broad enough to prevent reversible energy exchange. In this section, one examines the effect of «*form resonances*» (or morphology-dependent resonances), induced by particular microstructure or nanostructure shapes which behave like *microscopic dielectric cavities*. This situation is of interest in a number of nanotechnologies, where emitters or absorbers are located in the near field of microstructures (near field scanning optical microscopy, atom nanooptics, atom channeling and focussing...). Such a resonant behaviour exists for dielectric structures like planar dielectric waveguides [47], optical fibers [48], silica microspheres [11].

Resonant modes are caracterized by an e.m. field undergoing repeated total internal reflections at the surface of the dielectric structure, and then confined in an evanescent wave propagating along the interface. Highest spatial confinement of the electric field is obtained for dielectric microspheres, in which Mie resonances, or « Whispering Gallery Modes » (WGM) [49] can exhibit very large quality factors, $Q > 10^9$, corresponding to a photon lifetime in the WGM mode, Q/ω, in the μs range. WGMs correspond to an e.m. field localized along an equatorial circle of the microsphere - by total internal reflection at the sphere surface - and are analogous of the circular Rydberg states in atoms. Their resonance frequency is approximately given by the phase matching relation, $2\pi a \sqrt{\varepsilon} = \ell\lambda$ (a sphere radius ; ε, dielectric constant), the integer ℓ being the equivalent of the orbital quantum number in the Bohr atom. There are two others quantum numbers, n and m, in atomic physics notations : n gives the number of radial lobes of the electric field, and m is equivalent to the magnetic (azimuthal) quantum number : $\ell - |m| + 1$ gives the number of field lobes along a sphere meridian (*). The smallest mode volume V corresponds to $n=1$ and $|m| = \ell$. For large ℓ, $V \approx 2\pi a \lambda^2$, leading to very large vacuum field in the mode, $(\hbar\omega/V)^{1/2}$.

(*) Here, we closely follow the description given by Raimond in Ref [5].

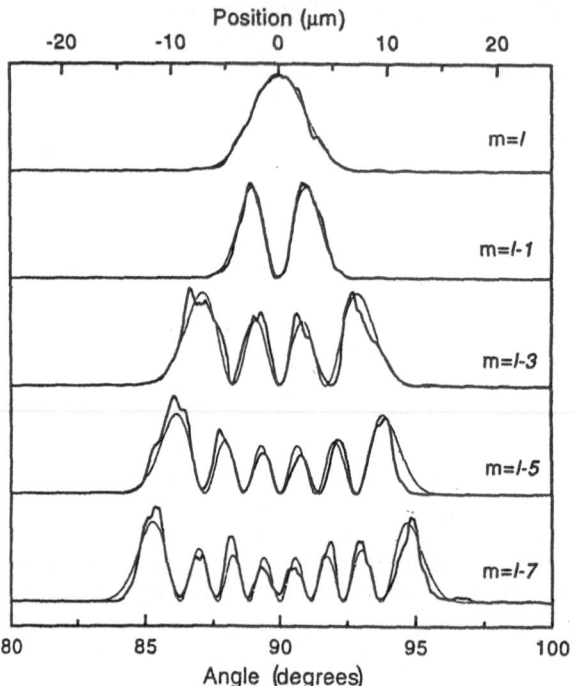

Fig. 4 : *SNOM mapping of WGM angular patterns along a microsphere meridian (from Knight et al, Ref [50])*

WGM's experimental characteristics have been thoroughly investigated by Haroche and collaborators at ENS [11, 50]. The silica microspheres (radius $\approx$ 50 µm) are fabricated by melting the end of an optical fiber [11]. Microsphere modes are excited via evanescent-wave coupling to a glass prism irradiated, under total internal reflection incidence, by a semiconductor laser (prism-microsphere gap $\approx$ 1 µm). To probe the WGM geometry and its spatial extent, the ENS group used scanning near-field optical microscopy (SNOM). They have measured the number of e.m. field lobes both along the equator, and across a meridian (see fig. 4) [50], determining in this way the WGM's quantum numbers. By tuning the incident laser frequency, they also observed the lifting of the Kramers degeneracy [51] which is induced by the weak backscattering coupling between the two identical modes counterpropagating along the sphere equator (differing only by the « azimuthal » quantum number, m, which is reversed). In this way, the ENS group has been able to assign the full set of WGM's quantum numbers.

Since WGMs with $m \approx \ell$ and small n's are characterized by a very small mode volume, they can be used to devise *microlasers* with very low thresholds. Sandoghdar et al [52] made Nd^{3+}-ion-doped silica microspheres, which lase at 1.06 µm when they are resonantly pumped by a 807 nm laser with threshold power as low as 200 nW. Also Chang and coworkers have used Mie resonances in liquid microdroplets to perform nonlinear optical processes like stimulated Raman scattering, or third-order sum frequency generation at low intensity level [53]

Due to the WGM small volume and correspondingly large vacuum field, vacuum Rabi splitting should be easily observed on the emission frequency of an atom near a dielectric microsphere. A theoretical study has been performed by Klimov et al, in which the atom is described as a classical oscillator of frequency ω_o and decay rate γ_o, placed inside [54], or outside a dielectric microsphere [55]. The influence of the microsphere on both atom's frequency and linewidth is thus analysed in the framework of radiation reaction of the reflected field at the location of the dipole. They are obtained as the solution of a dispersion equation [54] :

$$\omega^2 + i\omega\gamma_0 - \omega_0^2 = 2\omega_0 D_0 E_R(r,\omega) \tag{9}$$

(D_o : atomic dipole amplitude). E_R (r, ω), which is the e.m. field back-reflected from the microsphere on the atomic dipole at position r, has to be expanded in series of spherical Bessel and Hankel functions.

When the atomic emission frequency is far detuned from the microsphere resonances - or when these resonances are very broad - a linear response approach [equivalent to setting ω = ω_o in E_R (r, ω)] is well adapted to predict the main changes of the atomic properties : red or blue lineshift, lifetime decrease or increase (by up to one order of magnitude [54]). In the long wavelength limit, the theoretical predictions reduce to the electrostatic approach. For instance, the red lineshift of an atom confined at the center of an empty spherical microcavity (radius a), inside a dielectric medium (constant ε_d) is given by

$$\frac{\omega - \omega_0}{\gamma_0} = -\frac{3}{2}(ak_0)^{-3}\frac{\varepsilon_d - 1}{\varepsilon_d + 2} \tag{10}$$

(k_o, wavevector).

Perturbation theory fails when the atom frequency gets close to a high-Q microsphere resonance. Thus, one has to look for exact *non-perturbative* solutions of the dispersion equation (9), in which E_R now depends on the perturbed frequency ω. A resonance approximation [12] yields complex solutions for the atom-microsphere eigenfrequencies, - the imaginary part of which giving the relaxation rates. As an example, Fig 5 shows the variations of oscillation frequencies and linewidths with atom-dielectric microsphere distance, r (free-space atomic frequency is taken slightly below the l = 9 WGM resonance frequency) [12]. At large distances $(r \rangle 1.07a)$, overlapping between the evanescent field of the microsphere and the atom dipole is very weak. Long range coupling slowly starts to mix the linewidths. At $r \approx 1.07a$, *strong coupling* appears, characterized both by a r-dependent symmetrical frequency splitting (vacuum Rabi splitting $\Omega_v(r)$, proportional to

248

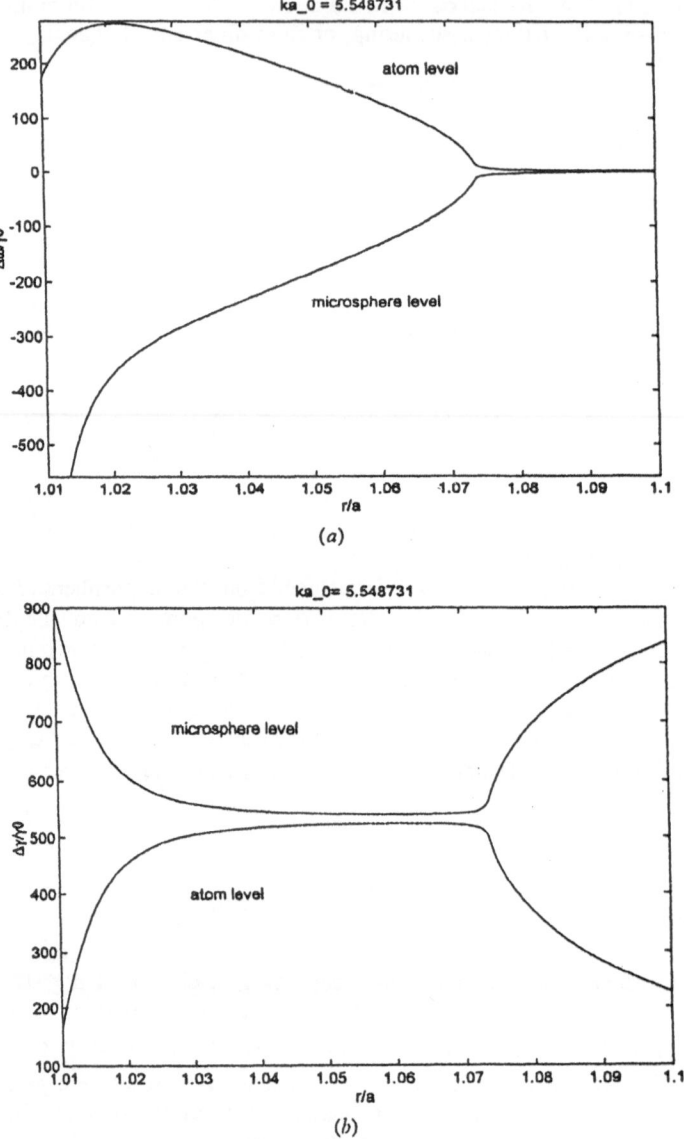

Fig 5 : (a) Relative frequency shift $(\omega - \omega_0)/\gamma_0$ and (b) relative linewidth γ/γ_0 as a
function of the location (r/a) of a radially oscillating atom relative to a
dielectric microsphere ($\omega_0/\gamma_0 = 10^7$; atomic frequency slightly below the WGM
resonant mode ; from Klimov et al, Ref [12]).

the local amplitude of the evanescent vacuum field), and by a complete mixing of the relaxation rates. This can be interpreted in the model of two coupled oscillators, in which $\Omega_V(r)$ yields the frequency of *reversible* energy exchange between the atomic oscillator and the WGM microsphere oscillator. Since the excitation is shared half-time between the two oscillators, its relaxation is governed by $\Gamma = \frac{1}{2}\left(\gamma_0 + \gamma_{WGM}\right) \approx \gamma_{WGM}/2$ (with the chosen parameters, $\gamma_{WGM} = \omega_0/Q \approx 10^3 \gamma_0$). The line splitting resolution increases proportionally to the WGM vacuum em field, when the distance is reduced. However for very small atom-microsphere separations, $r \langle 1.02a$, the non-retarded short-range vW attraction induced on the atom takes over, red-shifting the atomic frequency. The coupling weakens as can be observed on the relaxation rates which split again and approximately return to their free-space values (Fig. 5). The effect of atomic dispersion upon the microsphere resonance frequency is also visible at very small distances.

Several difficulties may hinder the experimental observation of such a vacuum Rabi splitting : inhomogeneous broadening due to atomic spatial distribution, or Doppler broadening (which could be reduced by laser cooling). Also high-Q WGM resonances produce very small evanescent field *outside* the microsphere. Strong coupling should be easier to observe on atoms or molecules implanted at the surface of, or inside the microsphere [52]. This could pave the way to an eventual single-atom microlaser, with a few circulating photons, thus generating highly non-classical e.m. fields. Also noticeable are recent proposals that atoms could orbit around microstructures - like in a planetary system - by using WGM light-induced forces [56], or be diffracted by microsphere in discrete directions reflecting the quantum nature of the WGM field [57] or be trapped in the WGM vacuum field [58].

To end this section, we would like to underline the important investigations performed on foreign atoms implanted in liquid Helium [59]. For those atoms embedded in microbubbles inside liquid He, some of the optical properties are closely related to the behaviour of atomic systems confined inside a microcavity in a dielectric medium (e.g. emission redshift like in Eq. (10), etc.). However, due to close contact between implanted atoms and Helium, the vW interactions are not sufficient to describe the physical situation, and short range repulsion due to exchange forces must be taken into account.

6. Conclusion

The cross-fertilizing interactions between fast-developing micro/nanotechnologies, and fundamental investigations in Quantum Optics is leading to novel multi-disciplinary fields, involving atomic physics and solid-state physics. Cavity QED - the simple concept of a quantum mechanical oscillator interacting with a few confined photons, or with the confined vacuum - is now finding applications in rapidly emerging fields like near-field optical technologies, atom lithography, nanoscale atom channeling and focussing [60], and represents a model system for nanoscience. In this field, dielectric microstructures, with their remarkable optical properties, are expected to play an increasing role in the future.

250

7. Acknowledgments

The author wishes to thank D. Bloch, M. Fichet, F. Schuller, V.S. Letokhov, V.V. Klimov, J. Baudon, M. Boustimi and V. Lefèvre-Seguin for fruitful discussions during the preparation of this report. This article follows in part the presentations and discussions held during a NATO shool organized in Les Houches, by D. Bloch and the author [2].

References

[1] Pohl D. and Courjon D. (Eds) « Near-field Optics »
     (NATO ASI Series E 242 ; Kluwer, Dordrecht 1993)

[2] Ducloy M. and Bloch D. (Eds) « Quantum Optics of Confined Systems »
     (NATO ASI Series E 314 ; Kluwer, Dordrecht, 1996)

[3] Rarity J. and Weisbuch C. (Eds) « Microcavities and Photonic Bandgaps : Physics
     and Applications » (NATO ASI Series E 324 ; Kluwer, Dordrecht, 1996)

[4] Haroche S., in « Fundamental Systems in Quantum Optics », Dalibard J., Raimond
     J.M. and Zinn-Justin J., Eds (North Holland, Amsterdam, 1992), pp 771

[5] Raimond J.M., « Basics of Cavity QED », in Ref [2], pp 1-46

[6] Drexhage K.H., in « Progress in Optics XII », Wolf E., ed. (North Holland,
     Amsterdam, 1974) pp 163

[7] Goy P., Raimond J.M., Gross M. and Haroche S., Phys. Rev. Lett. $\underline{59}$, 1903 (1983);
     Hulet R.G., Hilfer E.S., and Kleppner D., ibid $\underline{55}$, 2137 (1985) ; Meschede D.,
     Walther H., and Klein N., ibid $\underline{54}$, 551 (1985) ; for a general review, see Ref. [5]

[8] Brune M. et al, Phys. Rev. Lett. $\underline{77}$, 4887 (1996)

[9] Raizen et al, Phys. Rev. Lett. $\underline{63}$, 240 (1989) ;
     Thomson R.J., Rempe G. and Kimble H.J., ibid, $\underline{68}$, 1132 (1992).

[10] Yablonovitch E, Gmitter T.J. and Bhat R., Phys. Rev. Lett. $\underline{61}$, 2546 (1988) ;
     Weisbuch C., Nishioka M., Ishikawa A., and Arakawa Y., ibid, $\underline{69}$, 3314 (1992) ;
     Yamamoto Y., in Ref [2], pp 201-281; Yamamoto Y. and Slusher R.E.,
     Physics Today, $\underline{46}$ (6), 66 (1993).

[11] Braginsky V.B., Gorodetsky M.L. and Ilchenko V.S., Phys. Lett. $\underline{A137}$, 393 (1989)
     Collot L., Lefevre V., Brune M., Raimond J.M., and Haroche S.,
     Europhys. Lett. $\underline{23}$, 327 (1993).

[12] Klimov V.V., Ducloy M. and Letokhov V.S., J. Mod. Optics $\underline{44}$, 1081 (1997)

[13] See e.g. Zangwill A., «Physics at Surfaces» (Cambridge University Press, Cambridge, 1988).

[14] Lennard-Jones J.E., Trans. Faraday Soc., 28, 334 (1932).

[15] See, e.g., Jackson J.D. « Classical Electrodynamics » (Wiley, New York, 1975), pp 147-149.

[16] Raskin D. and Kusch P., Phys. Rev., 179 (1969) 172 ;
Shih A., Raskin D. and Kusch P., Phys. Rev. A 9 (1974) 652 and 1507.

[17] Anderson A., Haroche S., Hinds E.A., Jhe W. and Meschede D.
Phys. Rev. A 37 (1988) 3594 ; Sukenik C.I., Boshier M.G., Cho D., Sandoghdar V. and Hinds E.A., Phys. Rev. Lett. 70 (1993) 560.

[18] Landragin A., et al Phys. Rev. Lett. 77, 1464 (1996)
Feron S., Thèse, Université Paris-Nord (1994)

[19] Oria M., Chevrollier M., Bloch D., Fichet M. and Ducloy M., Europhys. Lett. 14 (1991) 527 ; Chevrollier M., Bloch D., Rahmat G. and Ducloy M., Opt. Lett. 16 (1991) 1879 ; Chevrollier M., Fichet M., Oria M., Rahmat G., Bloch D. and Ducloy M., J. Phys. II (France) 2 (1992) 631, and references therein.

[20] Heinzen D.J., and Feld M.S. Phys. Rev. Lett. 59, 2623 (1987)

[21] Casimir H.B.G. and Polder D., Phys. Rev. 73 (1948) 360.
For a general review, see, e.g., Hinds E.A., in « Advances in Atomic, Molecular and Optical Physics » 28, 237 (1991) and references therein ; also Spruch L., Physics Today, 39, 37 (1986) ; Tikochinsky Y. and Spruch L., Phys. Rev. A48, 4223 (1993).

[22] Mavroyannis C., Mol. Phys. 6 (1963) 593 ; Mc Lachlan A.D., Proc. Roy. Soc (London), A271 (1963) 387.

[23] Fichet M., Schuller F., Bloch D.and Ducloy M., Phys. Rev. A51, 1553 (1995)

[24] Wylie J.M. and Sipe J.E., Phys. Rev. A 30 (1984) 1185 ; A 32 (1985) 2030.

[25] Chance R.R., Prock A. and Silbey R., J. Chem. Phys. 62 (1975) 2245.

[26] Barnett S.M., Huttner B. and Loudon R., Phys. Rev. Lett. 68, 3698 (1992) ;
J. Phys. B29, 3763 (1996) ; Juzeliunas G., Phys. Rev. A55, R 4015 (1997)

[27] Girard C., Martin O.J.F. and Dereux A., Phys. Rev. Lett. 75, 3098 (1995) and references in.

[28] Sandoghdar V., Sukenik C.I., Hinds E.A. and Haroche S., Phys. Rev. Lett. 68, 3432 (1992)

252

[29] Wood R.W., Philos. Mag. 18, 187 (1909)

[30] Nienhuis G., Schuller F.and Ducloy M., Phys. Rev. A38 (1988) 5195

[31] Cojan J.L., Ann. Phys. (France) 9 (1954) 385

[32] Woerdman J.P. and Schuurmans M.F.H., Opt. Commun. 16, 248 (1975) ;
Schuurmans M.F., J. Phys. (Paris) 37 (1976) 469

[33] Akul'shin A.M., Velichanskii V.L., Zibrov A.S., Nikitin V.V., Sautenkov V.V.
Yurkin E.K. and Senkov N.V., Pis'ma Zh. Eksp. Teor. Fiz 36 (1982) 247
[J.E.T.P. Lett. 36 (1982) 303]

[34] Ducloy M. and Fichet M., J. Phys. II (France) 1 (1991) 1429

[35] Maki J.J., Malcuit M.S., Sipe J.E. and Boyd R.W., Phys. Rev. Lett. 67, 972
(1991) ; Vuletic V., Sautenkov V.A., Zimmermann C. and Hänsch T.W., Opt.
Commun. 99, 185 (1993) ; Papageorgiou N. et al Laser Physics 4, 392 (1994) ;
Wang P., Gallagher A. and Cooper J., Phys. Rev. A56, 1598 (1997)
and references in.

[36] Vuletic V. et al, Ref [35]

[37] Gorris-Neveux M. et al, Opt. Commun. 134, 85 (1997) ; Wang P. et al, Ref [35]

[38] Schuller F., Nienhuis G. and Ducloy M. Phys. Rev. A43, 443 (1991).

[39] Schuller F., Gorceix O. and Ducloy M. Phys. Rev. A47, 519 (1993)

[40] Rabi O. Amy-Klein A., Saltiel S. and Ducloy M. Europhysics Letters 25,579 (1994)

[41] Gorris-Neveux M. et al, Phys. Rev. A54, 3386 (1996)

[42] Failache H. et al, to be published

[43] Papargeorgiou N. et al, Ann. Physique (France) 20, 611 (1995)

[44] This surface-induced birefringence is also accompanied by an important anisotropy of
the spontaneous emission rate which can be enhanced or inhibited according to the
dipole orientation and the nature of the surface : Lukosz W. and Kunz R.E., Opt.
Commun 20, 195 (1997) ; Jhe W. et al, Phys. Rev. Lett. 58, 666 (1987).

[45] Similar TOF studies of inelastic processes in crossed-beam collisions have been
performed. See e.g., Robert J., Bocvarski V., Colomb de Daunant I., Vassilev G. and
Baudon J., J. Physique (France) 45, 225 (1984).

[46] Boustimi M. et al., private comunication.

[47] Levy Y., Zhang Y. and Loulergue J.C., Opt. Commun. 56, 155 (1985); Rigneault H. and Monneret S., Phys. Rev. A54, 2356 (1996) and references in; see also Ref [10]

[48] Knight J.C., Driver H.S.T., Hutcheon R.J., and Robertson G.N., Opt. Lett. 17, 1280 (1992); Rippin M.A. and Knight P.L., J. Mod Optics 43, 807 (1996) and references therein; Nha H. and Jhe W., Phys. Rev. A56, 2213 (1997).

[49] Nussenzveig H.M. « Diffraction effects in semi-classical scattering » (Cambridge University Press, 1992) ; Barber P.W. and Chang R.K. (Eds) « Optical effects associated with small particles » (World Scientific, Singapore, 1988); Chang R.K. and Campillo A.J. (Eds) « Optical processes in micro-cavities » (World Scientific, Singapore, 1996).

[50] Knight J.C., Dubreuil N., Sandoghdar V., Hare J., Lefèvre-Seguin V., Raimond J.M. and Haroche S., Optics Letters 20, 1515 (1995) ; 21, 698 (1996). A detailed presentation of the ENS work is given by Lefèvre-Seguin V. et al, in Chang and Campillo, Ref. [49], pp 101-133 (1996).

[51] Weiss D.S. et al, Opt. Lett. 20, 1835 (1995)

[52] Sandoghdar V. et al, Phys. Rev. A54, R1777 (1996).

[53] Chang R.K. Chen G. and Mazumder M.M., « Nonlinear Optics of micrometer sized droplets » in Ref. [2], pp. 75-99.

[54] Klimov V.V., Ducloy M. and Letokhov V.S., J. Mod. Optics 43, 549 (1996).

[55] Klimov V.V., Ducloy M. and Letokhov V.S., J. Mod. Optics 43, 2251 (1996)

[56] Mabuchi H. and Kimble H.J., Opt. Lett. 19, 749 (1994); see also Dowling J.P. and Gea-Banacloche J., Adv. At. Mol. Opt. Phys. 37, 1 (1996).

[57] Treussart F. et al, Opt. Lett. 19, 1651 (1994).

[58] Klimov V et al, to be published.

[59] Kanorsky S.I. and Weis A., « Atoms in nanocavities », Ref [2], pp. 367-393.

[60] Klimov V.V. and Letokhov V.S., Laser Physics 6, 475 (1996).

OPTICAL PROPERTIES OF InP QUANTUM WIRES GROWN IN HOLLOW CYLINDRICAL CHANNELS OF CHRYSOTILE ASBESTOS

S. G. Romanov[1,2] and C. M. Sotomayor Torres[3]

1. A. F. Ioffe Physical Technical Institute, Politekhnicheskaya st. 194021 St.Petersburg, Russia.

2. Department of Electronics and Electrical Engineering, University of Glasgow, Glasgow G12 8QQ, UK

3. Institute of Materials Science and Department of Electrical Engineering, University of Wuppertal, 42097 Wuppertal, Germany

1. Introduction

The behaviour of a large ensemble of nanostructures is of fundamental and practical interest since it cannot be simulated with bulk materials [1]. The discreteness of the energy spectrum added to its low density of states and controllable anisotropy, features inherent to each individual nanostructure, make these ensembles very attractive for application in, for example, non-linear optics. However, to meet a practical demand such nanostructures should be assembled in large arrays. The classical example of such ensembles is an array of quantum dots in a glass matrix, but this approach cannot provide neither ordered nor dense arrays and so far has been shown to work well only for II-VI semiconductors [2]. A self-organised strategy allows the growth of dense well-correlated arrays of quantum wires (QWRs) or dots on the crystal surface. Recently this method was extended to sequential growth of multiple dot-containing layers in order to construct three dimensional (3D) ensembles [3]. However, ordering in such arrays poses a significant challenge for five or more layers of quantum dots due to the extremely subtle nature of the interaction inducing the ordering. In contrast to the above techniques the template method based on structural confinement offers an opportunity to prepare very large 3D ensembles of nanoparticles with crystal-like ordering. A template in this case is a crystalline dielectric with structural voids, e.g. a zeolite or an opal. Voids in a porous dielectric lend themselves for infilling with a «guest» material [4]. The nano-scale size of these voids forces the "guest" material to follow their size and strictly limit the size distribution of nanostructures. Moreover, the crystallinity of a template forces the ensemble of nanostructures to reproduce this ordering. Obviously, this technique brings their own problems due to the increased role of incommensurability of crystal structures between the template and nanoparticles and the interaction of nanoparticles with the each other in the lattice and the dielectric confinement [5]. Nevertheless, this method currently permits the preparation of well ordered 3D lattices of nanoparticles with size from 1 to 100 nm and density of unit cells from 10^{12} to $10^{20} cm^{-3}$.

N. García et al. (eds.), Nanoscale Science and Technology, 255–270.

To obtain nanoparticles with a non-spherical shape, specific means are necessary. In particular, to prepare QWRs by structural confinement the use of a channel template is required. Among zeolites with structural channels, diameter of 0.5-1.5 nm can be found. Another interval of channel diameters from 2 to 20 nm can be can be found in chrysotile asbestos (CA). This asbestos template allows the study of individual properties of QWRs since channels in CA are well separated in contrast to the case of zeolite embedded "guests", where the mutual interaction of QWR affects dramatically the appearance of optical spectra of zeolite-based composites [5].

Chrysotile Asbestos is a type of naturally occurring fibrous minerals with $Mg_3Si_2O_5(OH)_4$ composition, which consists of nanotubes up to a few cm long. Each tube is a spiral roll of double silica-magnesium planes with an empty channel of 2-20 nm internal and of 30-50 nm external diameter [6]. Tubes are arranged hexagonally and stuck together in bundles with amorphous silica. Natural asbestos possess channel of different diameter and spacing, however there is less than 10% variation in channel diameter within a bundle of several mm diameter and a length of a few cm. A schematic diagram is shown in Fig.1. Note, that the internal surface of the channel is the natural crystallographic SiO_2- plane where all chemical bonds are saturated.

The first experiments took place in early 70s led by V. N. Bogomolov, who proposed to use CA as templates for embedding metals (e.g. In, Hg, Bi) from the melt under the high hydrostatic pressure [4]. This method results in the complete filling of nanotubes in CA or in the formation of array of collinear monolithic QWRs. The conductivity of 1D wires in CA has revealed a variety of effects at temperatures close to the superconducting phase transition as well as shown the competition of superconducting ordering with the Peierls transition and weak localisation [7]. Semiconductors such as Se, Te, CuCl, GaAs and InSb were embedded in CA nanotubes from the melt similar to metals or under the force of pressurised hydrogen [8,9,10]. Transmission electron microscopy (TEM) has shown that the semiconductor filling in each nanotube is a collection of ellipsoidal beads with the long axis along the channel direction. All of these QWRs exhibited the guest material characteristic Raman spectrum, high anisotropy of the optical absorption and evidence of size quantization. The optical anisotropy has been treated in the frame of the extended Maxwell-Garnett theory for metal ellipsoids embedded in an insulating medium. Using depolarisation factors to account for the shape of a particle and its orientation with respect to an electrical field E, the model produced a good correlation with experiments. As for size quantization, GaAs QWRs is the most interesting example, having exhibited a blue-shift of the absorption edge by 0.36 eV for 6 nm diameter wires. This shift was accounted for by quantum confinement using the effective mass approximation [8]. Unfortunately the melting temperature of GaAs is above the temperature of CA stability (~500 °C) and samples lost their bulk appeerence while keeping the form of nanotubes in a microscale. Time-resolved measurements revealed bleaching bands under high optical pumping power and a strong third-order optical non-linearity [11]. Another filling approach, proposed for low-temperature multiple-step in-void synthesis of semiconductors [12], was used to infill just a 0.1 vol.% of CA with CdSe [13]. TEM inspection and the shift of the absorption edge both agree in that CdSe forms rare beads in CA channels. In principle the behaviour of the latter material does not differ from other monolithic QWRs. The main lesson from the above description is that CA provides a good template for QWRs and for the study of physical properties of individual nanostructures by arranging them in macroscopically large arrays.

In what follows we shall concentrate on the analysis of optical properties of a qualitatively different configuration of QWRs templated in channels of CA. This material was synthesised by low-temperature metallorganic chemical vapour deposition (MOCVD) growth of thin layers on InP on the inner surface of CA nanotubes [14]. The peculiarity of this process is the formation of QWRs as thin-wall cylinders in contrast to the monolithic QWRs. There are two factors making optical properties of this material interesting for an investigation: (i) the strong effect of template-to-QWR interaction and (ii) the hollow cylinder configuration for confinement of electronic excitations.

2. Material and Experimental Technique

InP was deposited in CA with channels of $d = 6$ nm diameter as follows: first, trimethylindium (TMIn) was adsorbed on the inner surface of the CA channels; second, phosphine was passed through with a flow of nitrogen. The adsorption occurs due to the inhomogeneity of the potential relief caused by defects of the crystalline plane forming the channel surface in CA. Ideally, one may approach homogeneous loading with a semiconductor along the surface of tubes since the average distribution of defects is very smooth. Electron microprobe analysis showed that the InP content averaged over a bundle of several tens of tubes is nominally around 1 volume %. Since the free volume of channels in CA is around 3-5vol.%, the nominal thickness of InP layer is up to approximately 2 monolayers for the sample with the highest loading. Therefore, taking into account the nature of the deposition process we adopt the model of InP QWRs in CA as thin wall cylinders. The above consideration does not eliminate the possibility that InP forms islands on the channel surface, however, TEM demonstrated the homogeneity of InP loading with a resolution of ~10 Angstrom [15]. X-ray diffractometry of CA-InP cannot resolve the characteristic pattern of InP because of the low semiconductor content and only the CA fingerprint was observed.

Diffuse reflectance (DR) spectroscopy was used to examine qualitatively the absorption spectrum. Room and low temperature photoluminescence (PL) spectra of nanotubes of asbestos bundles both empty and filled with InP were recorded at ~ 45° geometry under excitation of 514.5 or 457.9 nm lines from an Ar-ion laser. The E_L vector of the laser beam was rotated with a $\lambda/2$ plate to analyse the PL dependence upon the polarisation of the exciting light and a linear polariser was used to examine the anisotropy of the emitted light E_{PL}. At the entry slit of a spectrometer the light arrived circularly polarised, achieved by keeping 45° difference between the axis of the $\lambda/4$ plate and a linear polariser. The variation of the excitation power P_e is given below in units of minimum $P_e = 25$ mW from the laser. The same set up was used to measure Raman scattering in the backscattering configuration with the spectral resolution around 2 cm^{-1}.

The Raman spectrum of CA-InP shows two lines at 302 and 339 cm^{-1}, which are close to the TO and LO phonon modes of a bulk InP at 304 and 345cm^{-1} (see Fig. 2). The softening of TO and LO lines from their positions in bulk InP indicates strained In-P bonds and the width of the Raman lines strongly suggests a wide distribution of strain over In-P bonds in the layer.

258

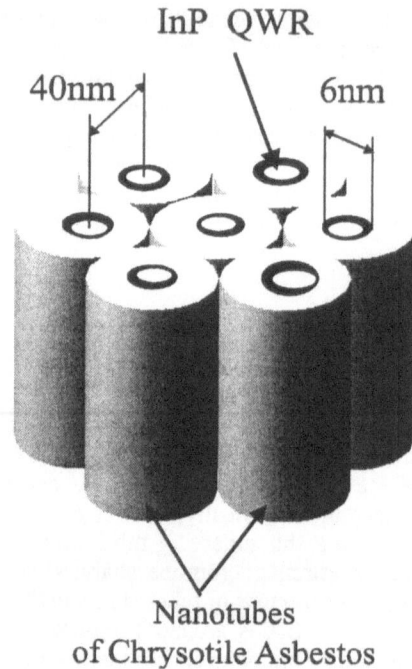

InP QWR

40nm

6nm

Nanotubes
of Chrysotile Asbestos

Figure 1. Schematic picture of hollow cylindrical QWRs in nanotubes of asbestos template

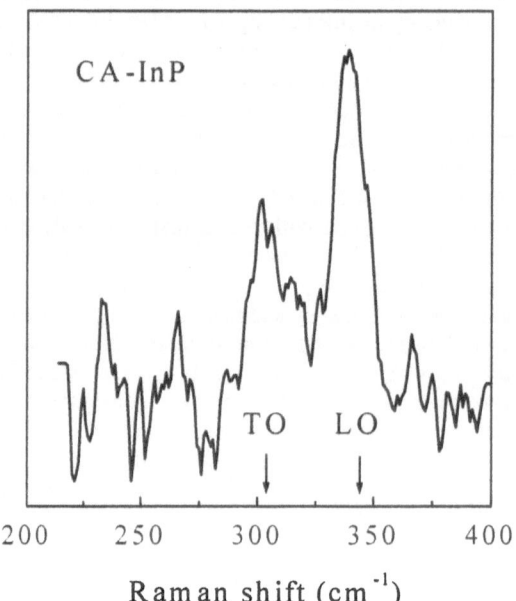

CA-InP

TO   LO

200   250   300   350   400

Raman shift (cm^{-1})

Figure 2. Raman spectrum of CA-InP. Arrows indicate the position of LO and TO modes in the crystalline InP

3. Optical absorption

Absorption spectra are plotted for three CA-InP samples with different InP loading and for bare CA (Fig.3). Comparing the bare template absorption to that of CA-InP in the range below 4 eV shows that below that energy the absorption is due to the InP component. The absorption edge of CA-InP was found at 1.6 eV, which is far above the edge of bulk InP at 1.32 eV (at T = 300 K). This confirms the quantum-effect size of InP fragments in CA. There is not much to be gained from an estimate of the edge blue shift within the effective mass approximation because the InP bonds are strained and it is not clear whether the actual QWR size is its diameter or the thickness of the InP layer. The effect of stress may exceed the contribution from confinement, but both are also affected by inhomogeneity of the InP network.

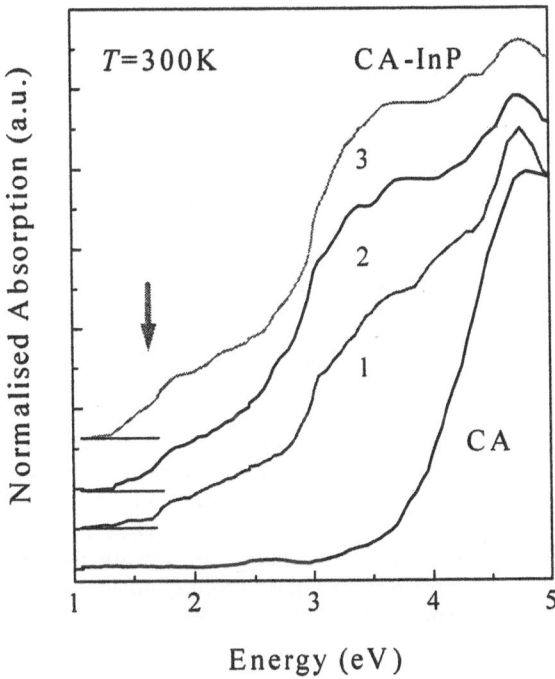

Figure 3. Absorption spectra normalised to their maximum value of CA and CA-InP with different InP loading 0.05 (curve 1), 0.2 (2) and 0.5 (3) at.%. Curves are shifted vertically. Note the region under the arrow. T=300K

In amorphous semiconductors selection rules do not apply and the absorption spectrum reflects a combined density of states from the valence and conduction bands. In crystalline InP the first maximum of electron density lies 2-3 eV below the Fermi level and originates from p-type bonding of In-P orbitals [16]. The distortion of a crystal lattice smears out this maximum and shifts it towards the fundamental gap. The 3.5 eV maximum of the CA-InP absorption corresponds well to this picture with the gap width taken into account. The absorption increases rapidly with loading in the

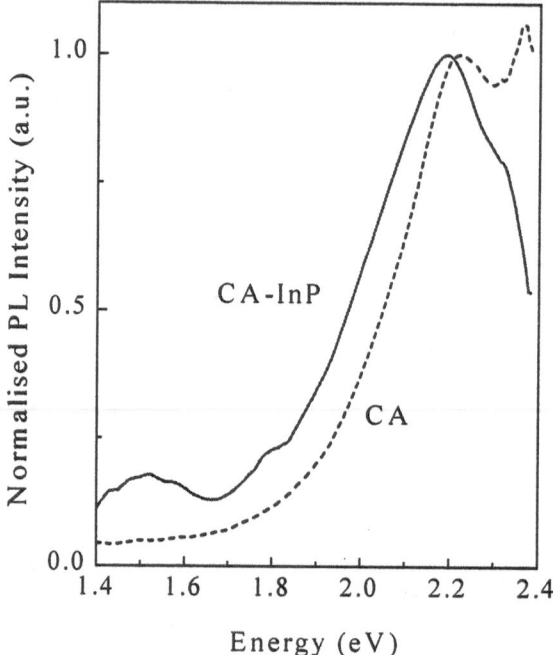

Figure 4. Comparison of PL spectra of bare and infilled CA, normalised to their magnitude at 2.2 eV. *T*=5K.

vicinity of the band edge (Fig.3). We interpret this as the result of an extended InP lattice formation which increases the density of states at the edge. Thus the absorption spectrum suggests the relative isolation of InP molecules from each other in the coating as well as random intermolecular distances compared to the InP lattice spacing. Accordingly, in the thin monolayer coating the InP-InP interaction is much weaker than that in the bulk and therefore the InP-to-template interaction contribution is important in the balance of forces in this composite.

4. Photoluminescence : General Discussion

PL spectra of CA-InP have two main bands at 1.5 and 2.2 eV if excited with 514.5 nm line (Figs.4 and 5) and at 1.5 and 2.4 eV if excited with 457.9 nm line. In bare CA a similar 2.2 eV band was observed, which also shifts upwards with decreasing wavelength of excitation [17]. This allow us to establish that the 1.5 and 2.2 eV bands have different origins. The extended study of a number of ensembles of QWRs confined within the different hosts demonstrate similar behaviour independently of the template used [5]. The low energy band was interpreted as the band edge recombination in InP and the high energy band as the wide band relaxation of excited defects states (oxygen vacancies) in the template. We note that the shape of the 2.2 eV PL band for bare and loaded CA (Fig.4 looks very similar, except for a distinct red shift in the PL spectrum

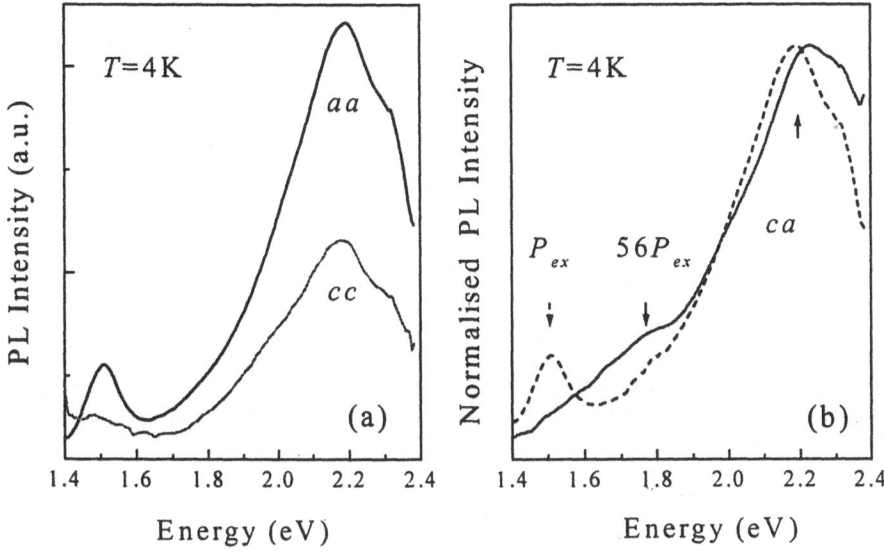

Figure 5. Anisotropy of low-level pumping PL spectra (a) and changing of the InP PL band with pumping power(b). *T*=5K.

for CA-InP in the range 1.7-2.2 eV. Based on these observations we consider both paths in the relaxation process to be responsible for this band.

The 1.5 eV PL band corresponds to the 1.6 eV width of the fundamental gap and may be assigned to an interband relaxation in the InP layer. We studied the anisotropy dependence of the PL upon the incident laser beam E_L electrical field orientation and upon the electric field orientation in the emitted light E_{PL}. The band at 1.5 eV (Fig.5a) appears when $E_{PL} \parallel a$. *aa* and *ca* denote polarisation configurations, where the first letter labels the orientation of E_L and the second that of E_{PL}, *c* is the direction along and *a* across the channel, respectively. The intensity the of 1.5 eV peak is practically independent from the orientation of E_L [18] and its polarisation rate is $P = (I_1 - I_2)/(I_1 + I_2) = 0.31$ for *aa/cc*, where I_1 and I_2 are the intensities of the recorded PL under crossed polarisations. The weakness of this band in the $E_{PL} \parallel c$ configuration contradicts common experience for 1D structures and the particular results for monolithic QWRs in asbestos. The 2.2 eV band is also polarisation-dependent with approximately the same values.

Two main contributions to the optical anisotropy of QWRs come from quantum mechanics, via the dependence of matrix elements of optical transition upon the confinement-influenced asymmetry of the wave functions along and across QWRs and classical electrodynamics, via the difference of depolarisation factors for cylindrically shaped particles. For monolithic QWRs in CA the second contribution was proved to be dominant [8]. Dielectric confinement may be another source of anisotropy for nanoparticles with dimensions smaller than the wavelength of the light λ and with a dielectric constant very different from that of the template. The latter is true for lattices of QWRs with $d/\lambda \ll 1$ and spacing $a \gg d$. Based on consideration of dielectric confinement it was expected that if $E_L \parallel c$, the field in the lattice would be homogeneous whereas if $E_L \perp c$, the electric field would be highly modulated, moreover, the PL itself was assumed to be depolarised [19]. Consequently, coupling of the light to QWRs is more efficient for $E \parallel c$. Therefore, the PL intensity should depend on the angle between E_L and c in an ensemble of collinear QWRs. However, none of these mechanisms can explain the observed anisotropy of the PL of CA-InP since all of them predict higher PL intensity for $E \parallel c$.

To understand the behaviour of a cylindrical shell QWR, the orientation of electrical dipoles should be taken into account. In the dipole approximation the PL amplitude is proportional to the square of the scalar product of the local electric field and the interband dipole momentum. Usually, the dependence upon the dipole momentum may be omitted in QWRs [19]. However, if one configures the array of dipoles such that the majority of them is parallel to one plane, the total anisotropy may approach 100% since the scalar product of two perpendicular vectors is zero. In-P bonds are known to be highly polar and a corresponding molecule may be easily oriented by the local electrostatic profile at the site where it is attached. A defect on the inner surface of CA nanotube, usually corresponds to an electron deficit in place of a missing oxygen in the Si-O-Si-O-Si network resulting in the distortion of the local electrical field. These sites are favourable for adsorbing TMIn molecules. As discussed above, the InP deposit inside CA is formed by weakly interacting InP molecules. Thus it is reasonably to assume, that in the monolayer-thin coating these molecules are aligned preferentially in the radial direction. The polarizability (dielectric constant) of the thin-wall InP QWRs is smaller than that of the monolithic QWR with the same diameter and depends upon the coating thickness [20]. Consequently, the dielectric inhomogeneity (depolarisation factors) and the dielectric confinement are both much weaker in the lattice of thin wall cylindrical QWRs. We believe, that this may lead to the opposite limit of the PL anisotropy, where the anisotropy due to the electric field redistribution loses out to the anisotropy due to the dipole orientation.

5. Photoluminescence : Relaxation Processes

A clear demonstration of the impact of relaxation mechanisms on PL properties of templated QWRs was obtained from the analysis of PL under high excitation power. With an increase of excitation power the transition giving rise to 1.5 eV band is saturated and replaced with the band at 1.78 eV at the low energy side of the spectrum (Fig.5b). Saturation of the low energy emission line is a common effect for the low-dimensional structures resulting from the reduced density of states. At $T = 300$ K no peak at 1.5 eV was observed at any level of pumping, whereas the 1.75 eV band became well resolved into two others and the 2.2 eV band split with growing excitation power (Fig.6a). Thus to observe the 1.5 eV band , low pumping power and low temperatures are needed.

Without translation symmetry in the InP layer, we can no longer use assume that bulk optical transitions in the crystalline lattice InP will also occur in CA-InP. Therefore, another process must be found involving a density of states maximum responsible for the 1.75 eV band. This has led us to suggest a redistribution of relaxation processes between to components of the CA-InP composite. On the one hand, 1.5 and 1.75 eV emission bands are closely related since one transforms into the other depending on excitation power and both arise from the recombination of

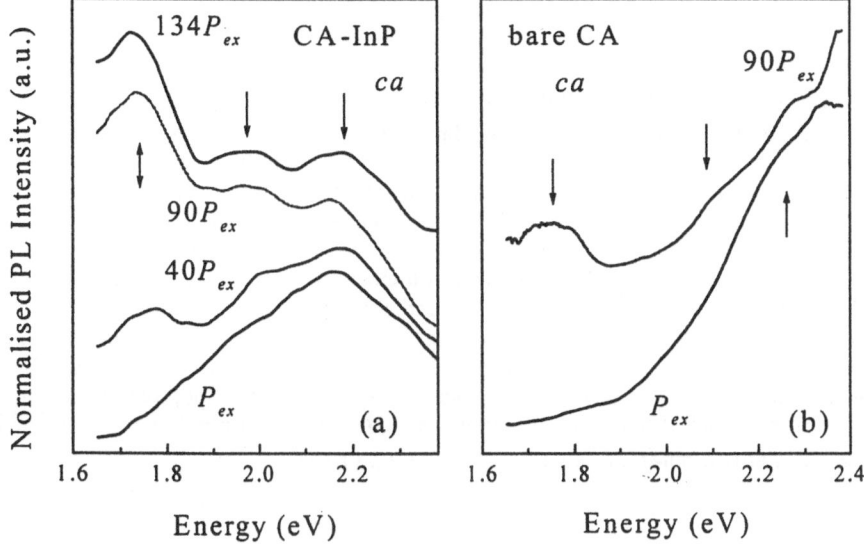

Figure 6. Dependence of the shape of CA-InP PL spectrum in *ca* configuration upon the pumping (a) and the same dependence for bare CA template (b). *T*=300K. Arrows indicate the similar bands observed in these spectra. To make these curves comparable the were normalised to the value at 2.4eV and offset vertically for clarity.

photoexcited carriers in the InP component of composite. On the other hand, the same 1.75 eV feature is present in CA-InP and in bare CA (Fig.6). It would appaear that, because of the limited efficiency of the 1.5 eV transition, relaxation takes place involving other paths with the corresponding dependence of PL spectra on excitation power as shown in Fig.6. With increasing pumping the 1.75 eV band grows very rapidly up to 90P_{ex}, then it saturates and two other bands grow with further pumping level increments (curve at 134P_{ex}). Therefore, we interpret the 1.78 eV band as arising from a low efficiency transition.

The first result from comparison of panels in Fig.6 is that there is a one-to-one correspondence of emission bands observed in bare CA and CA coated with InP. The main difference is the rapidly increasing background for bare CA onto which these bands are superimposed. Taking into account the minor effect of the selection rules on the radiative recombination from the defect-induced impurity band within the forbidden gap of bare CA, it is reasonably to ascribe the shoulders observed in the PL spectrum of the bare CA, to peaks in the density of states of this band. Therefore, the PL intensity from these maxima is directly related to their population with photoexcited carriers. Normally, defects form their own states in the continuum and the three weak bands seen under low-pumping conditions in the PL from bare CA reflect the weak modulation of the energy distribution of the density of states. The enhancement of PL from these bands of bare CA compared to the background emission after deposition of InP means that the CA defects couple selectively to InP. This is possible only if there is a spatial correlation between these states. Moreover, different defect states couple with different strengths. Fig.7a illustrates quantitatively the selectivity of pumping different defect state maxima. The relative PL intensity (the ratio

$I_{CA-InP}(\hbar\omega) / I_{CA}(\hbar\omega)$)should be the same if the population of all bands in bare CA and CA-InP changes by the same factor with increasing pumping. However, at low-level pumping the 1.78 eV is less populated relative to the 2 eV band, whereas at high pumping level the population of the 1.78 eV band increases by a factor of five. In other words, the relaxation path of photoexcited carriers from InP to the template defect states depends on the type of defect as well as on the relaxation efficiency of it.

For the analysis of PL curves at different pumping levels the distribution of photoexcited carriers participating in the radiative recombination process, has to be between the density of state maximum of the coupled InP-defect state band in CA-InP and the same maximum of defect states in bare CA, since all curves are shown for the same *ca* configuration. Fig.7b gives an idea of the difference in population increments of the different defect bands in bare CA and CA-InP by displaying the ratio of PL intensities at two pumping levels $I_{P_1}(\hbar\omega) / I_{P_2}(\hbar\omega)$.

The intensity of the 1.78 eV PL band in CA-InP increases linearly with pumping, whereas for bare CA all bands rise at nearly the same rate, showing a slight excess at 1.75 eV. It would appear that the 1.75 eV PL band corresponds to the lowest transition energy between an impurity-like band and the valence band in the template

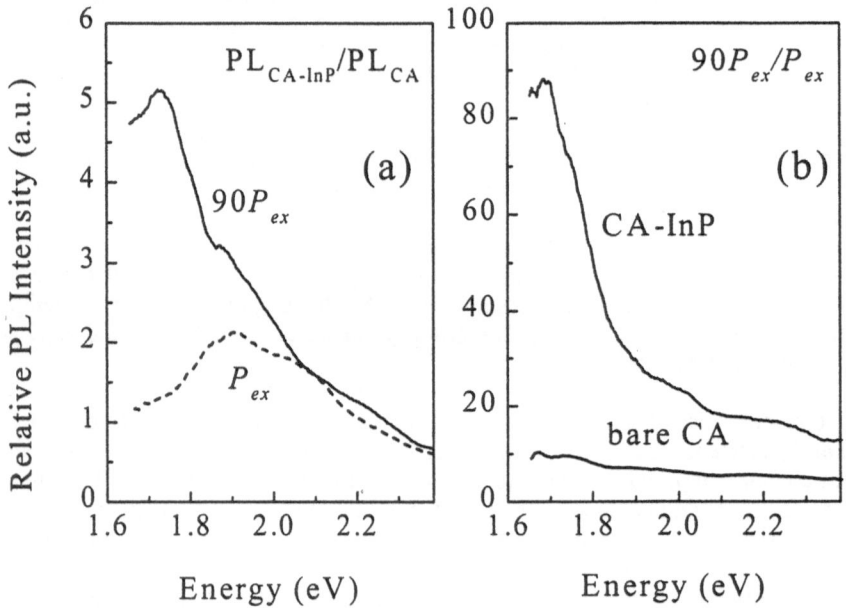

Figure 7. (a) Spectral dependence of the relative intensity of PL in QWRs with respect to the PL of the CA template shown for low and high-level pumping levels. Ratio (PL intensity CA-InP)/(PL intensity CA). (b) Spectral dependence of the increase of PL intensity for QWRs and template with increased pumping level.

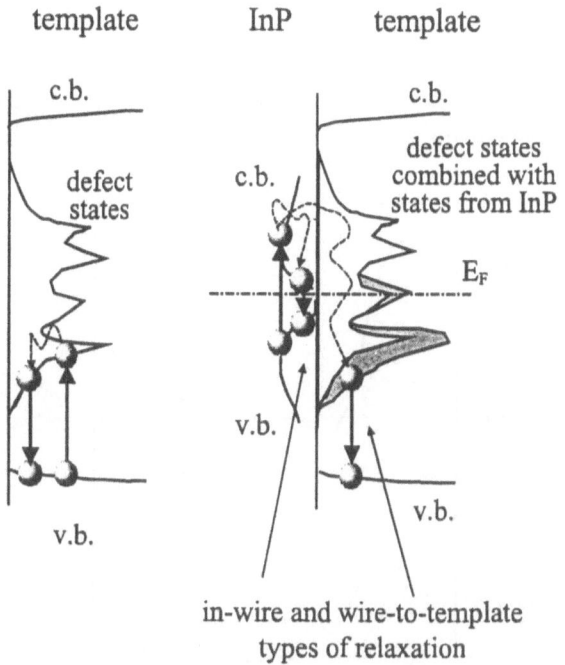

template InP template

in-wire and wire-to-template
types of relaxation

Figure 8. Schematic diagrams of excitation and relaxation processes in the bare template (left panel) and in the coated template (right).

energy structure. Photoexcited carriers in InP component filling the template defect band are distributed inhomogeneously. The reason for this selectivity is the interplay between the radiative and non-radiative relaxation of these carriers in the impurity-like band, resulting in preferential occupation of the lowest possible states or a selective pumping of the 1.75 eV band. The proposed scheme of photoexcited carrier relaxation processes is shown in Fig.8 in agreement with PL spectra shown in Fig.4. This mechanism can also explain the increase of PL intensity at 1.75 eV with increasing InP in CA [17].

6. Photoluminescence: Anisotropy of Coupled States

The PL of CA-InP in the range from 1.6 to 2.4 eV is also highly anisotropic (Fig.9). In the same manner as for the low temperature PL signal at 1.5 eV, a higher intensity is seen when both exciting and emission beams are polarised across the QWR axis (*ca* and *aa* configurations).

In Fig.10 polarised PL spectra of CA and CA-InP are shown. In contrast to CA-InP, weak PL bands of bare CA for *cc* and *ca* configurations are superimposed on the high featureless background which rapidly increases with energy. In both, the emission from bare CA (with subtracted background) and of CA-InP, there are different trends of the intensity dependence for *ca* and *cc* configurations. In the *ca* configuration, the 1.75 eV band has the highest intensity whereas the *cc* PL has the weakest.

Comparing the polarisation for each of these three bands, we find that the polarisation decreases as $P = 0.65$, 0.45 and 0.28 for CA-InP and as 0.51, 0.17 and -0.01 for bare CA, respectively, from the red to the blue. The trends are reverted since the 2.25 eV band of bare CA does not exhibit any polarisation and this is the band with the weakest polarisation in CA-InP.

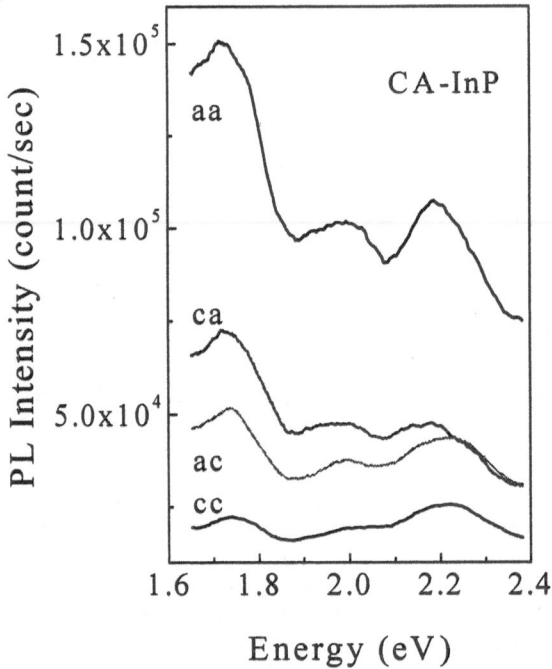

Figure 9. PL anisotropy under several polarization configurations of the exciting and emitted light in the array of QWRs under $134P_{ex}$. $T=300$K.

If the model of PL polarisation for the dipole orientation is valid, the polarisation of the PL in bare CA means that corresponding defects possess an ordering along their oriented local electrical fields. This is not surprising since the surface of asbestos channels is always coated with OH⁻-groups and organic molecules which are adsorbed from air in order to compensate the imbalance charge induced by the missing oxygen. These sites induce an anisotropic electrical field perpendicular to the channel axis. Correspondingly, in bare asbestos the surface defects with absorbed molecules are responsible for the anisotropy of PL, whereas the defects in the bulk of CA contribute to the isotropic background PL. Obviously, defects of different type possess their own charge distribution and the field-induced orientation-dependent polarisation is not the same for all cases. Attaching highly bipolar InP molecules to these defects improves the directionality of the local electrical field as in the case of 2.25 eV defect coupled with OH⁻-group or InP molecule. Therefore we conclude that this development results in the enhancement of the polarisation rate of the PL.

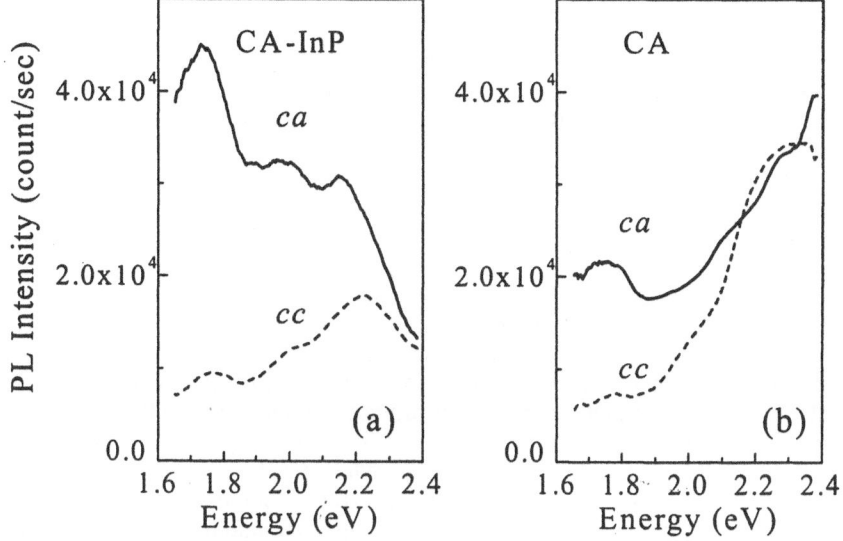

Figure 10. Comparison of PL anisotropy at $E_L \parallel c$ and high power excitation ($\sim 90 P_{ex}$) for an array of QWRs in the CA template (a) and bare template (b). T=300K

Thus, the deposition of InP only enhances selectively the inhomogeneity of defect-related features in the PL spectra of bare CA in the range 1.6-2.4 eV, but does not introduce a new transition. Since the energies of defect states do not change, this suggests the absence of chemical bonding between InP and the template. However, a more precise comparison yields a small but systematic red shift by ~ 0.02-0.03 eV of the band associated to CA-InP compared to those of CA for both *cc* and *ca* configurations. This effect corresponds most likely to an improved energy balance arising from the interaction between InP molecules with the template.

7. Summary

A template method was used to prepare arrays of collinear hollow cylindrical QWRs. It was based upon MOCVD growth of monolayer-thin InP coatings on the innerwalls of channels in chrysotile asbestos. The aim of this study was to design a material with a reduced power threshold for optical non-linearity.

It was deduced from Raman and absorption spectra, that the layer of InP in CA forms angstrom-size fragments and a highly distorted network of bonds, therefore its electronic structure is subjected to both quantum confinement and strain. Quantum confinement increases the energy of the ground state and induces the discreteness of the energy spectrum of electrons in QWRs. Disorder smears out this discreteness but does not cancel out the density of states associated with reduced dimensionality. Therefore, basic features of QWRs in asbestos are the reduced density of states and their directionality.

The interaction of InP molecules with surface defects of the template results in ~0.03 eV gain in electrostatic energy of the composite per molecule. This is not enough to form chemical bonds but it is sufficient to align the InP dipoles along the radial direction. This alignment is caused by the directionality of the local electrical field at the sites of defects on the channel surface and the highly polar nature of InP molecules themselves. This array of QWRs exhibited an unusual optical anisotropy, where a higher PL intensity was observed for the emission polarised across the QWRs. This anisotropy was explained within the dipole approximation as the result of the preferential alignment of InP molecules in the coating layer along the radial direction.

We differentiate two types of the PL bands by their origin accordingly to co-existence of InP and CA components in the composite. The PL band of InP itself may be easily suppressed by increasing the excitation power or temperature because of the limited amount of states available for radiative transitions. A sufficient number of states belonging to the InP layer appears correlated with the states belonging to the defects of template. The InP-to-template correlation is provided in real space by the adsorption of metal-organic components at the defect sites and, in energy space, by the alignment of electrochemical potentials between the template, where the level of the electrochemical potential is captured by the defect states band, and InP. With this consideration taken into account, we consider another channel of radiative recombination of photoexcited carriers in InP, namely, relaxation via the defect states band. This additional relaxation path changes remarkably the appearance of the PL in the energy range far above the band-to-band transitions in InP. In particular, the InP component transfers its own polarisation to the PL via the defect states band and induces a selectivity in the response of different defects to the excitation power. It became possible to observe these effects because of the increasing role of QWR-to-template interaction in the hollow cylinder geometry of QWR compared to monolithic QWRs.

Apart from being of interest to the field of nanostructure physics, we expect another useful application of the physical phenomena discussed in this paper. Owing to the common character of the energy exchange for a majority of the lightly infilled microporous materials, the deposit-induced enhancement of the inhomogeneity of the template energy structure may be used as a measure of the deposit-to-template interaction and directionality of the energy relaxation processes between its components. This provides a valuable tool to study, for example, the origin of catalytic activity of deposits in microporous materials.

8. Acknowledgements

The authors are grateful to their collaborators: V. Y. Butko and V.V. Tretiakov from the A.F.Ioffe Institute of St.Petersburg; H. M. Yates and M. E. Pemble from the University of Salford; J.R. Agger and A.R. Peaker from UMIST and E. Zhukov from the Moscow State University. This work was partly supported by the Russian Foundation for Basic Research (grant 96-02-17963), the UK EPSRC (grant GR/J90718), the EU ESPRIT (project SOLDES 7260), the Leverhulme Trust (grant F/179/AK) and the Russian Programme on the Physics of Solid State Nanostructures (grant 2-026/4).

9. References

1. C. Flytzanis, F. Hache, M.C. Klein, D. Richard, Ph. Roussignol, *Nonlinear optics in composite materials*, in *Progress in Optics*, vol.29, ed. E. Wolf, Elsevier, Amsterdam, 1991, pp.322-411
2. U. Woggon, *Optical Properties of Semiconductor Quantum Dots*, Springer Tracts in Modern Physics, v.136, Springer, Berlin, 1997
3. see, for exampl, N.N.Ledentsov, (1997) Submonolayer insertion in wide bandgap matrices: new principles for optoelectronic, Proc. Int. Symp. 'Nanostructures: Physics and Technology' St.Petersburg, Russia, 374-379
4. V. N. Bogomolov, (1978) Liquids in ultrathin channels, *Sov.Phys.Uspekhi*, **21**, 77-82
5. S.G. Romanov, H.M. Yates, M.E. Pemble, D.R .Agger, M.W. Anderson, C.M. Sotomayor Torres, V.Y. Butko, Y.A. Kumzerov. Interface phenomena and apperance of the optical properties of ensembles of structurally-confined InP quantum wires (1997) *Fizika Tverdogo Tela*, **39**, 727-734
6. K. Yada, (1967) Study of chrysotile asbestos by high resolution electron microscope, *Acta Cryst.*, **23**, 704-707
7. see for example V.N. Bogomolov, E.V. Kolla, Y.A. Kumzerov. (1983) One-Dimensional Effects In Low-Temperature Conductivity Of Ultrathin Metallic Filaments, *Solid State Communications*, **46**, 383-384
8. V.V. Poborchii, M.S. Ivanova, I.A. Salamatina, (1994) Cylindrical GaAs Quantum Wires Incorporated Within Chrysotile Asbestos Nanotubes - Fabrication And Polarized Optical-Absorption Spectra, *Superlattices and Microstructures*, **16**, 133-135
9. V.V. Poborchii, (1994), Optical-properties of the cylindrical quantum wires in the chrysotile asbestos channels, *Jpn. J. Applied Physics, Suppl.34-1*, **34**, 271-274
10. M.S. Ivanova, Y.A. Kumzerov, V.V. Poborchii, Y.V. Ulashkevich, V.V. Zhuravlev (1995) Ultrathin wires incorporated within chrysoile asbestos nanotubes: optical and electrical properties, *Microporous Materials*, **4**, 319-322
11. V. Dneprovskii, N. Gushina, O. Pavlov, V. Poborchii, I. Salamatina, E. Zhukov (1995) Nonlinear-Optical Absorption Of GaAs Quantum Wires, *Physics Letters A*, **204**, 59-62
12. S.G.Romanov, A.V.Fokin, V.V.Tretijakov, V.Y.Butko, V.I.Alperovich, N.P.Johnson, C.M.Sotomayor Torres (1996) Optical properties of ordered 3-dimensional arrays of structurally confined semiconductors. *J.Cryst.Growth*, **159**, 857-859
13. V.V. Poborchii, V.I. Alperovich, Y. Nozue, N. Ohnishi, A. Kasuya, O. Terasaki, (1997) Fabrication and optical properties of ultrathin CdSe filaments incorporated into the nanochannels of fibrous magnesium silicate, *J.Phys.: Condens.Matter*, **9** 5697-5695
14. H.M. Yates, W.R. Flavell, M.E. Pemble, N.P. Johnson, S.G. Romanov, C.M. Sotomayor Torres, (1997) Novel quantum confined structures via atmospheric pressure MOCVD growth in asbestos and opals, *J. Cryst.Growth,* **170**, 611-615
15. E. Zhukov, private communication.
16. M.L Theye., A. Gheoghiu, D. Udron, C. Senemaud, E. Bellin, J. Von Bardeleben, S. Squelard, J. Dupin, (1987) Defect States n Amorphous InP *J.Non-cryst.Solids* , **97&98**, 1107-1110

17. S.G. Romanov, C.M. Sotomayor Torres, H.M. Yates, M.E. Pemble, V. Butko, V. Tretijakov, (1997) Optical properties of self-assembled arrays of InP quantum wires confined in nanotubes of crysotile asbestos, *J. Appl. Phys.* **82,** 380-385

18. S.G. Romanov, N.P. Johnson, C.M. Sotomayor Torres, H.M. Yates, J. Agger, M.E. Pemble, M.W. Anderson, A.R. Peaker, V. Butko (1995) Self-assembled 3-dimensional arrays of InP quantum wires: imapact of the template geometry upon the optcal properties, *Proc. 3 Int. Symp. on Quantum Confinement: Physics and Applications,* v.**95-17,** ed. M. Cahay, S. Bandyopadhyay, J.P. Leburton, M. Razeghi, Electrochemical Society, Pennington, USA, pp.14-30.

19. N.A. Gippius, S.G. Tikhodeev, A. Forchel, V.D. Kulakovskii, (1994) Polarization-dependent optical effects in open quantum well wires, *Superlattices and Microstructures*, **16**, 165-167

20. H. C. Van De Hulst, *Light Scattering By Small Particles*, J.Wiley&Sons, New York, 1957

METAL CLUSTERS AND ATOMIC NUCLEI

T.P. MARTIN, I.M.L. BILLAS, W. BRANZ, M. HEINEBRODT AND
F. TAST
Max-Planck-Institut für Festkörperforschung
Heisenbergstr. 1, 70569 Stuttgart, Germany

AND

N. MALINOWSKI
Central Laboratory of Photoprocesses
Bulgarian Academy of Sciences, 1040 Sofia, Bulgaria

1. Introduction

It is not obvious that metal clusters should behave like atomic nuclei –
but they do. Of course the energy and distance scales are quite different. But aside from this, the properties of these two forms of condensed
matter are amazingly similar. The shell model developed by nuclear physicists describes very nicely the electronic properties of alkali metal clusters.
The giant dipole resonances in the excitation spectra of nuclei have their
analogue in the plasmon resonances of metal clusters. Finally, the droplet
model describing the fission of unstable nuclei can be successively applied
to the fragmentation of highly charged metal clusters. The similarity between clusters and nuclei is not accidental. Both systems consist of fermions
moving, nearly freely, in a confined space.

Many years ago it was noticed that atomic nuclei containing either 8,
20, 50, 82 or 126 protons have very long lifetimes. It was a challenge for the
nuclear physicists back in the forties to explain these so-called magic numbers. Since physicists tend to see most objects as perfectly round, it should
come as no surprise that they assumed atomic nuclei are spherically symmetric. Under this assumption they had to solve only a radial Schrödinger
equation.

$$\left[-\frac{d^1}{dr^2} + \frac{\ell(\ell+1)}{r^2} + V(r) \right] P_{n\ell}(r) = E_{n\ell} P_{n\ell}(r) \tag{1}$$

N. García et al. (eds.), Nanoscale Science and Technology, 271–295.

where ℓ is the angular momentum quantum number and V(r) is the radial dependence of the potential in which the nucleons move. They assumed further that the potential could be described by a simple potential well. Some confusion can arise because nuclear physicists and atomic physicists use slightly different definitions for the principal quantum number n,

$$n(\text{atomic}) = n(\text{nuclear}) + \ell \tag{2}$$

Throughout this lecture we will use the principal quantum number from nuclear physics, i.e., n denotes the number of extrema in the raidal wavefunction.

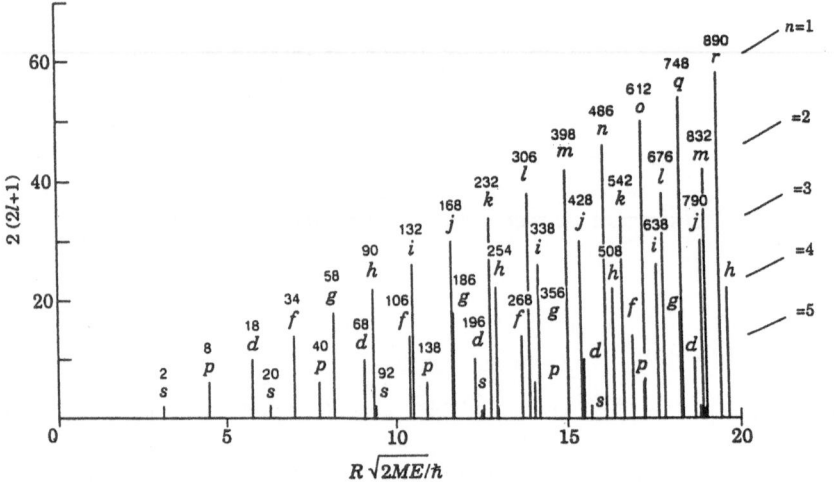

Figure 1. The degeneracy of states of the infinitely deep spherical well on a momentum scale. The total number of fermions needed to fill all states up to and including a given subshell is indicated above each bar.

Eigenstates of the radial Schrödinger equation are often called subshells. The subshells of the infinite spherical potential well are shown ordered according to momentum in Fig. 1. The lowest energy state is 1s then comes 1p, 1d, 1f, 2p ..., etc. This is, with 2, 8, 18, 20, 34, 40, 58, 90 ... nucleons, subshells are completely filled and the corresponding nuclei could be expected to be exceptionally stable. However, these are not the observed magic numbers.

In 1949 Maria Goeppert-Mayer [1] and Haxel, Jensen and Suess [2] came up with a modified model which yielded the obseved magic numbers. Their idea was that the spin-orbit interaction is unusually strong for nucleons. Subshells with high angular momentum split and the states rearrange themselves into different groups. As we shall see the original shell model, which the nuclear physicist had to discard, describes very nicely the electronic states of metal clusters [3-18].

2. Subshells, Shells and Supershells

If it can be assumed that the electrons in metal clusters move in a spherically symmetric potential, the problem is greatly simplified. Subshells for large values of angular momentum can contain hundreds of electrons having the same energy. The highest possible degeneracy assuming cubic symmetry is only 6. So under spherical symmetry the multitude of electronic states condenses down into a few degenerate subshells. Each subshell is characterized by a pair of quantum numbers n and ℓ. Under certain circumstances the subshells themselves condense into a smaller number of highly degenerate shells. The reason for the formation of shells out of subshells requires more explanation.

The concept of shells can be associated with a characteristic length. Every time the radius of a growing cluster increases by one unit of this characteristic length, a new shell is said to be added. The characteristic length for shells of atoms is approxiamtely equal to the interatomic distance. The characteristic length for shells of electrons is related to the wavelength of an electron in the highest occupied energy level (Fermi energy). For the alkali metals these lengths differ by a factor of about 2. This concept is useful only because the characteristic lengths are, to a first approximation, independent of cluster size.

The concept of shells can also be described in a different manner. An expansion of N, the total number of electrons, in terms of the shell index K will always have a leading term proportional to K^3. One power of K arises because we must sum over all shells up to K in order to obtain the total number of particles. One power of K arises because the number of subshells in a shell increases approximately linearly with shell index. Finally, the third power of K arises because the number of particles in the largest subshell also increases with shell index. Expressing this slightly more quantitatively, the total number of particles needed to fill all shells, k, up to and including K is

$$N_K = \sum_{k=1}^{K} \sum_{\ell=0}^{L(k)} 2(2\ell = 1) \sim K^3 \tag{3}$$

where $L(k)$ is the highest angular momentum subshell in shell k.

Shell structure is not necessarily an approximate and infrequent bunching of states as in the example of the spherical potential well, Fig. 1. Clearly, almost none of the subshells occur exactly at the same energy for this potential. Shell structure can be the result of exactly overlapping states. Such degeneracies signal the presence of a symmetry higher than spherical symmetry. Subshells of hydrogen for which $n + \ell$ have the same value, have exactly the same energy. This additional degeneracy in the states of hydrogen is a result of the form of its potential, $1/r$, which bestows on hydrogen $0(4)$ symmetry. Subshells of the spherical harmonic oscillator for which $2n + \ell$ have the same value also have exactly the same energy due to the form of the potential, r^2, and the resulting symmetry, $SU(3)$. For this reason it is said that these systems, hydrogen and oscillator, have quantum numbers $n + \ell$ and $2n + \ell$ that determine the energy. We have shown that $3n + \ell$ is an approximate energy quantum number for alkali metal clusters [16]. As the cluster increases in size, electron motion quantized in this way would finally be described as a closed triangular trajectory [19].

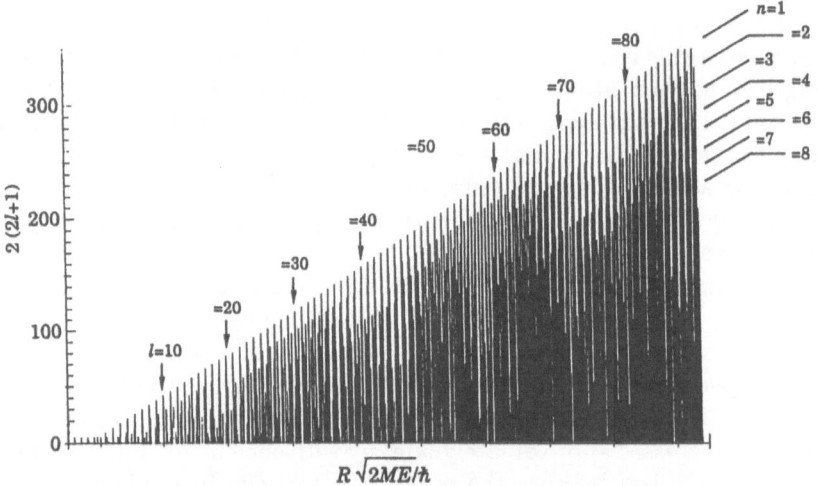

Figure 2. The states of the infinitely deep sphereical well for very large values of ℓ. Notice the periodic bunching of states into shells. This periodic pattern is referred to as supershell structure.

The grouping of large subshells into shells is illustrated in Fig. 2 for the spherical potential well. Here, it can again be seen that in certain energy or momentum regions the subshells bunch together. However, the states are so densely packed in this figure that the effect is perceived as an alternating light-dark pattern. That is, for the infinite potential well, bunching of states occurs periodically on the momentum scale. The periodic appearance of shell structure is referred to as supershell structure [20, 21]. Although supershell structure was predicted by nuclear physicists more than 15 years ago, it has never been observed in nuclei. The reason for this is very simple. The first supershell beat or interference occurs for a system containing 800 fermions. There exist, of course, no nuclei containing so many protons and neutrons. It is possible, however, to produce metal clusters containing such large numbers of electrons.

3. The Experiment

The technique we have used to study shell structure in metal clusters is photoionization time-of-flight (TOF) mass spectrometry, Fig. 3. The mass spectrometer has a mass range of 600 000 amu and a mass resolution of up to 20 000. The cluster source is a low-pressure, rare gas, condensation cell.

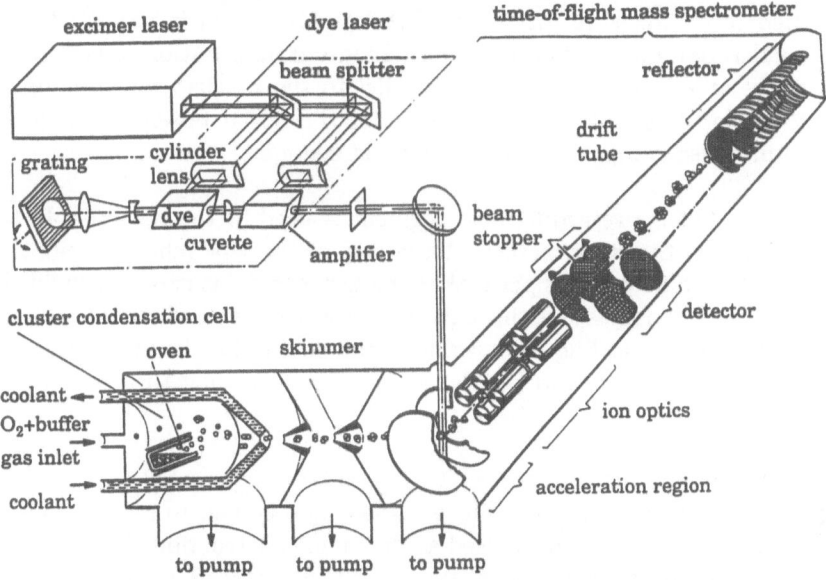

Figure 3. Apparatus for the production, photoionization and time-of-flight mass analysis of metal clusters.

Sodium vapor was quenched in cold He gas having a pressure of about 1 mbar. Clusters condensed out of the quenched vapor were transported by the gas stream through a nozzle and through two chambers of intermediate pressure into a high vacuum chamber. The size distribution of the clusters could be controlled by varying the oven-to-nozzle distance, the He gas pressure, and the oven temperature. The clusters were photoionized with a laser pulse.

Since phase space in the ion optics is anisotropically occupied at the moment of ionization, a quadrupole pair is used to focus the ions onto the detector. All ions in a volume of $1mm^3$ that have less than 500 eV kinetic energy at the moment of ionization are focused onto the detector [22].

The reflector consists of two segments with highly homogeneous electric fields, separated by wire meshes. The first segment, which is twice completely traversed by the ions, is called the retarding field, and the other segment is called reflecting field. This two-stage reflector allows a second-order time focusing of ions [23]. Two channel plates in series are used to detect the ions. The secondary electrons are collected on a metal plate and conducted to the electronics. The following main design features of the instrument are necessary to achieve such a resolution [24]:

1. The ions are accelerated at right angles to the neutral cluster beam. If clusters are ionized by a laser pulse from the gas phase, there will always be a distribution of initial *potential* energies. The reflector is used to compensate for these. If the neutral beam is parallel to the acceleration direction, there is also an initial distribution of *kinetic* energies or velocity components parallel to the acceleration diretion. If the reflector is used to compensate for the initial potential energy, it cannot also compensate for the kinetic energies.

2. A long (29cm) retarding field segment is used in the reflector. In the vicinity of the wire meshes at the end of the two reflector segments the electric field is not perfectly homogeneous. This causes a slight deflection of ions passing through them and thus a small time error. By using along retarding field segment, the field in he vicinity of the wire meshes is lowered, and the deflection of ions passing through them is reduced.

3. A reflector design is used which guarantees very stable and homogeneous electric fields. To achieve a good rigidity and to reduce temperature sensitivity, we have used as the main construction element of the reflector a 60 cm–long glass tube with a diameter of 15 cm. This glass tube holds all the field-defining elements. Since field errors as small as 10^{-4} can cause significant time-of-flight errors, great care has to be taken to acheive the necessary field homogeneity.

The mass spectra which will be displayed in this paper cover a large range of masses. For this reason it will not be possible to distinguish the individual mass peaks. For example, at the top of Fig. 4 we have reproduced a mass spectrum of Cs-O clusters which appears to be nothing more than a black smudge. How do we know how many oxygen atoms th clusters contain? This can be seen by graphically expanding the scale by a factor of 100, Fig. 4. Because of the high resolution of our mass spectrometer, we are quite certain about the composition of the clusters examined.

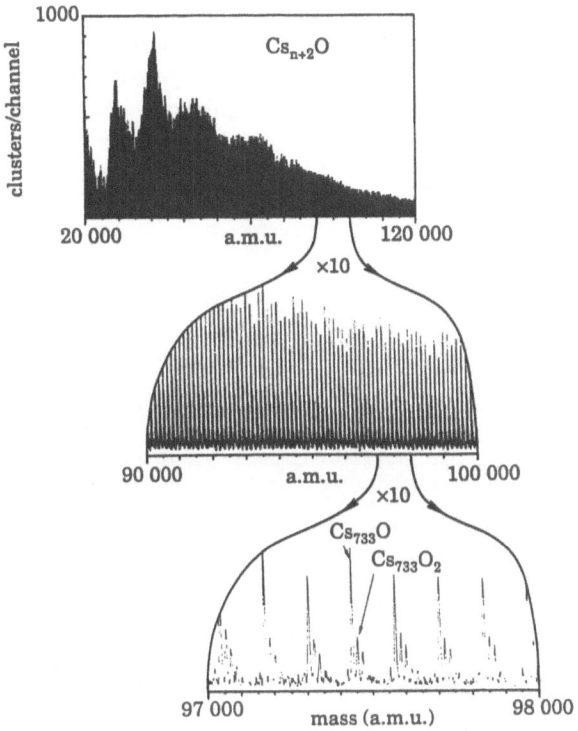

Figure 4. Mass spectrum of Cs-O clusters. Notice that the exact composition can be determined on an expanded mass scale.

4. Observation of Electronic Shell Structure

Knight, Clemenger, de Heer, Saunders, Chou and Cohen [3] first reported electronic shell structure in sodium clusters in 1984. Electronic shell structure can be demonstrated experimentally in several ways: as an abrupt decrease in the ionization energy with increasing cluster size, as an abrupt increase or an abrupt decrease in the intensity of peaks in mass spectra. The first type of experiment can be easily understood. Electrons in newly opened shells are less tightly bound, i.e., have lower ionization energies. However, considerable experimental effort is required to measure the ionization energy of even a single cluster. A complete photoionization spectrum must be obtained and very often an appropriate source of tunable light is simply not available. It is much easier to observe shell closings in photoionization, TOF mass spectra. However, depending upon the intensity and wavelength of the ionizing laser pulse, the new shell is announced by either an increase or a decrease in mass peak height.

For high laser intensities, multiple-photon processes cause the mass spectra to be less wavelength sensitive and also cause considerable fragmentation of large clusters. The resulting mass spectrum reflects the stability of cluster ion fragments. Clusters with newly opened shells are less stable and are weakly represented in the mass spectra. Notice in Fig. 5 that

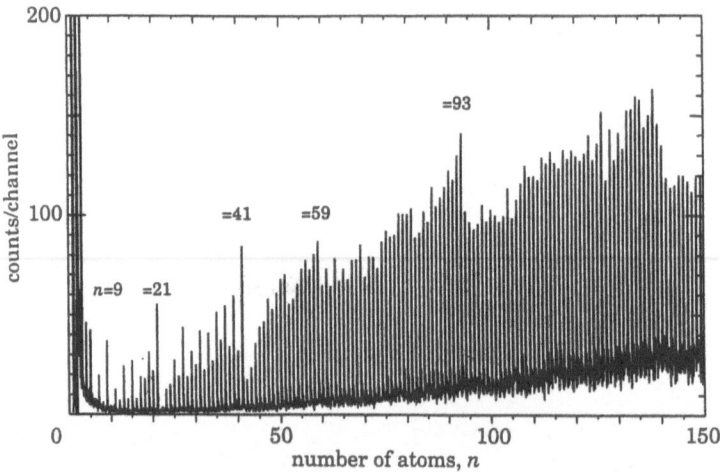

Figure 5. Mass spectrum of $(Na)_n^+$ clusters ionized with high-intensity, 2.53 eV light. The clusters are fragmented by the ionizing laser. Fragments having closed-shell electronic configurations are particularly stable.

as each new shell is opened there is a sharp step downward in the mass spectrum. Remember that cluster ions containing 9, 21, 41, 59, ... sodium atoms contain the magic number (8, 20, 40, 58, ...) of electrons.

For low laser fluence and wavelengths near the ionization threshold the mass spectra have a completely different character. As each new shell is opened there is a sharp step upward in the mass spectra, Fig. 6 (top). Open shell clusters have low ionization thresholds which fall below the energy of the incident photons, while closed shell clusters remain unionized.

Finally, for low laser fluence and wavelengths well above the ionization threshold, it is possible to observe the neutral distribution of cluster sizes. If the source conditions are appropriately chosen, this distribution can peak at sizes corresponding to closed electronic shells, Fig. 6 (bottom). Cluster intensities can sometimes be increased by a factor of ten by using a seed to nucleate the cluster growth. For example, by adding less than 0.02% SO_2 to the He cooling gas, Cs_2SO_2 molecules form which apparently promote further cluster growth. Mass spectra of $Cs_{n+2}(SO_2)$ clusters obtained [15] using four different dye-laser photon energies are shown in Fig. 7. Although it is not possible to distinguish the individual mass peaks in this condensed plot, it is evident that the spectra are characterized by steps. For example,

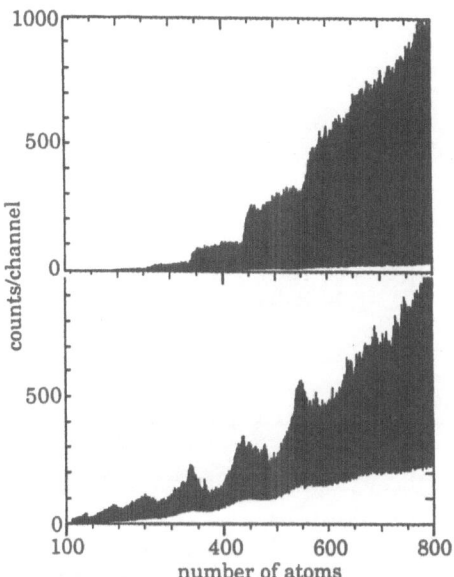

Figure 6. Mass spectra of $(Na)_n$ clusters obtained using ionizing light near the ionization threshold (top) ($\hbar\nu = 3.1$ eV) and well above the ionization threshold (bottom) ($\hbar\nu - 4.0$ eV). In both cases the neutral cluster beam was heated with 2.54 eV and 2.41 eV laser light.

a sharp increase in the mass-peak intensity occurs between n = 92 and 93. This can be more clearly seen if the mass scale is expanded by a factor of 50 (Fig. 8).

Notice also that the step occurs at the same value of n for clusters containing both one and two SO_2 molecules. In addition to the steps for n = 58 and 92 in Fig. 7, there are broad minima in the 2.53 eV spectrum at about 140 and 200 Cs masses. These broad features become sharp steps if the ionizing photon energy is decreased to 2.43 eV. By successively decreasing the photon energy, steps can be observed for the magic numbers n = 58, 92, 138, 198±2, 263±5, 341±5, 443±5, and 557±5 [15, 17]. However, the steps become less well defined with increasing mass. We have studied the mass spectra of not only $Cs_{n+2}(SO_2)$ but also $Cs_{n+4}(SO_2)_2$, $Cs_{n+2}O$, and $Cs_{n+4}O_2$. They all show step-like features for the same values of n.

First, we would like to offer a qualitatiave explanation for these results and then support this explanation with detailed calculation. Each cesium atom contributes one delocalized electron which can move freely within the cluster. Each oxygen atom, and each SO_2 molecule, bonds with two of these electrons. Therefore, a cluster with composition $Cs_{n+2}(SO_2)$, for example, can be said to have n delocalized electrons. The potential in which the electrons move is nearly spherically symmetric, so that the states are char-

280

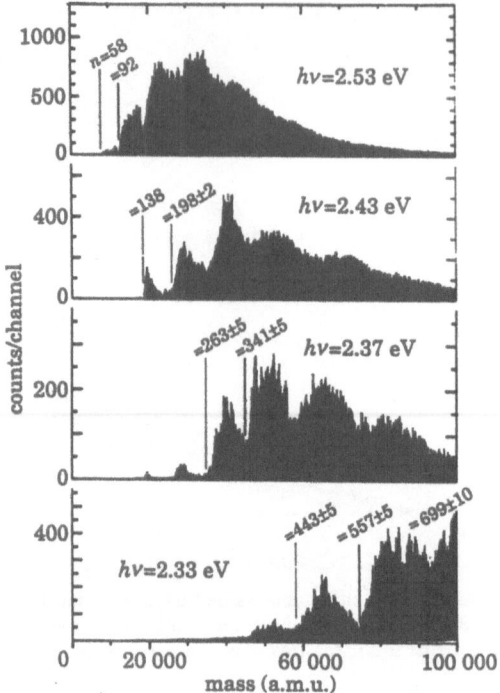

Figure 7. Mass spectra of $Cs_{n+2}(SO_2)$ clusters with decreasing photon energy of the ionizaing laser from 2.53 eV (top) to 2.33 eV (bottom). The values of n at the steps in the mass spectra have been indicated (Ref.[15]).

acterized by a well-defined angular momentum. Therefore, the delocalized electrons occupy subshells of constant angular momentum which in turn condense into shells. When one of these shells is fully populated with electrons, the ionization energy is high and the clusters will not appear in mass spectra obtained using sufficiently low ionizing photon energy.

In other experiments [9] the closing of small subshells of angular momentum was shown to be accompanied by a sharp step in the ionization energy for Cs-O clusters having certain sizes, namely for $Cs_{n+2}O_z$ with n= 8, 18, 20, 34, 58 and 92. The closing at n = 40 seen in all other alkali-metal clusters could not be observed, neither in the experiments nor in the calculations. The steps were observed for clusters containing from one to seven oxygen atoms.

5. Density Functional Calculation

Self-consistent calculations have been carried out applying the density functional approach to the spherical jellium model [10, 11]. We used an exchange

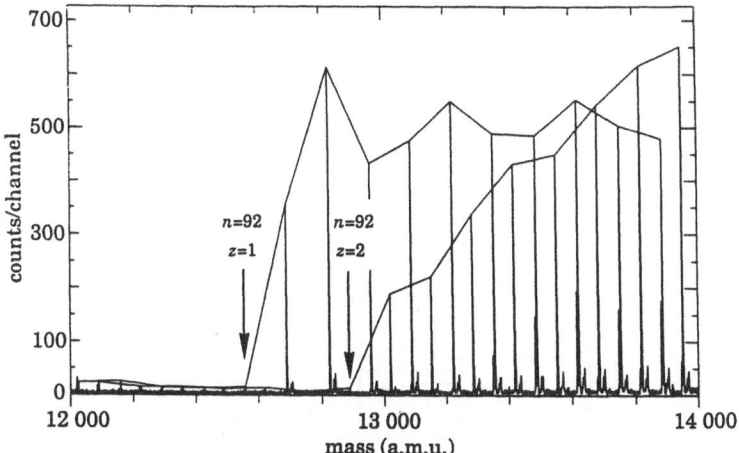

Figure 8. Expanded mass spectra of $Cs_{n+2z}(SO_2)$ clusters for an ionizing photon energy of 2.48 eV. The lines connect mass peaks of clusters containing the same number z of SO_2 molecules. Notice that the steps for clusters containing (SO_2) and $(SO_2)_2$ are shifted by two Cs atoms (Ref.[15]).

correlation term of the Gunnarsson-Lundqvist form and a jellium density $r_S = 5.75$ corresponding to the bulk value of cesium. This model implies two improvements over the hard sphere model discussed earlier. Firstly, electron-electron interaction is included. Secondly, the jellium is regarded to be a more realistic simplification of the positive ion background than the hard sphere. The O^2 ion is taken into account only by omitting the cesium electrons presumably bound to oxygen. The calculations were performed on Cs_{600} clusters [25].

We found, that if a homogeneous jellium was used, the grouping of subshells was rather similar to the results of the infinite spherical potential well. However, a nonuniform jellium yielded a shell structure in better accordance to experimental results. We found that the subshells group fairly well into the observed shells only if the background charge distribution is slightly concentrated in the central region. This was achieved, for example, by adding a weak Gaussian (0.5% total charge density, half-width of 6 a.u.) charge distribtuion to the uniform distribution (width 48 a.u.). Figure 9 shows the ordering of subshells obtained from this potential. This leads to the rather surprising result that the Cs^+ cores seem to have higher density in the neighborhood of the center perhaps due to the existence of the O^{2-} ion. All attempts to lower the positive charge density in the central region led to an incorrect ordering of states.

The first calculation addressed the problem of the grouping of low-lying energy levels in one large Cs_{600} cluster. However, in the experiment the

main quantum number

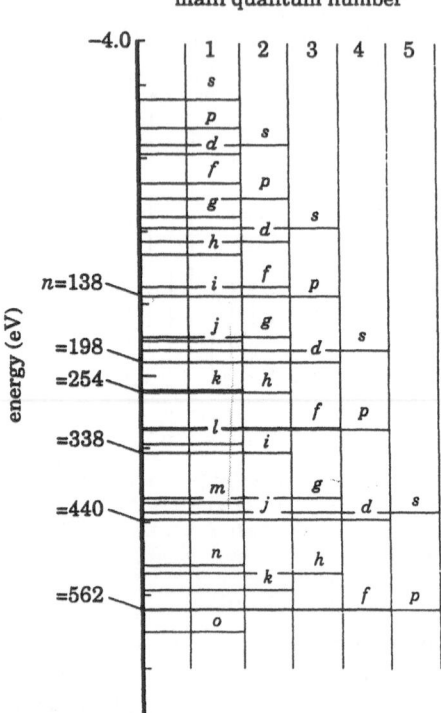

Figure 9. The self-consistent, one-electron states of a 600 electron cesium cluster calculated using a modified spherical jellium background (Ref.[25]).

magic numbers were found by a rough examination of ionization potentials of the whole distribution of cluster sizes. A more direct way to explain magic numbers is to look for steps in the ionization potential curve of Cs–O clusters. Therefore, we calculated the ionization potentials of $Cs_{n+2}O$ For $n \leq 600$ and of $(Na)_n$ for $n \leq 1100$ using the same local-density scheme described above, Fig. 10. Starting from a known closed-shell configuration for n = 18, electrons were successively added. Three test configurations were calculated for each cluster size testing the opening of new subshells. The configuration with minimum total energy was chosen for the calculation of the ionization potential.

We found that the lower magic numbers n = 34, 58, 92 were well reproduced. For higher n distinct steps in the ionization potential were observed for n = 138, 196, 268, 338, 440, 562, 704, 854 and 1012. The absolute values of the calculated ionization potentials can be brought into better agreement

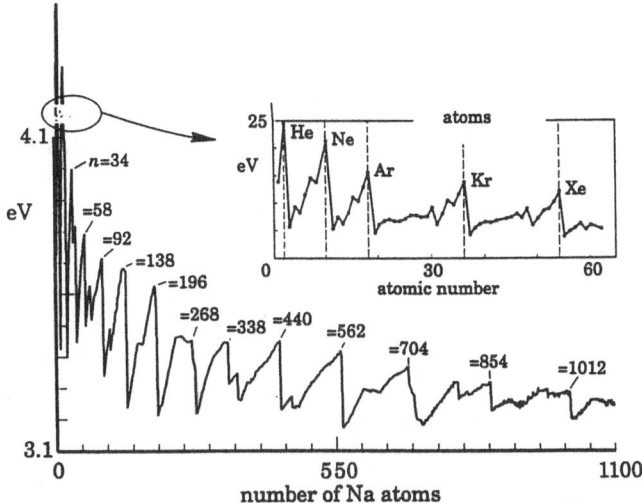

Figure 10. Ionization potentials calculated as a function of n for $(Na)_n$ clusters. A positive background charge distribution slightly concentrated in the central region has been used. Notice the similar behavior of the ionization energies of the chemical elements (inset) (Ref.[25]).

with experiment by assuming that clusters have a 10–15% lower electron density than is fouond in the bulk. Magic number clusters exhibit unusually high ionization energies for the same reason rare gas atoms do: they possess a closed shell electronic configuration, Fig. 10. In this sense the metallic clusters behave like giant atoms.

6. Observation of Supershells

Although nuclear physicists speculated on the possible existence of supershells several decades ago, the phenomenon has never been observed in atomic nuclei for a very simple reason. No nucleus contains enough fermions to allow supershell formation. However, there is almost no limit to the number of electrons that can be contained in metal clusters.

Supershells are the periodic appearance and disappearance of shell structure in the energy density of states of a fermion system. In order to make clear the physical origin of supershells, it is necessary to go back one step to the semiclassical description of shells. Shells are associated with a charactristic length. Each time an integral number of fermi wavelengths fit into this length, a new shell has formed. The systems that we are studying are so large, that the classical picture of an electron bouncing back and forth inside a metal cluster is not completely without meaning. The character-

284

istic length associated with a set of shells is just the length of a <u>closed</u> electron trajectory within the clusters. For spheres, two closed trajectories with almost the same length turn out to be the most important – a triangualr path and a square path. This leads to two sets with nearly the same energy spacing. These two contributions interfere with one another to produce a beat pattern known as quantal supershells. The first attempts

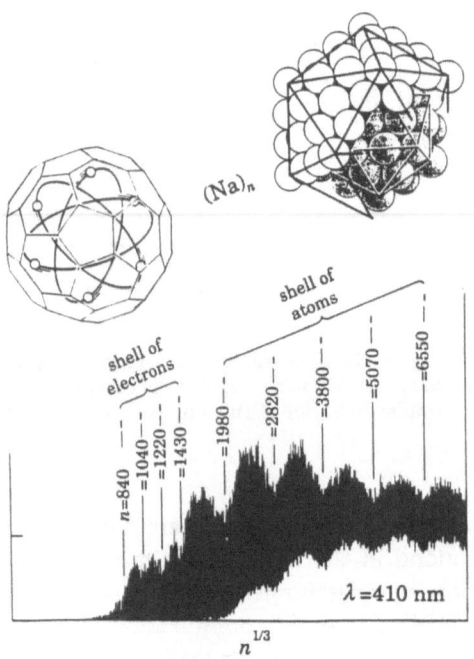

Figure 11. Mass spectrum of $(Na)_n$ clusters photoionized with 3.02 eV photons. Two sequences of structures are observed at equally spaced intervals on the $n^{1/3}$ scale – an electronic shell sequence and a structural shell sequence.

to observe supershell structure in our laboratory were hindered by the unexpected appearance of a second set of shells in clusters containing more than 1500 atoms, Fig. 11. These proved to be geometric shells of atoms that masked the weaker electronic shell structure. In the new experiments the geometric shell structure was surpressed by "melting" the clusters through heating with a continuous laser beam tuned to the plasmon frequency of the electron system.

The clusters were warmed prior to ionization with a continuous Ar-ion laser beam running parallel to the neutral cluster beam. The laser light entred the ionization chamber through a heated window, passed through the ionization volume, through 2.2 mm Ø and 3.0 mm Ø skimmer aperatures, through a 3.0 mm nozzle, throught the oven chamber and finally exited through a second window where the laser intensity was recorded. Short wavelength light was found to warm much more efficiently. Using the 458 nm (2.71 eV) laser line, 10 mW proved sufficient to appreciable alter the neutral size distribution.

The size distribution obtained with ionizing photons having energy well above threshold are quite different from the spectra discussed in section 5 Without the warming laser the mass spectra are without structure, i.e. the size distribution of the cold clusters emerging from our source is smooth, Fig. 12. If the warming laser is turned on we obtain not steps but peaks

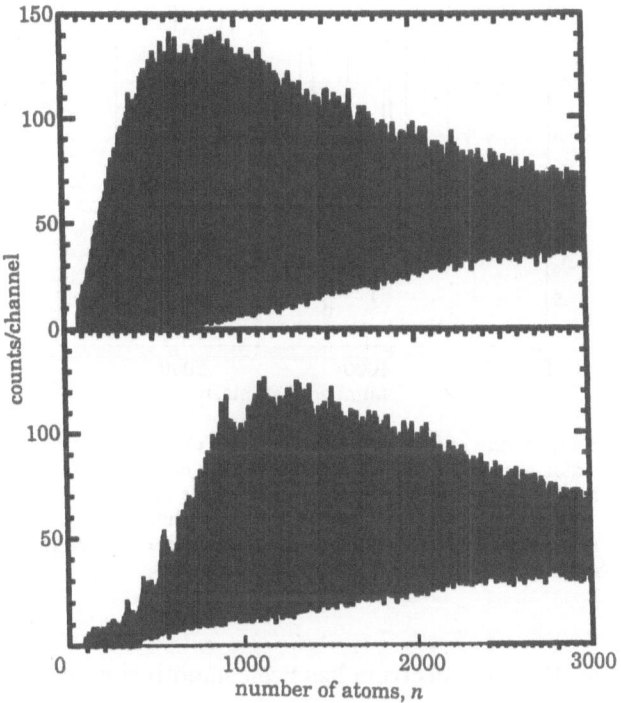

Figure 12. Mass spectrum of (Na)$_n$ clusters using 4.0 eV ionizing light. The top spectrum shows the size distribution of cold clusters produced in the source; the bottom spectrum, after heating with 2.54 eV laser light.

as seen in the bottom of Fig. 12. We believe these peaks reflect the neutral size distribution of the laser-warmed clusters. It appears that it is usually possible to correlate a <u>falling</u> edge of the size disbribution with a step in the threshold ionization spectrum. Because of this correlation, we will characterize mass spectra obtained using excimer light by the number of atoms at steep negative slopes. A more extended mass spectrum of laser-warmed sodium clusters obtained with 4.0 eV ionizing photons is shown at

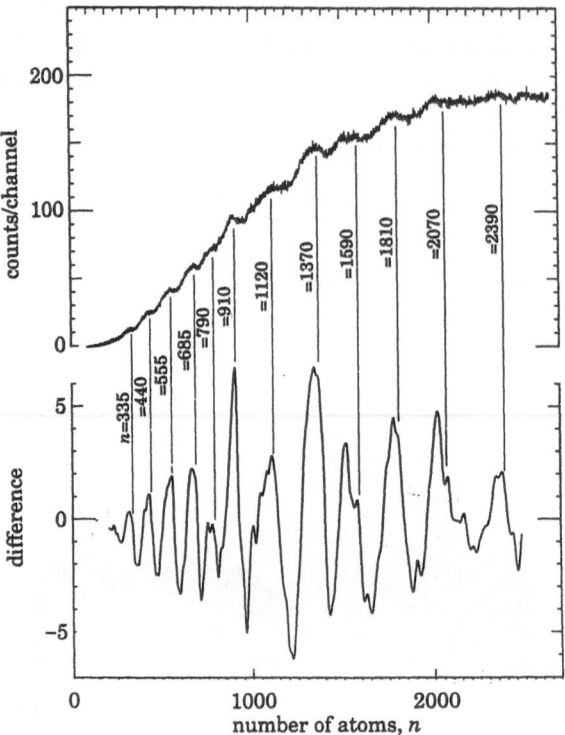

Figure 13. Mass spectrum of (Na)$_n$ clusters using 4.0 eV ionizing light and (458nm) 2.71 eV continuous axial warming light having an intensity of 500 mW/cm^{-2}. The spectrum has been smoothed over one-hundred 16 ns time channels (top). In order to emphasize the shell structure an envelope function (obtained by smoothing over 20 000 time channels) is subtracted from a structural mass spectrum (smoothed over 1500 time channels). The difference is shown in the bottom spectrum.

the top of Fig. 13. This spectrum has been smoothed with a spline function extending over one-hundred 16ns time channels. Notice that the structure observed does not occur at equal intrvals on a scale linear in mass. In order to present this structure in a form more convenient for analysis, the data have been procesed in the following way. First, the raw data is averaged with a spline function extending over 20 000 time channels. The result is a smooth envelope curve containing no structure. Second, the raw data is averaged with a spline over 1500 channels. Finally, the two avearges are subtracted. The result is shown in the bottom of Fig. 13. Five independent measurements were made under the same experimental conditions. The positions, relative heights and widths of features in the mass spectra were well reproducible.

The clusters in this experiment have been warmed with a continuous laser beam running parallel to the neutral cluster beam. But what is implied by "warming"? consider the fate of a typical 500-atom cluster as it moves from the nozzle to the detector.

It leaves the nozzle with the temperature of the He carrier gas ($\sim$100K) traveling at a velocity of about 350 m/s. during its 1 ms flight to the ionization volume it undergoes no further collisions but does begint o absorb photons. We don't really know the absorption cross-section of this cluster at the warming laser wavelength (458nm). However, 1Å^2/atom is a typical upper limit for smaller clusters. It can be expected that the cross-section will be cluster size dependent. This size dependence will be reflected in the final mass distribution. The cluster absorbs the first 25 photons without evaporating any atoms, gaining an excess energy of about 70 eV and reaching a temperature of about 500 K. This all takes place in the first 450 μs. The temperture of the cluster remains rather constant for the last half of its journey to the ionization volume. It continues to absorb photons, of course, but after each absorption it evaporates 2 or 3 atoms returning to its original temperature before absorbing the next photon. It loses a total of 80 atoms, i.e. 16% of its original mass. It appears that this repeated heating and cooling through the "critical temperature for evaporation" on this time scale favors the evolution of a size distribution with relataively strong peaks near sizes corresponding to closed electronic shells.

The photon energy (4.0eV) of the ionizing laser has been chosen so that it is well above the ionization threshold (3.0 eV) of the sodium clusters investigated. The excess energy (1eV) insufficient to cause only one atom to evaporate. This is a neglibible loss on the mass scale we will be considering. For this reason, we believe that the magic numbers obtained reflect variations inthe size disbtribution of the neutral clusters induced by the warming laser.

The concept of shells can be associated with a characteristic length. Every time the radius of a growing cluster increases by one unit of this characteristic length, a new shell is said to be added. A good rough test of whether or not shell structure has been observed can be quickly carried out by plotting the shell index as a function of the radius or $n^{1/3}$. If the points fall on a straight line, the data is consistent with shell formation. That this is indeed the here, can be seen in Fig. 14. However, an even better fit can be obtained using two straight lines with a break between shell 13 and 14. This too can be interpreted in an interesting way.

It has been suggested [19-28] that shell structure might periodically appear and disappear with increasing cluster size. Such a supershell structure can be understood as a beating pattern created by the inteference of two nearly equal periodic contributions. Quantum mechanically the contribu-

288

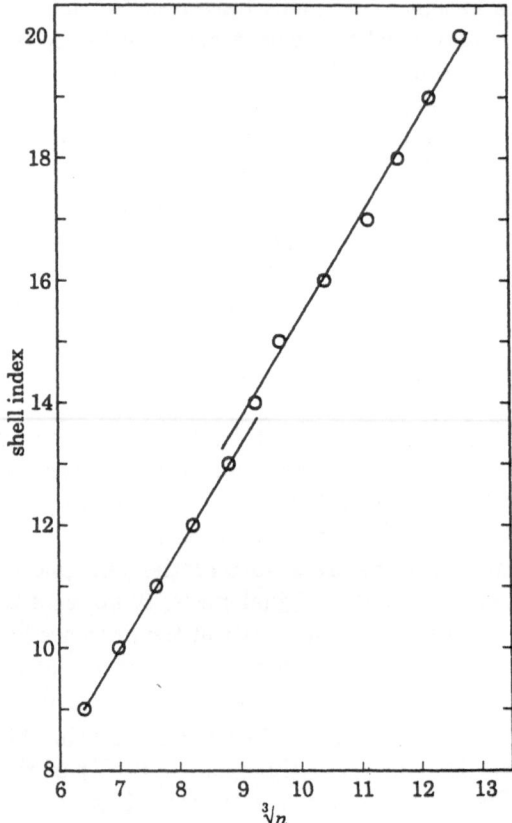

Figure 14. The electronic shell closing falls approximately on a straight line if plotted on an $n^{1/3}$ scale. An even better fit is obtained using two straight lines with a break between shells 13 and 14. Such a break or "phase change" would be an indication of supershell structure.

tions can be described as arising from competing energy quantum numbers. Classically, the contributions can be described as arising from two closed electron trajectories within a spherical cavity. One trajectory is triangular, the other square.

7. Fission

The fission of clusters was one of the first subjects [29-43] to be investigated in the newly developing field of cluster research. It is often referred to as Coulomb explosion, since the fission is caused by the Coulomb repulsion of like charges concentrated in a cluster smaller than a critical size. The kinetic energy that the charged fragments acquire can be as high as several eV.

Most of these studies have delt with the fission of doubly or triply charged clusters. Recently we have shown that it is possible to induce charges as high as +14 on large Na clusters by photoionization [44]. In this section we will discuss fission in these highly charged clusters.

The technique we have used to study fission in sodium clusters is photoionization time-of-flight (TOF) mass spectroscopy. The cluster source is a low pressure, inert gas, condensation cell. The clusters were photoionized with a 50 mJ, 15ns, 193nm (6.4eV) excimer laser pulse focussed onto the neutral cluster beam with a 150 cm focal length quartz lens. The ionized clusters were heated 30 ns later with a second 5 mJ/mm^2, 470 nm (2.6eV) laser pulse.

The energy (I) required to remove an aditional electron from a cluster that already has charge +z can be written

$$I(z, R) = W + (\alpha + z)e^2/r \qquad (4)$$

where W is the bulk work function, e is the electronic charge, and R is the radius of the cluster. Clearly, we have assumed that the cluster can be modelled as a conducting sphere. Various values of α have beenused in the literature. We will assume α is 0.5 and point out that for larg values of z the value of α used becomes unimportant. Since the radius of the cluster can be related to the number of electrons (or in our case atoms) through the Wigner-Seitz radius, $R^3 = r_s^3 n$, Eq. 4 can be rewritten as

$$I(z, n) = W + (\alpha + z)e^2/r_s n^{1/3}). \qquad (5)$$

It can be seen from this expression that a larage amount of energy is required to remove electrons from small, highly charged clusters. If the amount of energy available is limited to that in one photon, then the maximum charge attainable for a cluster of a given size is

$$z_{max} = 1 - \alpha + (h\nu - W)r_s n^{1/3}/e^2. \qquad (6)$$

This means that a two-dimensional cluster space (n,z) can be divided by a line into clusters that can be formed with, for example, an ArF excimer laser and those that cannot, Fig. 15. Also indicated in this figure is a line dividing the space into stable and unstable clusters. Notice that all of the values of n and z accessible with the ArF photons characterize stable clusters. The unstable clusters which we would like to investigate cannot be produced by direct multi-step ionization with this laser. There is, however, a way out of this dilemma.

The ArF laser can be used to prepare a stable, highly charged, large cluster and then this large cluster can be reduced in size by heating and

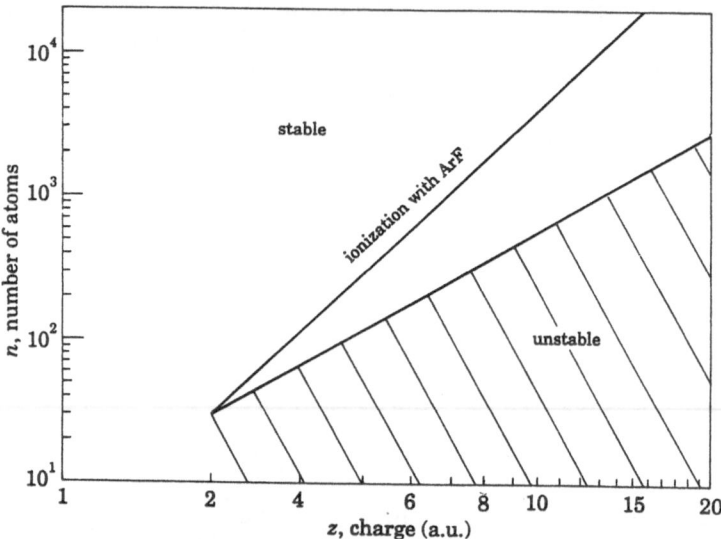

Figure 15. The two-dimensional cluster space (n,z) can be divided by straight lines into Na$_n^z$ clusters which can (cannot) be produced with an ArF laser and into clusters which are (are not) stable against Coulomb explosion. Notice that all clustrs (except for z = 2) that can be ionized with 6.4 eV photons are stable.

subsequent evaporation. A second laser pulse, containing photons with energies near the plasmon resonance of the sodium clusters, is used for heating. The clusters shrink down in size without charge until they reach a critical size at which they undergo fission. A mass spectrum, or better said, an n/z spectrum for Na$_n^z$ clusters produced in this way is shown in Fig. 16. The log scale emphasizes, perhaps even overemphasizes, the effect we wish to show.

The highest set of mass peaks belongs to singly-charged sodium clusters. The peaks which occur exactly half-way between the Na$_n^+$ peaks are due to Na$_n^{2+}$ clusters. Notice that new sets of peaks appear in the spectrum at various critical values of n/z. This is perceived as a step-wise darkening of the mass spectrum. In the lower part of Fig. 16 we see the threshold region for the appearance of Na$_n^{5+}$ on an expanded scale. Another segment of the spectrum on an expanded scale is shown in Fig. 17. This segment is near the threshold for the appearance of Na^{6+}n.

In this way, by careful examination of the fine structure in the mass spectra, it is possible to determine that the critical sizes for z = 1,2,3,4,5,6 and 7 are 27±1, 64±1, 123±2, 208±5, 321±5 and 448±10 atoms, respectively. These values are plotted on a double log scale in Fig. 18. They lie on a straight line with slope 2. This means that the critical condition for

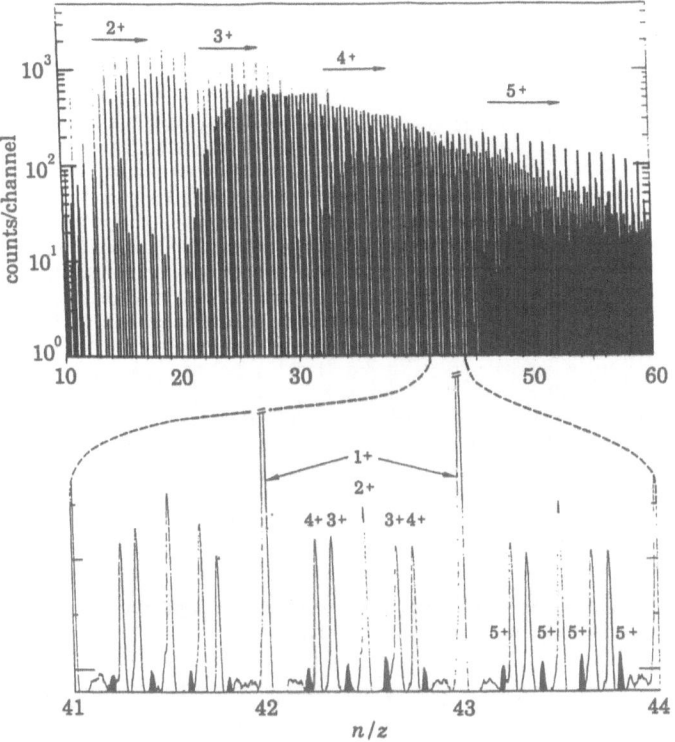

Figure 16. An n/z spectrum of Na_n^z clusters. large clusters were first charged by multistep ionization using a high-fluence, arF laser. The clusters were then heated to reduce their size by evaporation. A portion of the spectrum is expanded to show the appearance threshold for Na_n^{5+} clusters (black filled).

stability is

$$z^2/n \leq 0.125 . \tag{7}$$

z^2/n is proportional to the so-called fissility parameter used in nuclear phyciscs as a measure of stability. It has been shown inthe past that this parameter is also useful for clusters with small total charge [40, 41]. Here we see that it continues to be applicable for values of z up to 7.

The results of an extensive theoretical investigation of fission in Na clusters have recently been published [45, 46]. an important assumption made inthis work was that the fission is symmetric, i.e. the mass and charge of the original cluster are divided nearly equally between the fission productes. Unfortunately, we have no evidence at this time to either support or to challenge this assumption. Still, it is useful to compare the results of this calculation with our experiment, Fig. 18. Here, cluster space (n,z)

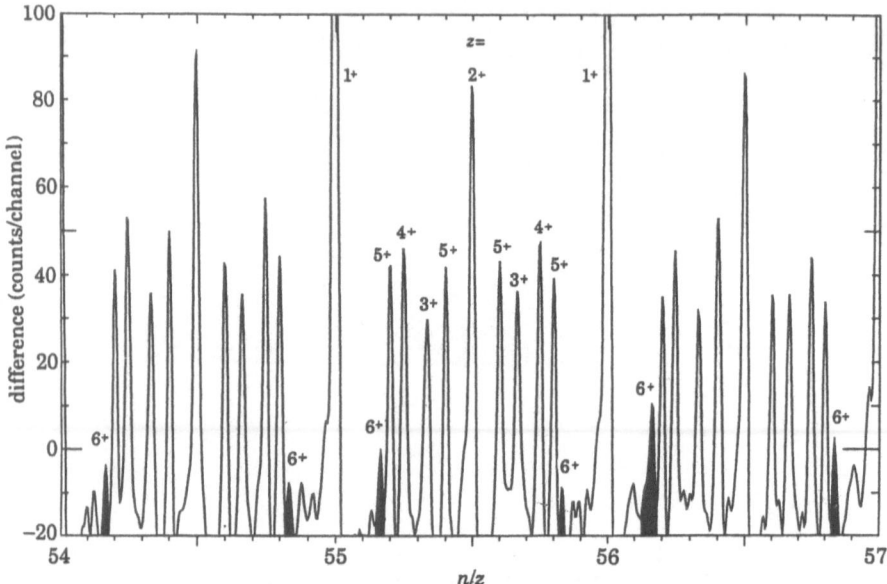

Figure 17. An expanded portion of Fig. 2 showing the first appearance on Na_n^{6+} clusters. Smaller clusters in this charge state are not stable.

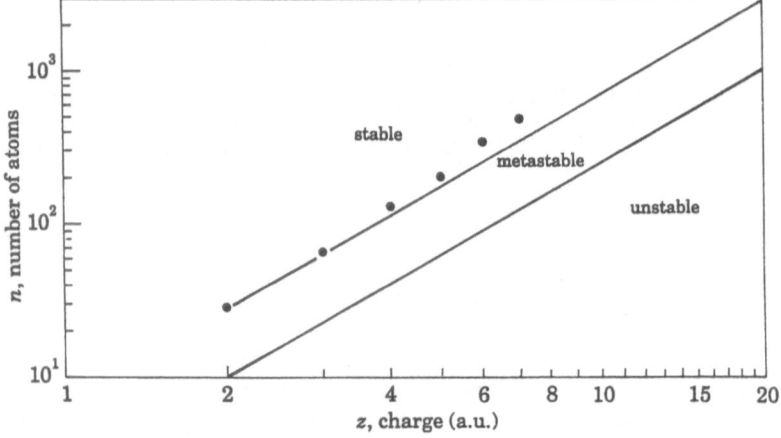

Figure 18. The number of atoms in the smallest experimentally observed Na_n^z clusters (filled circles). The cluster space can be divided into stable, metastable and unstable regions, according to Ref.[45], using the symmetric liquid-drop model.

has been divided into stable metastable and unstable regions according to the tunneling criteria appropriate in nuclear physics. The fission process for nuclei can be described qualitatively in terms of three energies; the initial energy (E_i) of the charged, nondeformed clusters, the final energy (E_f) which is the sum of the energies of the noninteracting fission products. The third energy necessary to characterize fission products. The third energy necessary to characterize fission is the energy (E_b) of the lowest barrier separating the initial and final states. If $E_b < E_i$, the nucleus is unstable. If $E_f < E_i < E_b$, the nucleus is metastable to fission by tunneling through the barrier. Finally, if $E_f > E_i$, then the nucleus is stable. since tunneling for clusters has negligible probability, it is more accurate to say at zero temperature clusters are either stable or unstable, depending on whether there is a barrier or not. At finite temperature clusters can be classified as either unstable $(E_B < E_i)$ or as metastable $(E_B > E_i)$ to thermal hopping over the barrier. In practice it is useful to further subdivide the set of metastable clusters [36-38]. At finite temperatures clusters can lose mass and thermal energy by the evaporation of neutral atoms. Evaporation will always compete with fission and will, in fact, dominate if E_B is greater than E_v, the energy needed to evaporate an atom. For this reason, we have the conditions

a) unstable to fission, $\quad\quad\quad E_B < E_i$;

b) metastable to fission, $\quad\quad E_i < E_B < E_v$;

c) metastable to evaporation, $\quad E_i < E_v < E_B$.

Since previous experiments [36-38] on doubly-charged Na and K clusters indicate that fission is strongly asymmetric, it would be appropriate now to consider this alternative. Even using the droplet model it is not easy to calculate the height of the barrier if the mass and charge can be distributed arbitrarily between the fission products. For this reason we will consider an energy which does not exactly characterize a real system, but it is trivial to calculate and therefore useful. It is the energy at the instant of scission, E_s. That is, starting from the final state we merely bring the fission products together until they just touch. Of course the energy increases monotonically from E_f to E_s according to Coulomb's Law. We assume that E_s is nearly equal to the barrier height for the fission process. There is no unique value of E_s for a cluster in initial state n and z. Rather, a whole set of values exist corresponding to the various ways of distributing charge and mass between the fission products. However, one value of E_s has special significance and that is the minimum value. If this minimum value of $E_s < E_i$ the cluster is unstable and will spontaneously fission, even at zero temperature. If $E_s > E_i$ the cluster is stable against fission. Strictly speaking, one shuld say metastable because at finite temperatures the final state can be reached by jumping over the barrier, no matter how high. The results of these calculations are summarized in Fig. 19. The initial and final energies are

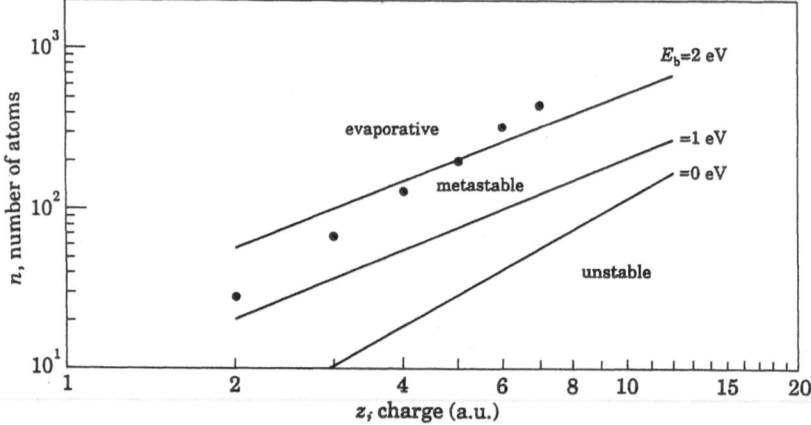

Figure 19. Cluster space divided into evaporative and metastable und unstable regions. The barrier height has been obtained using an oversimplified model (see text). The number of atoms in the smallest experimentally observed Na_n^{z+} clusters is shown by the filled circles.

determined using the sphereical droplet model assuming only two fission fragments with arbitrary size and charge and assuming a surface tension parameter $\sigma = 200$ dyne/cm appropriate for sodium. One might expect that the experimental points would fall on the line corresponding to $E_B = E_v$. Clearly, this is not the case since E_v is known [36-38] to have a value of about 1 eV. That is, this rough model overestimates the barrier height by a factor of two. Various refinements are clearly needed; proper treatment of the Coulomb energy allowing for electron redistribution as the fragments move away from one another [48], a description of the asymmetric fission before scission, shell effects [49, 50] and entropy effects. Also needed are experiments demonstratinghow mass and charge are distributed between the fission fragments.

8. Concluding Remarks

Clearly, cluster science has greatly benefitted from the inspired work carried out by nuclear physicists decades ago. The shell model, the liquid droplet model, and the theory of giant dipole resonances have provided a ready and appropriate framework for understanding the properties of metal clusters. Hopefully, in the future, the exchange between nuclear science and cluster science will not be so one-sided, because metal clusters offer us a unique opportunity to study well-characterized, large fermion systems.

References

1. Goeppert-Mayer, M. (1949) *Phys.Rev.* **75**, 1969L.
2. Haxel, O., Jensen, J.H.D., Suess, H.E. (1949), **75**, 1766L.
3. Knight, W.D., *et al.* (1984) *Phys. Rev. Lett.*
4. Kappes, M.M., Kunz, R.W. Schumacher, E. (1982), *Chem. Phys. Lett.* **91**, 413.
5. Katakuse, I., *et al.* (1985) *Int. J. Mass Spectrom. Ion Processes* **67**, 229.
6. Bréchignac, C., Cahuzac, Ph., Roux, J.-Ph. (1986) *Chem. Phys. Lett.* **127**, 445.
7. Begemann, W., *et al.* (1986) *Z. Phys.* **D3**, 183.
8. Saunders, W.A., *et al.* (1986) *Phys. Rev.* **B32**, 1366.
9. Bergmann, T., Limberger, H., Martin, T.P. (1988) *Phys. Rev. Lett.* **60**, 1767.
10. Martins, J.L., Car, R., Buttet, J. (1981) *Surf. Sci.* **106**, 265.
11. Ekardt, W., (1984) *Ber. Bunsenges. Phys. Chem.* **88**, 289.
12. Clemenger, K. (1985) *Phys. Rev.* **B32**, 1359.
13. Ishii, Y. Ohnishi, S., Sugano, S. (1986) *Phys. Rev.* **B33**, 5271.
14. Bergmann, T., Limberger, H. (1989) *J. Chem. Phys.* **90**, 2848.
15. Göhlich, H., *et al.* (1990) *Phys. Rev. Lett.* **65**, 748.
16. Martin, T.P., *et al.* (1991) *Chem. Phys. Lett.* **72**, 209.
17. Bjørnholm, S. ,*et al.* (1990) *Phys. Rev. Lett.*, **65**, 1627.
18. Persson, J.L., *et al.* (1990) *Chem. Phys. Lett.*, **171**, 147; Honea, E.C., *et al.* (1990) *Chem. Phys. Lett.*, **171**, 147; Lerme, J., *et al.* (1992) *Phys. Rev. Lett.* **68**, 2818.
19. Balian, R. and Bloch, C. (1971) *Ann. Phys.* **69**, 76.
20. Bohr, A. and Mottelson, B.R. (1975) *Nuclear Structure* Benjamin, London).
21. Nishioka, H., Hansen, K., Mottelson, B.R. (1990) *Phys. Rev.* **B42**, 9377.
22. Bergmann, T. , *et al.* (1990) *Rev. Sci. Instrum.* **61**, 2585.
23. Mamyrin, B.A., *et al.* (1973) *Sov. Phys. JETP* **37**, 45.
24. Bergmann, T., Martin, T.P. and Schaber, H. (1990) *Rev. Sci. Instrum.* **61**, 2592.
25. Lange, T. *et al.*, (1991) *Z. Phys. D* **19**, 113.
26. Martin, T.P., *et al.* (1991) *Chem. Phys. Lett.* **186**, 53.
27. Pedersen, J., *et al.* (1991) *Nature* **353**, 733.
28. Bréchignac, C. *et al.* (1993) *Phys. Rev. B* **47**, 2271.
29. Kreisle, D., *et al.* (1986) *Phys. Rev. Lett.* **56**, 1551.
30. Echt, O. (1987) Physics and Chemistry of Small Clusters, P. Jena, B.K. Rao and S.N. Khanna (eds.), Plenun Press, New York.
31. Echt, O. (1988) *et al.*, *Phys. Rev.* **A38**, 3236.
32. Märk, T.D., *et al.* (1989) *Z. Phys.* **D12**, 279.
33. Gotts, N.G., Lethbridge, P.G. and Stace, A.J. (1992) *J. Chem. Phys.* **96**, 408.
34. Kandler, O, *et al.* (1991) *Z. Phys.* **D19**, 151.
35. Sattler, K., *et al.* (1985) *Phys. Rev. Lett.* **47**, 160; Sattler, K. (1985) *Surf. Sci.* **156**, 292.
36. Bréchignac, C., *et al.* (1990) *Phys. Rev. Lett.* **64**, 2893.
37. Bréchignac, C., *et al.* (1991) *Z. Phys.* **D19**, 1.
38. Bréchignac, C., *et al.* (1991) *Phys. Rev.* **B44**, 11386.
39. Katakuse, I., Itoh, H., Ichihara, T. (1990) *Int. J. Mass. Spectrum. Ion Proc.* **97**, 47.
40. Saunders, W.A. (1990) *Phys. Rev. Lett.* **64**, 3046.
41. Saunders. W.A. (1991) *Z. Phys.* **D20**, 111.
42. Schulze, W. (1987) *J. Chem. Phys.* **87**, 2402.
43. Rabin, I., Jackschath, C. and W. Schulze (1991) *Z. Phys.* **D19**, 153.
44. Näher, U. *et al.* (1992) *Phys. Rev. Lett.* **68**, 3416.
45. Sugano, S. (1991) *Microcluster Physics*, Springer, Berlin, Heidelberg.
46. Nakamura, M. *et al.* (1991) *Z. Phys.* **D19**, 145.
47. Lipparini, E. and Vittori, A. (1990) *Z. Phys.* **D17**, 57.
48. Garcias, F. *et al.* (1991) *Phys. Rev.* **B43**, 9459.
49. Rao, B.K. *et al.* (1987) *Phys. Rev. Lett.* **58**, 1188.
50. Barnett, R.N., Landman, U. and Rajagopal, G. (1991) *Phys. Rev. Lett.* **67**, 3058.

Nanoscale Probes of the Solid - Liquid Interface

H. SIEGENTHALER, E. AMMANN, P.-F. INDERMÜHLE[1], G. REPPHUN[+]
Universität Bern, Departement für Chemie und Biochemie, Freiestrasse 3,
CH-3012 Bern (Switzerland),
[1]*Université de Neuchâtel, Institut de Microtechnique, CH-2007 Neuchâtel (Switzerland)*
[+]*Present address: Paul Scherrer Institut, CH-5232 Villigen PSI (Switzerland)*

Abstract: In-situ nanoscale probing of the solid-liquid interface, in particular by electrolytic STM and SFM, has gained considerable importance in electrochemistry and has given rise to novel nanoscale applications in electrochemical systems, including first attempts for electrochemical nanostructuring methods. In this report, selected examples are presented to illustrate the present state and future potential of nanoscience in electrochemistry.

1.   Introduction and Scope

In 1986, Hansma et al. [1] reported the first STM images of graphite and gold surfaces immersed in water and in NACl-solution. These results confirmed the expectation that the principle of STM imaging was also applicable to the interface between a conducting solid and a liquid (including electrolyte) phase. The development of the potentiostatic STM technique by several groups [2-5] around 1988, involving independent potential control of the STM tip and the substrate by means of a bipotentiostatic circuitry, enabled a drastic reduction of interfering electrochemical currents at the STM tip together with the possibility of a defined control of electrochemical processes at the substrate surface. In combination with refined procedures for STM tip preparation, this method has now become the established technique in electrochemistry for STM investigations at the electrode - electrolyte interface in a wide resolution range from micrometer to atomic scale

297

N. García et al. (eds.), Nanoscale Science and Technology, 297–315.
© 1998 *Kluwer Academic Publishers.*

resolution. In 1990, Magnussen et al. [6] reported the first STM images of the atomic structure of a metal surface and a metal adsorbate in contact with an electrolyte solution during the adsorption of Cu at Au(111). This result demonstrated that a similar imaging resolution can be obtained at the metal-electrolyte interface as achieved in UHV or ambient gas atmosphere and was followed in 1991 by atomic resolution images of the same system in the first application of SFM in an electrolyte solution operated with potential control of the substrate [7]. The mentioned developments have stimulated and promoted in the last years numerous electrolytic STM and SFM studies of the surface structure and morphology of electrodes in their real electrolytic environment, as well as investigations of the local onset and propagation of electrochemical processes in relation with the local structural and electronic properties of the electrode surface. Representative surveys of this work have been given in recent reviews [8-10].

While most of the the initial nanoscale probing experiments of the solid-liquid interface were focussed on electrochemical systems with (mostly metallic) conducting solids and electrolytic liquid phases, especially the introduction and application of electrolytic SFM has greatly promoted the extension of electrolytic SFM- and STM probing in liquids onto biological systems in a growing field of applications [11] beyond the scope of this chapter.

This contribution covers some selected examples of nanoscale probing in electrochemical systems, based mostly on recent and ongoing work in the author's group.

2. Local Probing of Electrochemical Processes at Non-ideal Electrodes

A specially important aspect of STM and SFM probing techniques for the local characterization of electrode surfaces and processes is their capability to image also nonperiodic features at the electrode-electrolyte interface. This offers attractive possibilities for the investigation of "real" electrode systems, as applied widely in electrochemical technology (e.g. galvanic metal deposition and battery technology), and as encountered in corrosion problems. As shown schematically in Figure 1, such electrodes are usually characterized by pronounced structural and morphological heterogeneities (e.g. monoatomic or polyatomic steps, islands and pits, surface defects and dislocations, grain boundaries) and chemical heterogeneities (e.g. foreign adsorbates, heterogeneous alloy phases, passive layers), and a detailed elucidation of electrochemical processes in such systems implies the correlation of the global electrochemical response with the local monitoring of electrode properties and processes.

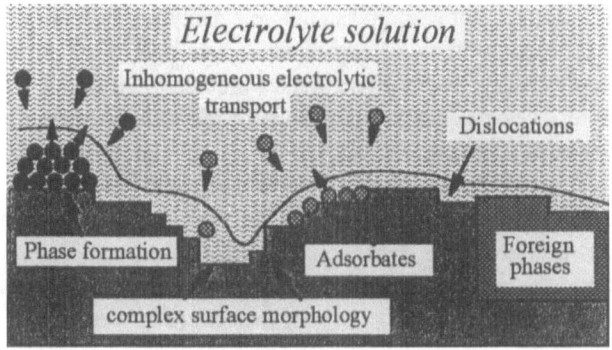

"Real" electrode

Figure 1: Schematic of a "real" electrode with structural, morphological and chemical heterogeneities

In the following, some specific aspects related with the application of local probes at non-ideal electrodes are illustrated in selected studies of the early stages of metal deposition and of a conducting polymer, and are supplemented by first results   of a newly developed thin-layer STM probing method.

2.1   STM-Investigation of Pb and Tl Underpotential Deposition at non-ideal Ag(111) electrodes

The use of "non-ideal" single-crystal electrodes with a relatively high density of stacked terrace domains, monoatomic pits and monoatomic islands has proved to be a very interesting tool for the elucidation of the influence of the atomic scale electrode morphology upon the local onset and progress of electrochemical reactions.

This is demonstrated in the following studies of the formation of Pb and Tl monolayers at a non-ideal Ag(111) electrode substrate during underpotential deposition of Pb^{2+} and Tl^+. More detailed reports are presented in [12,13].

300

2.1.1 Experimental methods and nanoscale surface morphology of the non-ideal Ag(111) electrodes

The STM measurements were performed by means of a commercial Nanoscope II instrument equipped with a homebuilt electrolytic 4-electrode cell [12] and operated in the potentiostatic mode with independent control of the substrate and tip potentials. Time-dependent local alterations were recorded either by evaluating the difference between several subsequently recorded scans of the same substrate domain, or by special line-scan techniques involving repetitive 1-dimensional scans of a selected surface region recorded as function of time [13]. The experiments were performed in 0.01M $HClO_4$ containing 0.005M Pb^{2+} or Tl^+. The non-ideal Ag(111) electrodes were prepared by mechanical and subsequent chemical (chromate) polishing, followed by transfer under electrolyte cover into the electrolytic STM cell.

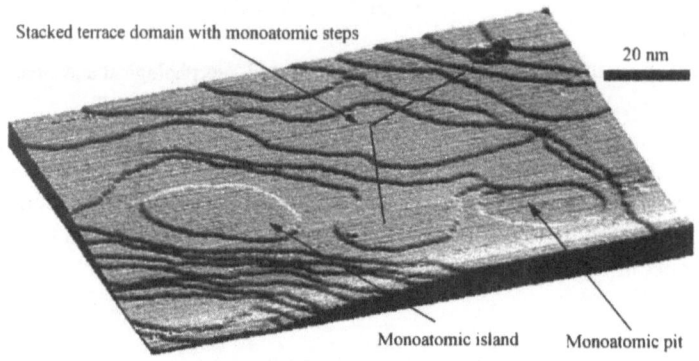

Figure 2: STM image of a chemically polished Ag(111) electrode in 0.01M $HClO_4$. From [13].

Figure 2 shows a typical example of a chemically polished Ag(111) electrode: The largest part of the surface consists of *stacked terrace domains* composed of stacks of monoatomic terraces, whose width varies between ca. 2nm and more than 20 nm. In some cases, terrace widths up to 100 nm have been observed. In addition to these terraced domains, *monoatomic islands* and *monoatomic pits* with typical average widths of ca. 25 nm are observed regularly.

2.1.2   Local Progress of Pb and Tl Adsorbate Formation

Adsorption and desorption of Pb

Figure 3 summarizes the voltammetric and STM results observed during the adsorption and desorption of Pb at a non-ideal Ag(111) electrode of the type shown in Figure 2:

The cyclic voltammogram (Figure 3a) indicates that the formation of the Pb monolayer occurs in 3 distinct potential intervals associated with the voltammetric adsorption/desorption peaks A1/D1, A2/D2 and A3/D3. From the results of the local STM imaging and linescan measurements performed at the stacked terrace domains, monoatomic pits, and monoatomic islands of the electrode substrate, the local progress of adsorbate formation at the morphologically different domains of the non-ideal Ag(111) substrate is presented schematically in Figure 3c and can be described as follows:

- In the initial adsorption stage, associated with the *voltammetric peak A1*, the monoatomic step edges at all different substrate domains are decorated by a spatially delimited adsorbate extending laterally no further than ca. 1-3 nm from the step edge. Although the lateral extension of this initial coverage is remarkably stable within time scales of several hours, it has not been possible up to now to image a stable atomic structure of this decorating coverage. It cannot be excluded that this adsorbate forms a locally delimited coverage with a temporally unstable (fluctuating) structure. An additional adsorption of Pb adatoms or small Pb clusters on the substrate terraces has not been observed up to now, but cannot be fully excluded.

- In the *voltammetric peak A2*, the adsorbate growth proceeds differently at the different morphological domains: In the monoatomic pits, the adsorbate coverage grows inwards from the decorated pit boundaries resulting in a complete coverage of the pit. At the stacked terrace domains the adsorbate propagates also from the decorated step edges. However, at the negative potential limit of peak A2, the resulting coverage only extends to within 1-3 nm from the peripherical terrace boundaries, thus forming a "partial" adsorbate coverage. On the monoatomic islands, no adsorbate layer growth has been observed, although there is evidence for occasional nucleation and subsequent disappearance of cluster-like adsorbate domains.

- Finally, in the most negative potential range of the *voltammetric peak A3*, the adsorbate-free peripherical parts of the stacked terrace domains are completely covered, and a complete adsorbate coverage is also formed on the monoatomic islands. No further adsorption occurs in the monoatomic pits that have been covered previously by a complete adsorbate layer within peak A2.

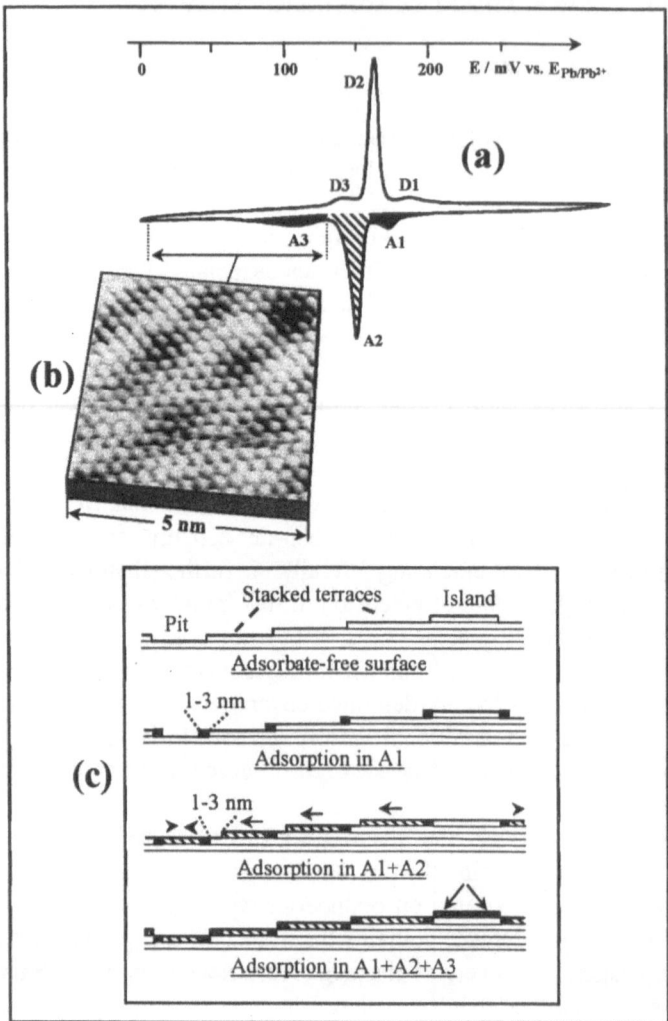

Figure 3: Summary of the adsorption/desorption behaviour of Pb at chemically polished Ag(111) electrodes in 0.01M HClO$_4$ + 0.005M Pb^{2+}: (a): Cyclic voltammogram of Pb adsorption and desorption with characteristic adsorption/desorption peaks A1/D1, A2/D2, A3/D3. (b) Atomic structure of the hexagonally close-packed adsorbate coverage observed at the negative limit of adsorption peak A2 and after adsorption in peak A3. (c): Schematic presentation of the local progress of Pb adsorption at monoatomic pits, stacked terrace domains and monoatomic islands, investigated by STM linescan techniques. From [13] (slightly modified).

As shown in Figure 3b, and presented previously [14], the Ag(111) electrode surface in contact with $HClO_4$ electrolyte is not reconstructed in the adsorbate-free potential range. The complete Pb monolayer prevailing between the negative limit of peak A3 and the onset of 3D Pb deposition is stable over several hours. It has a hexagonally close-packed structure with a higher-periodicity Moiré-pattern observed by *Müller et al.* [15] and interpreted in terms of the electronic or geometric superposition of an incommensurate Pb adlayer with the topmost substrate layer. A systematic study of the dependence of the periodicity of this Moiré superstructure on the electrode potential in the interval between peak A3 and the onset of 3D Pb deposition potential [16] has revealed a nearly linear decrease of the Pb-Pb nearest-neighbour distance in the Pb adlayer with decreasing undervoltage (i.e. the difference between the electrode potential in the adsorbate range and the equilibrium potential of a 3D Pb bulk phase). These results are in good agreement with an in-situ GIXS study [17]. On the monoatomic pits and on the stacked terraces the same hexagonally close-packed adlayer structure is also observed at the negative limit of peak A2, before the adsorption in peak A3.

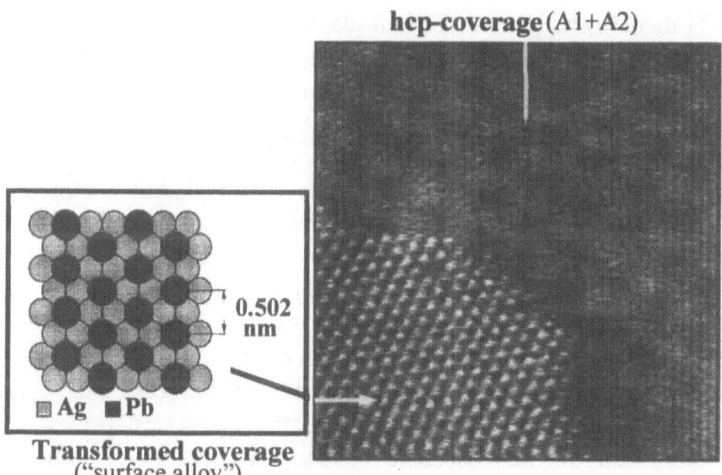

Figure 4: In-situ STM image of slow adsorbate-substrate rearrangement phenomena after adsorption of an incomplete Pb monolayer (peaks A1 + A2) on Ag(111) in 0.01M $HClO_4$ + 0.005M Pb^{2+}, and subsequent extended polarisation at constant potential between peaks A2 and A3. STM window size: 12x12 nm; grayscale range 0.07 nm. The upper right-hand part of the image shows the original hexagonally close-packed Pb monolayer, the lower left-hand part the rearranged "surface alloy" coverage with [√3 × √3]R30° - symmetry. The rearrangement has started near a monoatomic step, at the border to an adsorbate-free peripherical part on a stacked terrace. From [12] (slightly modified).

However, on the stacked terraces, the partial adsorbate coverage formed in peaks A1 + A2, with the observed adsorbate-free peripherical terrace boundaries, is not stable and undergoes interesting adsorbate-substrate rearrangement processes [12], if the electrode potential remains for extended time intervals (> ca. 600 s) in the potential range between peaks A2 and A3, or if it is cycled within the entire potential range of peaks (A1 + A2) / (D1 + D2). In this case, the originally formed hexagonally close-packed Pb monolayer is slowly converted into a "surface alloy" coverage with [$\sqrt{3} \times \sqrt{3}$]R30° - symmetry, where every third Ag atom of the topmost substrate layer is replaced by a Pb atom, and the excess Pb is desorbed into the electrolyte. An example of this "surface alloy" formation is shown in Figure 4.

Adsorption and desorption of Tl

In earlier voltammetric experiments it has been found that Tl underpotential deposition at chemically polished non-ideal Ag(111) substrates occurs in 2 distinctly separated potential intervals that have been associated with the consecutive formation of 2 Tl monolayers prior to Tl bulk deposition. As a remarkable feature, the voltammogram in the more anodic potential range (associated with the formation of the first monolayer) exhibits a very similar splitting into 3 distinct peaks A1/D1, A2/D2, A3/D3, as observed in the system Pb/Ag(111) (see Figure 3). These assumptions have been confirmed in an investigation by *Carnal et al.* [12] by the observation that a hexagonally-close packed Tl adlayer with slightly compressed Tl-Tl interatomic distances is formed at more anodic potentials, followed by the formation of a second hcp adlayer with slightly disordered domains at small undervoltages. Similarly to the system Pb/Ag(111), the formation of the first Tl adlayer at stacked terrace domains proceeds as follows:
- Peak A1: Decoration of the steps at a lateral width of ca. 1-3 nm.
- Peak A2: Formation of a hcp adlayer on the stacked terraces, except for the peripherical terrace boundaries that remain adsorbate-free over ca. 1-3 nm.
- Peak A3: Completion of the adsorbate coverage at the peripherical terrace boundaries.

The progress of Tl adsorbate formation in the monoatomic pits and at monoatomic islands has not been investigated yet.

A similar transformation of the original hcp adlayer to a surface alloy coverage with the same [$\sqrt{3} \times \sqrt{3}$]R30° - symmetry and Tl-Tl interatomic distances as shown in Figure 4 for the system Pb/Ag(111) has been observed in the system Tl/Ag(111) during extended polarisation of the incompletely formed first Tl adsorbate layer.

As in the system Pb/Ag(111), there is strong evidence that these adsorbate-substrate rearrangement phenomena proceed from the boundaries of the peripherical adsorbate-free domains on the stacked terraces . In contrast to the system Pb/Ag(111), however, the transformed coverages include also disordered domains, and their desorption is accompanied with the formation of monoatomic pits in the substrate with widths of ca. 3-10 nm. These pits diminish and finally vanish within the time scale of a few minutes by lateral displacement and coalescence at a rate that increases markedly upon positive shift of the electrode potential [12].

The presented results illustrate the relevance of nanometer-scale morphological heterogeneities at "non-ideal" electrode substrates not only for the local progress of adsorbate formation, but also for the long-time stability of the resulting adlayers. In the discussed model systems Pb/Ag(111) and Tl/Ag(111), the stepwise formation of the initial metal adlayers, combined with the observed slow formation of a "surface alloy", illustrate experimentally recently discussed thermodynamic and kinetic aspects of various growth modes of metal deposits [18]. In these 2 systems, the complete hcp monolayer coverage formed in the peaks A1+A2+A3 represents obviously a *metastable system*, whereas the rearranged "surface alloy" coverage resulting during extended polarisation of the incompletely formed adlayer can be considered as *thermodynamically stable coverage*. Further insight into the role of atomic-scale heterogeneities in the local progress of electrochemical reactions can be expected for example from the use of "custom-tailored" nanostructured model electrodes.

2.2   Local Probing of Conducting Polymers

In the field of chemically heterogeneous electrodes, the combined electrochemical and local probe investigation of conducting polymers has become an important technique for clarifying possible influences of electrolyte composition and polarisation dynamics upon the electropolymerization process, to investigate the dependence of the morphology of polymer films on the oxidation state of the polymer, and to study possible effects of morphological and electronic inhomogeneities in the polymer films upon their global electrochemical behaviour. Based on earlier STM work [19] demonstrating the application of in-situ STM for the investigation of the electrochemical growth and the morphology of the polymers in the polymerization electrolyte, more recent reports have also presented STM- and

SFM-based methods for measuring the thickness of polymer films in the electrolytic environment, and for investigating thickness changes resulting from electrochemical oxidation/reduction of the polymer [20].

As an example for the application of local probes to electropolymerized polymers, Figure 5 shows an STM probing result obtained in our group [21] at an electrochemically polymerized film of poly-hydroxy-phenazine (POPh). In difference to conducting polymers containing conjugate bonds with delocalized π-orbitals (e.g. polyaniline or polythiophene), the polymer chains of POPh films consist of monomer units bridged by ether bonds. This results in a conduction mechanism that is presumably controlled by electron hopping between fixed redox centers and is strongly affected by the potential-dependent oxidation state of the film. The dependence of the conductivity of the polymer bulk on its oxidation state in contact with the electrolyte can be studied with a potentiostatic STM assembly by recording tip current-distance-curves. Figure 5 shows such tip current-curves recorded at an electropolymerized POPh film of ca. 80 nm thickness in contact with 1M H_2SO_4: At positive potentials, in the fully oxidized state

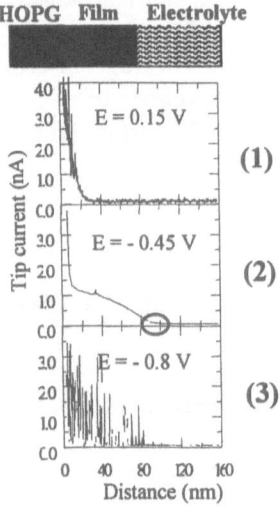

Figure 5: Tip current-distance-curves recorded with a potentiostatic STM assembly at an electropolymerized poly-hydroxy-phenazine film (ca. 80 nm thickness) in contact with 1M H_2SO_4. (1): fully oxidized film; (2): film at intermediate oxidation state; (3): fully reduced film. The schematic on top represents the position of the different phases HOPG (substrate), polymer film, and electrolyte solution, with regard to the tip position. The distance of the tip is defined versus the HOPG film substrate. From [21].

(curve 1), the tip penetrates into the film without any significant current flow, until a tunneling current is observed at a sufficiently small distance to the conductive HOPG film substrate. Obviously, in this case the film is non-conductive, and STM imaging of the interface polymer-electrolyte is impossible. At very negative potentials, in the fully reduced state (curve 3), the penetration of the tip into the polymer is accompanied with the onset of considerable current signals in the nA range. However, these current signals fluctuate without maintaining a stable level. Obviously, in this fully reduced state, the film cannot maintain a steady state charge transfer, expect at small distances to the HOPG substrate, where again tunneling transfer is observed. STM imaging of the film-electrolyte interface is equally impossible. A markedly different behaviour is observed, however, at intermediate potentials (curve 2), where the oxidation state of the polymer is characterized by the simultaneous presence of redox centers in the fully oxidized, fully reduced, and in an intermediate radicalic state. Upon penetration of the tip into the film, a stable current flow is maintained that indicates intrinsic film conductivity, presumably by electron hopping between the intermediate radicalic redox centers. At the tip position near the polymer-electrolyte interface, marked in curve 2 by a circle, stable and reproducible STM images can be obtained that resemble closely in-situ non-contact mode SFM images of the film surface. This suggests, that under these conditions tunneling may occur across an electrolytic barrier between the STM tip and the surface of the now conductive polymer.

Due to the observed changes in conductivity of the polymer at different oxidation state, it can be expected that the electrochemical film formation and growth is considerably affected by the polarisation routine applied during the electropolymerization of the film. This has indeed been observed.

3.   New Developments and Outlook

The increasing application of nanoscale probes in electrochemistry has promoted in the last few years a considerable effort for the development of *electrochemical or other electrolytic nanostructuring techniques* at the interface solid-electrolyte. A short survey on the main principles of such methods is given below and presented schematically in Figure 6:

308

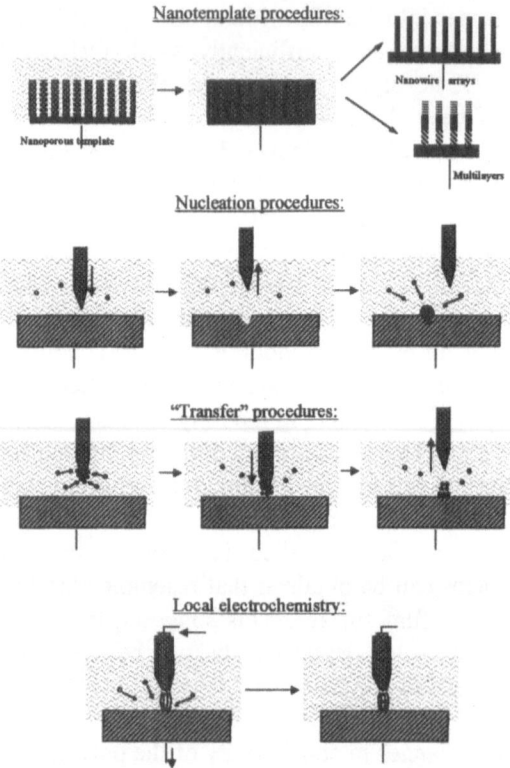

Figure 6: Basic principles for electrochemical or other electrolytic nanostructuring techniques

- In the case of *nanotemplate procedures*, conductive "nanowires" are deposited electrochemically into single or multiple electrolyte-filled nanopores of a template mounted on top of a conductive substrate. Even though the electrochemical growth and electrolytic transport mechanisms inside the nanopores is not fully clear yet, it has been possible to deposit not only single metal nanowires and nanowire arrays, but also nanowires consisting of multiple metallic components, and of conducting polymers [22]. Compared with other nanowire preparation methods, the electrochemical template technique is relatively simple. Possible applications of nanowire systems are focussed for example onto giant magnetoresistive devices and on nanosize electrochemical sensor systems.

- In the case of the *nucleation procedures*, conductive metal or polymer clusters are formed electrochemically on top of a conductive surface at nanoscale indentations induced previously by an STM or SFM probe and acting as nucleation centers. This technique has proved to produce cluster arrays with relatively uniform cluster sizes [23] and may become an attractive method for the production of nanoscale batteries and other cluster-based devices.

- Spatially defined arrangements of metal clusters have also been prepared recently by *transfer procedures*. In this case the clusters are first deposited electrochemically at an STM tip and are then transferred by potential-induced tip movement from the tip to the conductive substrate. The method allows an accurate tuning of the average cluster size, and in recent experiments [24] it has become possible to produce circular and rectangular densely packed multiple cluster arrays.

- Another very promising technique is based on concepts enabling the formation of nanosize features at an electrode surface by a highly *localized electrochemistry*. One principle is based on the use of an ultramicroelectrode (e.g. an STM tip) as current-carrying counter electrode in close vicinity to the conducting substrate, where nanostructuring is to take place. By using electrolyte solutions with sufficiently low conductivity, such a configuration results in a highly inhomogeneous electrochemical current density distribution, where the current flow is confined to the immediate domain between the current carrying ultramicroelectrode and the closest part of the electrode substrate. The method has already been successfully applied for the localized nanoscale etching of Ni surfaces [25], and a large variety of possible applications for both nanoscale phase formation and phase dissolution, as well as localized electrochemical intercalation processes can be anticipated.

Other interesting applications of electrolytic nanoscale probing methods can be anticipated, if the localized nanoscale probing of electrode surface domains can be combined with the simultaneous change or analysis of the electrolyte composition in the immediate electrolytic environment of the probed domain. First experiments with such an experimental concept have recently been started in our group with the design and test of electrolytic **Thin-layer Scanning Probe Microscopy** [26], whose principle is described in Figure 7: In this technique, the scanning probe consists of a microfabricated unit (Figure 7a) containing an STM tip that is surrounded, at lateral distance of ca. 20 µm, by an annular *Generator Electrode (or Sensor Electrode)*. If the probe is lowered under potentiostatic STM conditions onto the electrode substrate, until a tunneling mode ist established between the STM tip and the electrode substrate, it forms together with the electrode substrate a diffusionally shielded thin electrolyte layer ("thin layer"), whose thickness is defined by the

310

vertical protrusion of the etched STM tip and amounts to ca. 50 μm. This probe configuration is completed to a potentiostatic electrochemical setup by an additional reference and counter electrode placed in the electrolyte solution near electrode substrate and probe. The generator (or sensor) electrode potential E_G, the potential E_S of the electrode, and the potential of the STM tip can be controlled and varied independently by means of a tripotentiostatic control unit, and the currents flowing at the electrode substrate and at the generator (or sensor) electrode can be measured independently. The entire unit is integrated into an electrolytic STM cell. Depending on the choice and electrochemical control of the annular electrode, it can be designed either as an electrochemical generator, releasing a compound into the thin layer by a specific electrochemical reaction, or as a sensor electrode detecting specific compounds in the electrolyte that have been released at the electrode substrate. Due to the micrometer scale dimensions of both the thin layer thickness, as well as the radial distance between the annular electrode and the STM tip, the diffusional transport of the generated (or sensed) electrolytic compounds between the STM-probed domain of the electrode substrate and the generator (sensor) is expected to occur within time scales in the ms range.

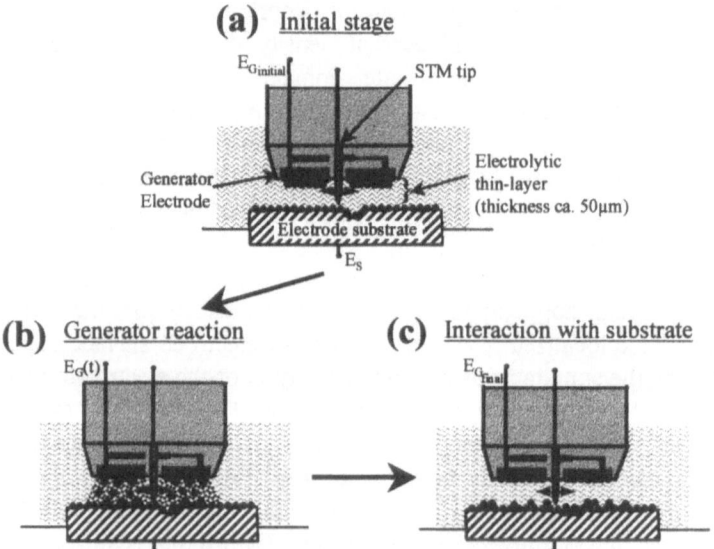

Figure 7: Schematic principle of Thin-Layer Scanning Probe Microscopy with a thin-layer generator electrode. Further explanations are given in the text

In the simplest operational concept, shown schematically in Figure 7, the annular electrode acts as a potential-controlled electrochemical generator of a specific electrolyte compound (e.g. a metal ion Me^{z+}), that is initially not present in the electrolyte, through a reversible faradaic reaction. In the *initial state* (Figure 7a), the generator potential is kept at a value $E_{G\ initial}$, where the corresponding equilibrium concentration of that specific compound is vanishingly low, and where the electrode substrate can be investigated locally by STM in absence of that compound. During the subsequent *generator reaction* (Figure 7b), a finite amount of the specific compound is generated electrochemically at the generator by an appropriate polarisation routine $E_G(t)$ (e.g. a pulse or cyclic polarisation), and will then diffuse within the thin layer to the electrode substrate, including the surface domain imaged by STM, where its interaction with the electrode substrate can afterwards be locally investigated (Figure 7c). During this entire sequence, the STM probe remains in the tunneling mode and enables the continuous monitoring of an electrode surface domain during the generation of the specific compound, and its possible interaction with the electrode substrate. A variety of possible applications can be anticipated, including the nanoscale generation and study of atomic-scale interactions within nonequilibrium mixed metal adlayers, the local change of pH by an electrochemical H^+-generator, or the characterisation of the onset of adsorbate- (e.g. Cl^--) induced corrosion reactions.

In a first experimental series, the principal operation concept of the thin-layer STM probe with an electrochemical generator has been tested with a microfabricated probe combining a polycrystalline Ag generator electrode with a Pt/Ir STM tip. This generator system can be used to inject electrochemically generated Ag^+-ions into the electrolyte. First experiments with this probe, shown in Figure 8, have been carried out at a chemically Ag(111) electrode substrate with an adsorbed Pb monolayer in an electrolyte solution of $0.01M\ HClO_4 + 0.005M\ Pb^{2+}$.

Figure 8a represents the first part of the experimental sequence: Image 1 shows a 100x100 nm surface area of the Ag(111) substrate at a potential E_S = 20 mV vs the Pb/Pb^{2+} equilibrium potential, where the Ag(111) substrate is covered by a complete hcp Pb monolayer. In this stage, the polycrystalline Ag generator is at a potential E_G = 350 mV vs. Pb/Pb^{2+}, where only a vanishingly low concentration of Ag^+ is present in the electrolyte. Afterwards, a finite amount of Ag^+ is injected electrochemically into the thin layer and deposited at the Pb adsorbate-covered Ag(111) electrode, by shifting the generator potential E_G linearly from 350 to 800 mV vs. Pb/Pb^{2+}, maintaining E_G for ca. 40 sec at 800 mV vs. Pb/Pb^{2+}, and then reversing E_G again to 350 mV vs. Pb/Pb^{2+}. The amount of Ag^+ produced at the generator during this sequence corresponds to a coverage of ca. 5 Ag monolayers (with respect to the entire surface area of the generator). On the imaged domain, which is not located

312

directly opposite the generator, it is estimated, however, that only ca. 1-2 monolayers of Ag are deposited during the injection experiment. Image 2 shows the same surface domain as in image 1 during Ag^+-injection, and during and after the reverse potential step, whereas image 3 shows the same surface domain ca. 50 sec after recording image 2. As seen in images 2 and 3, the injection of Ag^+ results in a noticeable change of the substrate morphology due to the deposition of the injected Ag^+. This change is caused by both, lateral terrace growth at the terrace edges, and by the formation of new islands on top of the terraces. Recently obtained additional results with atomic resolution suggest that, presumably, a relatively fast exchange occurs between the initially adsorbed Pb adlayer and the subsequently deposited Ag, whereafter the formation of a new Pb adsorbate layer can be expected.

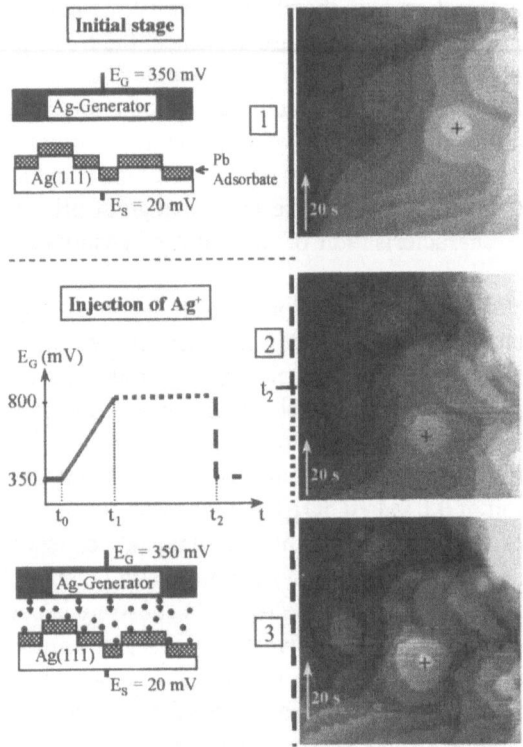

Figure 8a: Test experiment for thin-layer STM probe performance in the system Pb/Ag(111) with a thin-layer STM probe with Ag generator (part 1). Approximate thin-layer thickness: 40 μm. Electrolyte: 0.01M $HClO_4$ + 0.005M Pb^{2+}. STM scan size: 100x100 nm. The line patterns on the left-hand side of the STM images refer to the polarisation sequence diagram $E_G(t)$ during the injection of Ag^+. Further explanations are given in the text. From [26].

Figure 8b shows the subsequent behaviour of the system during and after an anodic linear sweep of the substrate potential Es from 20 mV vs. Pb/Pb^{2+} to 250 mV vs. Pb/Pb^{2+} in the adsorbate-free range. Starting with image 3, ca. 100 sec after the Ag deposition, the anodic potential shift is accompanied in images 4 and 5 by significant changes in the morphology of the substrate terraces, and even the formation of monoatomic pits. This behaviour contrast markedly to the simple desorption of a complete Pb monolayer, that would not be accompanied by changes in the terrace morphology. The terrace morphology resulting ca. 150 sec after the anodic potential shift of the substrate then remains stable for several hrs.

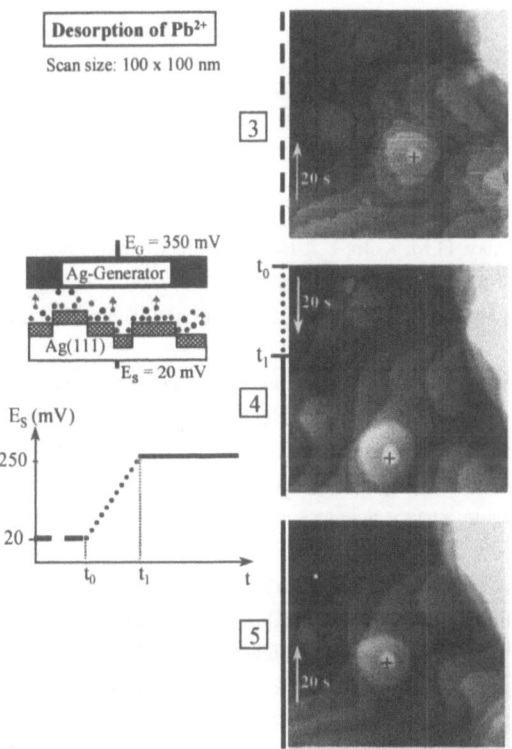

Figure 8b: Test experiment for thin-layer STM probe performance in the system Pb/Ag(111) with a thin-layer STM probe with Ag generator (part 2). Further explanations are given in the text. From [26].

The presented results in the 100x100 nm dimensional range give first clear evidence that the anticipated generator principle in the thin-layer STM probe works and enables the local observation of electrochemically induced changes in the electrolyte composition without loss of the stable STM imaging mode.

314

Another interesting combination of electrochemistry and nanoscience consists in methods for probing the *electron transfer behaviour of single fixed redox centers at the electrode/electrolyte interface*. First STM-based experimental concepts for such investigations have already been proposed [27].

Acknowledgements

The authors thank A. Daridon, P. Forrer, P.F. Indermühle, P. Häring, W.J. Lorenz, and F. Niederhauser for their cooperation and acknowledge gratefully financial support from the Schweiz. Nationalfonds

References

[1] R. Sonnenfeld, P.K. Hansma, Science **232**, 211 (1986).

[2] O. Lev, Fu-Ren Fan, A.J. Bard, J. Electrochem. Soc. **135**, 783 (1988).

[3] P. Lustenberger, H. Rohrer, R. Christoph, H. Siegenthaler, J. Electroanal. Chem. **243**, 451 (1988).

[4] J. Wiechers, T. Twomey, D.M. Kolb, R.J. Behm, J. Electroanal. Chem. **243**, 225 (1988).

[5] K. Itaya, E. Tomita, Surf. Sci. **201**, L507 (1988)

[6] O.M. Magnussen, J. Hotlos, R.J. Nichols, D.M. Kolb, R.J. Behm, Phys. Rev. Lett. **64**, 2929 (1990).

[7] S. Manne, P.K. Hansma, J. Massie, V.B. Elings, A.A. Gewirth, Science **251**, 183 (1991).

[8] A.J. Bard, H.D. Abruna, C.E.D. Chidsey, L.R. Faulkner, S.W. Feldberg, K. Itaya, M. Majda, O. Melroy, R.W. Murray, M.D. Porter, M.P. Soriaga, H.S. White, J. Phys. Chem. **97**, 7147 (1993).

[9] H. Siegenthaler, in "Scanning Tunneling Microscopy II" (R. Wiesendanger, H.-J. Güntherodt, Eds.), Springer Series in Surface Sciences, pp. 7-49 and 303-312, Springer-Verlag Heidelberg, 1995.

[10] "Nanoscale Probes of the Solid-Liquid Interface", H. Siegenthaler, A.A. Gewirth, Eds., NATO ASI Series E: Applied Sciences, Vol 288, Kluwer Academic Publishers, Dordrecht, 1995

[11] R. Guckenberger, T. Hartmann, W. Wiegräbe, W. Baumeister, in "Scanning Tunneling Microscopy II" (R. Wiesendanger, H.-J. Güntherodt, Eds.), Springer Series in Surface Sciences, pp. 51-98, Springer-Verlag Heidelberg, 1995.

[12] D. Carnal, P.I. Oden, U. Müller, E. Schmidt, H. Siegenthaler, Electrochim. Acta **40**, 1223 (1995).

[13] E. Ammann, P.I. Oden, H. Siegenthaler, in IUPAC Monograph "Local probe Techniques for Studies of Electrochemical Interfaces" (W.Plieth, W.J. Lorenz, Eds.), VCH, in press; E. Ammann, diploma thesis, University of Bern, 1993.

[14] W. Obretenow, M. Höpfner, W.J. Lorenz, E. Budevski, G. Staikov, H. Siegenthaler, Surf. Sci. **271**. 191 (1992).

[15] U. Müller, D. Carnal, H. Siegenthaler, E. Schmidt, W.J. Lorenz, W. Obretenow, U. Schmidt, G. Staikov, E. Budevski, Phys. Rev. B 46, 12899 (1992).

[16] U. Müller, D. Carnal, H. Siegenthaler, E. Schmidt, W.J. Lorenz, W. Obretenow, U. Schmidt, G. Staikov, E. Budevski, Phys. Rev. B 49, 7795 (1994).

[17] M.F. Toney, J.G. Gordon, G.L. Borges, O.R. Melroy, D. Yee, L.B. Sorensen, Phys. Rev. B 49, 7793 (1994).

[18] E. Budevski, G. Staikov, W.J. Lorenz, "Electrochemical Phase Formation and Growth – An Introduction to the Initial Stages of Metal Deposition", VCH, Weinheim, 1996.

[19] Y.T. Kim, H. Yang, A.J. Bard, J. Electrochem. Soc. 138, L71 (1991); R. Nyffenegger, C. Gerber, H. Siegenthaler, Synth. Metals 55-57, 402 (1993).

[20] H. Yang, Fu-Ren Fan, Sh.-L. Yau, A.J. Bard, J. Electrochem. Soc. 139, 2182 (1992); R. Nyffenegger, E. Ammann, H. Siegenthaler, R. Kötz, O. Haas, Electrochim. Acta 40, 1411 (1995).

[21] P. Forrer, G. Repphun, E. Schmidt, H. Siegenthaler, in "Solid-Liquid Electrochemical Interfaces" (G. Jerkiewicz, M.P. Soriaga, K. Uosaki, A. Wieckowski, Eds.), pp. 210-235, ACS Symposium Series, 1997; G. Repphun, Ph.D. thesis, University of Bern, 1997

[22] See for example Ch. Schönenberger, B.M.I. van der Zande, L.G. J. Fokkink, M. Henny, C. Schmid, M. Krüger, A. Bachtold, R. Huber, H. Birk, U. Staufer, J. Phys. Chem. B 101, 5597 (1997).

[23] See for example W. Li, G.S. Hsiao, D. Harris, R.M. Nyffenegger, J.A. Virtanen, R.M. Penner, J. Phys. Chem. 100, 20103 (1996).

[24] See for example D.M. Kolb, R. Ullmann, T. Will, Science 275, 1097 (1997).

[25] See Proc. of the Swiss-Japanese Workshop on Nanosciences, Ascona, 1996.

[26] E. Ammann, P. Häring, P.-F. Indermühle, R. Kötz, N.F. de Rooij, H. Siegenthaler, Proc. Of the 1997 Joint ECS/ISE International Meeting, Paris, 1997.

[27] W. Schmickler, C. Widrig, J. Electroanal. Chem. 336, 213 (1992).

THE FIRST 130 YEARS OF ELECTRON MICROSCOPY
Focusing, imaging, spectroscopy and lithography

A. Howie
University of Cambridge
Cavendish Laboratory, Cambridge CB3 0HE, UK.

1. Introduction

In the one hundred years which have elapsed since its discovery by J.J. Thomson, the electron has enjoyed exponential growth as an object for basic study, as a vital tool of science and engineering and as a potent agent for sociological change. It retains its status as the first elementary particle of physics and is the most efficient agent in the coupling of radiation to matter. The electron provides the basis for atomic structure, chemical bonding, photosynthesis, electronics, communication and computing. In this vast landscape, electron microscopy and the related topics included in the title of this article cover a terrain of comparatively limited extent although more appropriate for review in the present context. As is true for many other aspects of electron beam and current phenomena however, the exploration of the territory began long before 1897.

2. The propagation and focusing of cathode rays

The study of the conduction of electricity through low-pressure gas discharge tubes goes back to Faraday but gathered momentum when Geissler invented his mercury piston vacuum pump in 1860. The appearance in the glowing discharge of straight lines which seemed to emanate from the cathode led to the concept of cathode rays. In some striking observations in 1869, Hittorf [1] sketched the paths of the rays when the pole of an electro-magnet was brought up to the glass wall of the discharge tube. His drawings (see fig. 1) clearly show the bending of the ray path into a spiral about the magnetic field lines and the tendency for the rays to be concentrated or focused into a tighter spiral as they entered a region of increasing field strength. Hittorf even mentioned that, as the current in the magnet windings

N. García et al. (eds.), Nanoscale Science and Technology, 317–332.

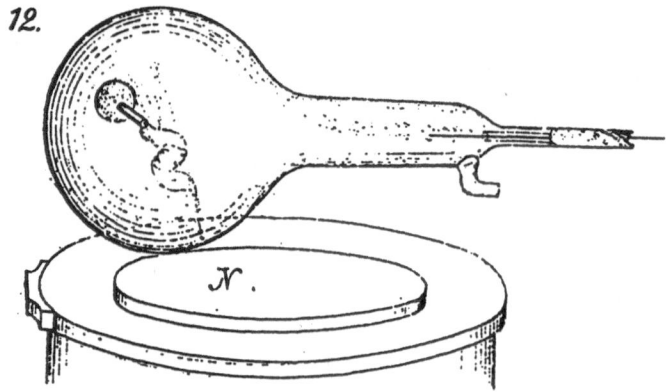

Figure1. Hittorf's drawing of cathode rays focused near a magnet [1].

was increased, the fluorescent circle produced by the rays shrunk almost to its centre point, becoming strongly luminous and heated and that, at still stronger currents, the glass would melt on the surface and partially volatilize, leading to the eventual destruction of the equipment. Although nowadays we would prefer to think of this as an ionisation damage rather than as a simple heating process, Hittorf should be credited with the first observations of e-beam lithography.

The ability of the cathode rays emitted normally from the planar cathode to cast on the wall of the tube sharp shadow images of objects placed in their path was first noticed by Goldstein and exploited by Crookes [2]. Using a more pointed cathode, Crookes could even obtain slightly magnified images of a star or Maltese cross and may thus perhaps be regarded as the first projection electron microscopist. Today's projection microscopes [3,4] use electrons of about the same energy but operate with atomically sharp field emitters and very much smaller specimen working distances to achieve a magnification of a million or more.

Although these early observations must have encouraged the idea that the cathode rays might be composed of charged particles, this perception was thrown into doubt by the experiments of Hertz and others who firstly failed to find any deflection of the rays when a transverse electric field was applied and secondly found that the rays could actually pass through very thin metal foils. Many German physicists then began to believe that the rays were some form of electromagnetic radiation.

3. J.J. Thomson and the discovery of the electron

Viewed as the initiating event in the momentous developments which have taken place in the subsequent century, the discovery of the electron and the discoverer have both been analysed extensively [5-10], particularly in recent months. Here we merely summarise some key conclusions.

J.J. Thomson's experiments, begun soon after his astonishing appointment as Cavendish Professor in 1884 when he was only 28 years old, were part of the long series of investigations by many workers into the conduction of electricity in low pressure gas discharge tubes. The discovery of X-rays in 1895 brought fresh impetus to the field as well as a more benign means of initiating the discharge. Perrin's (1895) experiment of magnetically deflecting the rays into a Faraday cup indicated that they were negatively charged. Thomson then realised that the failure of Hertz to observe any deflection of the cathode rays in an electric field was possibly due to the screening effect of the residual gas ions in the tube and set about repeating the experiments in a much better vacuum. From the deflections which he then obtained in both electric and magnetic fields he deduced a charge to mass ratio e/m for the "corpuscles" about 2000 times larger in magnitude than the value for a hydrogen ion. Similar, and indeed more accurate, values of e/m were actually measured by Wiechert and by Kaufmann a few months before Thomson but he went considerably beyond these two workers in appreciating the fact that he had found something fundamentally new and, by showing that the value of e/m was independent of the gas in the tube or of the material of the cathode used, that it must be a universal constituent of all matter.

Figure 2. J.J. Thomson in 1909 (Courtesy Cavendish Laboratory).

J.J. Thomson was an accomplished mathematical physicist as well as a remarkably insightful, if personally clumsy, experimenter. It would be wrong to regard Maxwell and Rayleigh as "loners" but, compared with these somewhat patrician predecessors in the Cavendish chair, Thomson had a more pronounced talent for running and inspiring a research team. From 1895 onwards, when the University first admitted research students, he was blessed with a remarkable array of junior colleagues. No less than eight Nobel prizewinners (E.V. Appleton, F.W. Aston, C.G. Barkla, W.L. Bragg, O.W. Richardson, E. Rutherford, G.P. Thomson and C.T.R. Wilson) were trained in Thomson's Cavendish between 1895 and 1919 when he resigned to become Master of Trinity College.

The word electron, coined originally in 1891 by Johnstone Stoney to describe the smallest (ionic) unit of electric charge involved in electrolysis, was applied almost immediately by other scientists to Thomson's corpuscle, but Thomson himself did not adopt the universal usage until almost twenty years later. Ten years later still, he had to admit a rather more fundamental inadequacy in the corpuscle concept when, independently of Davisson and Germer, his own son G.P. Thomson made the first observations of electron diffraction.

4. Transmission electron microscopy and atomic imaging

4.1 THE LONG HAUL TO ATOMIC RESOLUTION

The challenge of improving the performance of high voltage oscillographs, rather than any idea of the electron as a short wavelength particle with potential in microscopy, was what drove Gabor and Busch in the development of the iron-shrouded, cylindrical magnetic lens [11]. This lens however became the basis of Ruska's first transmission electron microscope [12] and indeed the key component in electron microscopy for the ensuing forty years which were required to achieve atomic resolution imaging. Fortunately for morale, there were several significant milestones on this long journey such as the diffraction contrast imaging of defects in thin crystals [13] (which sidesteps the need to image the atoms directly) as well as a progressive series of inorganic crystals with large unit cells in addition to many stained biological structures which could be usefully studied at better and better resolution. Incidentally, the diffraction contrast theory and the associated picture of the channeling of electron waves through crystals produced the first satisfactory explanation of the puzzling ability of cathode rays to penetrate thin films which was mentioned above.

High resolution electron microscopy (HREM) transmission images [14] are in fact a kind of in-line hologram, dominated by small-angle coherent, Bragg diffraction and phase contrast effects which depend sensitively on crystal thickness and defocus effects. They can be obtained and interpreted almost routinely, at least in two-dimensional structures such as thin crystal films where the incident electron beam can be aligned with a prominent crystal axis as well as with any linear or planar defects or interfaces of interest. Using ion milling techniques, it is often possible to produce thin slices meeting these requirements from more complex, bulk samples. Although the atomic-scale information which the images provide is mainly structural rather than chemical, since the atomic scattering in the small-angle range varies only as $Z^{1/3}$, small regions of different phases can often be identified from their interplanar spacings. The towering column of lenses with which these extremely useful images are obtained still embodies the original design of Ruska [12] and is a remarkable endorsement of his vision. Although there is now little obvious scope for significant improvements in the performance of the cylindrical lens itself, there has not been any perceptible slackening in the search for higher resolution. Super resolution *i.e.* resolution substantially better than the usual figure determined by diffraction and lens aberrations, has been demonstrated by electron holography [15] as well as by Fourier processing of image intensities due to electrons scattered at higher angles [16] but, in both cases, so far carried out only in periodic structures. An alternative approach, using octupole lens combinations to cancel spherical aberration, has now been made much more practical by the availability of sophisticated computer control of the many interacting lens current supplies. Preliminary results from a pilot project [17] look promising.

For the much more general three-dimensional specimen structures, the simple projection image becomes confused by overlap effects and is harder to interpret. Sometimes, as in the cases of nanotubes, or the multiply twinned structures adopted by small atomic clusters, the main features of the geometry can still be sorted out although it may be impossible fully to characterise internal strains or defects. For other problems, such as the core structure of dislocations which cannot be aligned with the beam or the medium range structure of "amorphous" materials, conventional transmission atomic imaging is much less useful and does not add anything to the low spatial resolution (>1.5nm) or even spatially averaged results of more traditional diffraction methods. Although the continuing drive towards improved instrumental resolution seems unlikely to be directly relevant in addressing many of the outstanding challenges which still face the electron microscopist, on past history it could well be extremely useful in unforeseen ways.

4.2 UNEXPECTED SPIN-OFFS

It must be admitted that the cylindrical lens, not readily compatible with good vacuum conditions, and depending on a second-order effect to achieve focusing inherently linked to significant spherical aberration, has presented a serious bar to progress in electron microscopy. Had these problems been less severe, the field would clearly advanced more rapidly. Had the problems been worse, progress might still have been faster since alternative solutions would have been pursued earlier and more energetically. As it is, the drive for higher resolution in electron optics presents no less than four magnificent examples of how the desperate hunt for remedies can throw up the most radical ideas capable of addressing problems much wider or even totally different than the original goal.

First of all the principle of scanned imaging [11] vastly increased the number of different image signals available to the electron microscopist and allowed bulk samples to be used. For much of this work, the small numerical apertures, to which aberrations restricted the round magnetic lens, were a positive advantage since they brought enormous depth of field, allowing comparatively large objects to be imaged as a whole in sharp focus.

Secondly, Muller's field emission tip [18] has not only become an electron source of unparalleled brightness but has also provided the basis of the field ion microscope (FIM) and the first atomic resolution images of surfaces without the aid of lenses. The atom probe is a still later modification of the FIM where tips of different materials can be used and, by application of successive voltage pulses, sequential layers of atoms can be removed from its surface for analysis in a mass spectrometer. Although it is completely destructive and restricted to the special geometry of a tip, this system represents the ultimate in three-dimensional mapping of atomic arrangements in a material.

The family of scanned probe microscopes, particularly the STM [19], is in a sense a fusion of Muller's tip with the scanning principle but is such a novel advance in the three hundred year-old field of microscopy that it deserves to be our third example of radical serendipity. New ideas in such a mature field cannot be expected to arise from any obvious quarter so it is perhaps not surprising that one of the best pre-existing uses of a scanning probe should by someone for whom microscopy would seem to have no relevance - the blind person sweeping and tapping the ground ahead with his stick.

The final and perhaps most spectacular example of spin-off from the travails of electron optical imaging is the invention of the holography principle by Gabor [20]. Even at its inception it was clearly easier to demonstrate holography with light waves rather than with electron waves and, with the advent of the laser, holography flowered so magnificently as an optical technique that most scientists are unaware of its origin. In more recent years, with the availability of highly coherent, field emission sources, holography has come back to electron optics. As remarked above, the conventional high resolution imaging process is a form of on-axis holography whereby the lens collects and deflects the various diffracted beams back to the axis to interfere with the strong, unscattered beam. Off-axis holography, using an electron biprism invented by Mollenstedt, has been pioneered by Pozzi, by Lichte and by Tonomura and is now available as an optional facility on commercial electron microscopes. So far it has found more application [21] in quantitative and sensitive phase contrast microscopy at moderate resolution such as the imaging of flux lines in superconductors than in producing higher resolution in non-periodic structures than can currently be achieved by various alternative methods.

5. Scanning transmission electron microscopy

5.1 "Z CONTRAST" IMAGING

The usefulness of Muller's high brightness, field emission source for practical atomic resolution scanning transmission electron microscopy (STEM), requiring reasonable scan times for relatively noise-free images, was first spectacularly demonstrated by Crewe and co-workers [22]. In principle a large number of image signals can be collected simultaneously with high spatial resolution in terms of the position of the focused incident probe. These include fast electrons scattered at both small and large angles, electrons of known energy loss, slow secondary electrons, X-rays, light. In particular, the fast electron signal collected by the annular dark field (ADF) detector, covering scattering angles in the range of 60 mrad to 300 mrad, depends on Rutherford scattering from the atoms and is much more sensitive to atomic number than the signal at lower angles. The usefulness of the ADF signal, particularly in imaging heavy atoms either individually or as small clusters such as catalyst particles on disordered, light atom supports, has been clearly demonstrated [23, 24]. More recently STEM ADF imaging has begun to present formidable competition to the conventional transmission electron microscope in atomic imaging of crystals and two-dimensional defect structures [25]. It has been shown [26] that the ADF images are essentially dependent on incoherent scattering and consequently rather simpler to

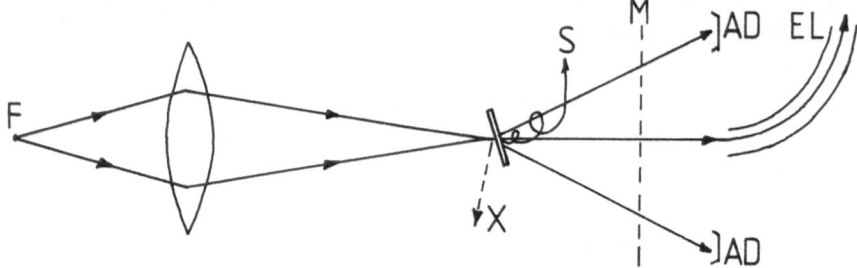

Figure 3. Schematic STEM operation with energy loss, secondary emission, annular dark field, microdiffraction and X-ray modes.

interpret than HREM images since they do not exhibit the same complex, phase contrast effects and sensitivity to specimen thickness and defocus. The ADF images also carry much more chemical contrast than is found in the traditional HREM images.

5.2 SPATIALLY LOCALISED ENERGY LOSS SPECTROSCOPY

As indicated in fig. 3, an energy loss spectrometer working in the small angle (< 15 mrad) scattering region is readily compatible with STEM. This facility was originally exploited not only for nanoanalysis by using the inner-shell excitation edges characteristic of individual atoms, but also, via the low loss, valence excitations, as a means of normalising the ADF signal to local variations in specimen thickness in order to achieve more accurate Z contrast imaging. Electron energy loss spectroscopy (EELS) has developed enormously in more recent years, particularly with the development of parallel recording systems allowing the whole loss spectrum to be collected simultaneously rather in a serial fashion.

Since the excitation processes depend on a Coulomb rather than on a contact interaction, the spatial resolution available in EELS may not be as good as the size of the focused probe might indicate. Using an old, impact parameter - excitation time argument due to Bohr, the typical interaction distance for an energy loss ΔE is $hv / \Delta E$ where v is the fast electron velocity. For 100 keV electrons, characteristic core loss excitations (150 eV < ΔE < 2000 eV) have typical interaction distances of below 1nm whereas valence losses are less highly localised. Core loss EELS has thus become a very powerful tool for chemical analysis on the atomic scale, particularly when applied to essentially two-dimensional structures in conjunction with ADF imaging which allows individual columns of atoms to be imaged and identified. Additional information about the local electronic structure can also be extracted from the shape of the loss spectra just above the absorption edge (the near edge region) since this is controlled by the density of empty electron states into which the excited core electron can be ejected. Recent applications of this approach have been to the silicon-silica interface [27] and to grain boundary embrittlement problems [28].

The much more intense, valence excitation ($0 < \Delta E < 50eV$) part of the loss spectrum has so far been less thoroughly exploited because it is less easily interpreted. Collective excitations such as plasmons as well as single electron excitations are a feature of this spectral region and in simple geometries such as thin slabs, planar interfaces and spheres the results have been successfully interpreted using a version of Fermi's dielectric excitation theory adapted for inhomogeneous samples (for references see [29]).

6. Electron beam lithography

The profoundly damaging capability of an intense electron beam is often a serious problem in electron microscopy but can be turned to advantage in the context of lithography which opens up exciting opportunities for the microscopist to play a less passive, more interventionist role in relation to the sample. Contrary to Hittorf's conjecture quoted above, electron beam heating is only rarely a significant cause of specimen damage in electron microscopy. At electron beam energies exceeding about 300 keV, atomic displacement damage can be apparent, even in metals, following high-angle scattering events associated with sufficiently energetic recoil of the struck atom. Much more important however, particularly in many organic materials and insulators, is atomic displacement following electron excitation i.e. ionisation damage. In many of the most sensitive materials, including many e-beam resists, damage can result from valence excitation which can be initiated by quite low energy electrons such as the secondary electrons generated in a cascade around the primary beam. As noted above, the fast primary electron can also directly induce such low energy excitations at some distance from its track. These proximity effects are a serious limitation in the context of e-beam lithography at high spatial resolution < 10nm.

In aromatic organic crystals [30] as well as in a number of insulators, the valence excitation process seems to be relatively ineffective in iniating damage compared with the far rarer core excitations where the energy and momentum transfer requirements mean that proximity effects are much less serious. The greater damage efficiency of the core excitation event is probably due to the subsequent de-excitation process which often leaves, via

326

an Auger transition, a double hole in the valence system. In insulators, this Knotek-Fiebelman mechanism has been studied in considerable detail in broad beam experiments with mass spectrometers to detect the desorbed atoms [31]. In the context of STEM however, the effects are rather more dramatic since the focused beam can sometimes drill extremely fine, 2nm diameter holes through a film of 100nm thickness [32,33]. The rate of hole drilling is a highly non-linear function of the beam current which suggests the presence of other processes favouring strong localisation and high writing resolution [34]. In favourable cases, pages of newsprint have been transcribed in a few seconds [35] and diagrams such as fig ? can be reproduced very easily. Although the spatial resolution is not quite at the atomic level achieved in STM atomic manipulation experiments, the speed of writing is considerably greater.

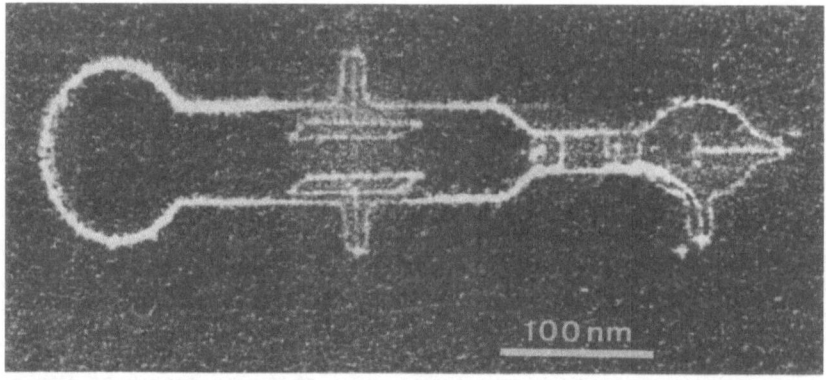

Figure 4. Nanoscale e-beam lithograph of J.J. Thomson's aparatus (created by S. Granleese and Y. Itoh, Cavendish Laboratory).

There is almost certainly an exciting future for e-beam lithography in the fabrication of one-off specialist devices. Whether it can play a significant role in the large-scale manufacture of devices with feature size too small to be made by the current optical lithography methods is much more problematical since enormously faster writing procedures will be needed to get the necessary throughput. Probably the best prospect at present is the SCALPEL system of Berger and Gibson [36] which uses parallel electron beam illumination of a thin mask as in transmission electron microscopy, with an angular filter in the back focal plane to boost the contrast to the required value for satisfactory copying. This system would presumably employ photoresists and is currently under active development. The hole drilling phenomenon has greater potential for high spatial resolution but will require large arrays of field emission tips operating in parallel if the speed is to be increased even part of the way to the level which could ultimately be needed.

7. Electron microscopy of electron behaviour

Electron microscopy provides most direct information about the atomic structure of materials but it is the electronic structure which is often of primary interest and in any case usually determines the atomic structure adopted.   Once known, the atomic structure can increasingly provide the basis for some understanding of electronic behaviour or even for a reliable computation of the electronic structure.   For example the opto-electronic properties of porous silicon appear to be largely understood on the basis of quantum confinement effects in the nanocrystals of Si which it has been shown to contain [37].   On a more complicated level, there is now a reasonably good picture [38] of the details of the photosynthesis process thanks to a whole range of optical and spectroscopic experiments in conjunction with detailed structure determinations by X-ray crystallography and supporting electron microscopy work.   In other cases however the situation is still confused.   More work remains to be done for instance to reconcile the electron microscopist's view of small (supported) clusters with optimal forms such as icosahedra, or decahedra with the cluster beam physicist's picture that they are quantum wells of quasi-spherical shape which can be distorted to accomodate the valence electrons in a minimal energy configuration [39].   These two approaches yield quite different magic numbers for cluster sizes of exceptional stability and a special interest could attach to their reconciliation since the particles concerned may span the region dividing quantum objects from macroscopic classical objects.

More direct information about the spatial variations in valence electron density can sometimes be obtained from imaging [40], or from accurately measured electron diffraction intensities, whilst (e,2e) spectroscopy [41] (a form of Compton scattering) provides both spatial and momentum data. Both methods are limited to thin single crystal regions of area exceeding about 10nm x 10nm however and the latter technique requires rather long exposure times. Electron scattering is sensitive to the presence of unpaired spins, because of the magnetic fields they generate. If some spatial resolution can be sacrificed by working outside the strong field of the objective lens, the deflection of the fast electron beam can then be used to image the structure and dynamics of domains in ferromagnets and of flux lines in superconductors [42].

As remarked above, core loss EELS gives information about the empty states just above the Fermi level via the electron energy loss near-edge structure (ELNES). Compared with XANES, the related technique in X-ray absorption, ELNES can be applied with close to atomic spatial resolution. In both techniques one should take account of the polarisation of the empty states by the core hole and, in the ELNES case, of the partially compensating effect of the passing fast electron [43]. In a relatively simple case of Si-Ge alloy multilayers, the band gap off-sets were successfully measured from the local Si core L-edge shifts [44].

Valence EELS can be used to measure the local, frequency-dependent, dielectric response function which describes many of the electronic response properties of interest in different contexts and depends on occupied as well as empty states. Although the dielectric function cannot be inverted in general to give complete electronic structure data,, useful quantities like the band gap can be measured fairly directly in local regions or small crystals, provided care is taken to subtract the effect of the strong zero-loss peak [45]. Dielectric data can be processed to yield the joint density of states function which can in turn be modelled in a critical point analysis of the type appropriate to crystals. Proceeding in this way, Mullejans and French have been able to conclude [46] that the bonding at a $\Sigma 11$ boundary in α-alumina is somewhat more ionic than in the bulk.

A number of SEM or STEM imaging modes depend on the transport, trapping or escape of excited electrons in the conduction band. Thus secondary electron (SE) imaging, until recently a very poor relation of photoelectron spectroscopy and used mainly as a source of topographic information, is begining, with improved sample surface cleanliness, to exhibit more clearly the expected dependence on band structure and surface barrier effects [47,48]. It is now being applied systematically to imaging on the 10nm scale, of the differently doped regions in semiconductor structures [49]. Electron traps and recombination centres in semiconductors can be imaged by electron beam induced conductivity (EBIC) mode or by cathodoluminescence. In the latter case, additional spectral information about the trapped level near an individual dislocation could be obtained [50].

The photon has many advantages over the electron as a probe of electronic structure. Its coupling is dominated by its interaction with the electrons of the sample and its capacity for spectroscopy with high energy resolution or femtosecond timing is unmatched. Despite the development of near-field microscopy, the electron however still offers the sometimes crucial advantage of structural sensitivity and high spatial resolution. A pioneering experiment

[51] has been conducted in picosecond pulsed electron diffraction of transient structural phenomena in a gas initially excited by a femtosecond laser pulse which was also used to trigger the electron pulse with a controlled delay. In this broad beam experiment however, each pulse could contain thousands of electrons. In the typical focused probe in the STEM, which would be needed if high spatial resolution were required, the average time interval between successive electron arrivals is about 100 picoseconds so that even if the timing were adequate, the statistics would be very poor. It remains however an extremely attractive proposition to devise some method of using focused electrons and photons together to combine the advantages of each. In principle for instance it would be possible to illuminate the sample in the STEM with a tuned laser beam and observe the effect on the EBIC or cathodoluminescence signal associated with particular traps as the laser frequency is adjusted either to fill or empty the traps. It might even be possible at very high levels of laser illumination to detect energy gain in EELS when an electron picks up energy from an excitation generated by the laser. In the meantime it must remain a challenge for the electron microscopist to extend to the subsurface regions the success in providing at high spatial resolution the electronic structure information that the STM achieves in the surface region.

8. References

1. Hittorf, W. (1869) Uber die elektrizitatsleitung der gase II, *Ann Phys. u. Chem.* **136**, 197-234.
2. Crookes W. (1879) On the illumination of lines of molecular pressure and the trajectory of molecules, *Phil. Trans. Roy. Soc.* **170**, 135-164.
3. Fink, H.W., Stocker, W. and Schmidt, H. (1990) Holography with low energy electrons, *Phys. Rev. Lett..* **65**, 1204-1206.
4. Binh, V.T., Semet, V. and Garcia, N. (1995) Nanometre observations at low energy by Fresnel projection microscopy, *Ultramicrosc.* **58**, 307-317.
5. Pais, A. (1986) *Inward Bound*, Oxf. Univ. Press, New York.
6. Pippard, A.B. (1997) J.J. Thomson and the discovery of the electron, in M. Springford (ed.) *Electron*, Camb. Univ. Press, Cambridge, pp. 1-24.
7. Squires, G.L. (1997) J.J. Thomson and the discovery of the electron, *Physics World* **10** (4) 33-36.
8. Weinberg, S. (1997) The first elementary particle, *Nature* **386**, 213-215.
9. Lord Rayleigh (1942)*The Life of Sir J.J. Thomson*, Camb. Univ. Press, Cambridge.

10. Davis, E.A. and Falconer, I.J. (1997) *J.J. Thomson and the discovery of the electron,* Taylor and Francis Ltd., London.

11. Hawkes, P.W., (1985) (ed) The beginnings of electron microscopy, *Adv. in electronics and electron physics, suppl.* **16,** Academic press, London.

12. Ruska, E. (1934) Progress in the construction and performance of the magnetic electron microscope, *Z. Phys.,* **87,** 580-602.

13. Hirsch, P.B., Howie, A., Nicholson, R.B., Pashley, D.W. and Whelan, M.J. (1977) *Electron microscopy of thin crystals,* Krieger, New York.

14. Buseck,P., Cowley, J.M.and Eyring, L. (1988*) High resolution electron microscopy,* Oxford Univ. Press, New York

15. Orchowski, A., Rau, W.D. and Lichte, H. (1995) Electron holography surmounts the resolution limit of electron microscopy, *Phys. Rev. Lett.* **74,** 399-402.

16. Nellist, P.D., McCallum, B.C. and Rodenburg, J. M. (1995) Resolution beyond the 'information limit' in electron microscopy, *Nature* **374,** 630-632.

17. Krivanek, O.L., Delby, N., Spence, A.J., Camps, R.A. and Brown, L.M. (1997) Aberration correction in the STEM in J. Rodenburg (ed*) Electron microscopy and analysis 1997* Inst of Phys., Bristol (in press)

18. Muller, E.W. (1937) Beobachtungen uber die felde emission, *Z. Phys.* **106,** 132-140.

19. Binnig, G., and Rohrer, H. (1983). Scanning tunneling microscopy, *Surface Sci.* **126,** 236-244.

20. Gabor, D. (1948) A new microscope principle, *Nature* **161,** 777-778.

21. Tonomura, A. (1993) *Electron holography,* Springer, Berlin.

22. Crewe, A.V., Langmore, J.P. and Isaacson, M.S. (1975) Resolution and contrast in the scanning transmission electron microscope, in B. Siegel and D. Beaman (eds.), *Physical aspects of electron microscopy and microbeam analysis,* Wiley, New York, pp 47-62

23. Treacy, M.M.J., Howie, A. and Wilson, C.J. (1978) Z contrast of platinum and palladium catalysts, *Phil. Mag.* A38, 569-585.

24. Treacy, M.M.J. and Rice, S.B. (1989) Catalyst particle sizes from Rutherford scattered intensities, *J. Microsc.* **156,** 211-234.

25. Pennycook, S.J. and Boatner, L.A. (1988) Chemically sensitive structure imaging with a scanning transmission electron microscope, *Nature* **336,** 565-567.

26. Jesson, D.E. and Pennycook, S.J. Incoherent imaging of crystals using thermally scattered electrons, *Proc. Roy. Soc A* **449,** 273-293.

27. Batson, P.E. (1993) Simultaneous STEM imaging and electron energy loss spectroscopy with atomic column sensitivity *Nature* **366,** 727-28.

28. Muller, D.A., Subramanian, S., Batson, P.E., Sass, S.L. and Silcox, J. (1995) Near atomic scale studies of electronic structure at grain boundaries in NiAl, *Phys. Rev. Letts.* **75**, 2744-47.

29. Walls, M.G. and Howie, A. (1989) Dielectric theory of localised valence energy loss spectroscopy, *Ultramicrosc.* **28**, 40-42.

30. Howie, A., Rocca, F.J. and Valdre, U. (1985) Electron beam radiation damage processes in p-terphenyl, *Phil. Mag.B* **52**, 751-757.

31. Miura, K., Suguira, K. and Siguira, S. (1991) F$^+$ - desorption mechanism from a CaF$_2$ (111) surface by low energy electron irradiation, *Surf. Sci. Letts.* **253**, L407-L410.

32. Broers, A.N. (1988) Limits of thin film microfabrication, *Proc. Roy. Soc. A* **416**, 1-42.

33. Berger, S.D., Salisbury, I.G., Milne, R.H., Imeson, D. and Humphreys, C.J. (1987) Electron energy-loss spectroscopy studies of nanometre-scale structures in alumina produced by intense electron beam irradiation, *Phil. Mag. B.* **55**, 341-358.

34. Berger, S.D., Macauley, J.M., Brown, L.M. and Allen, R.M. (1989) Hig current density electron beam induced desorption, *Mat. Res. Symp. Proc.* **129**, 515-520.

35. Berger, S.D., McMullan, D., Macauley, J.M. and Brown, L.M. (1987) A high density storage system based on electron beam writing, in L.M. Brown (ed.) *Electron microscopy and analysis 1987,* Inst of Physics, Bristol. pp 93-96.

36. Berger, S.D. and Gibson, J.M. (1990) New approach to projection electron lithography, *Appl. Phys. Letts.* **57**, 153-155.

37. Cullis, A.G., Canham, L.T. and Calcott, P.D.J. (1997) The structure and luminescence properties of porous silicon, *J. Appl. Phys.* **82**, 909-965.

38. Hoff, A.F. and Diesenhofer, J. (1997) Photophysics of photosynthesis, *Physics Rep.* **287**, 1-248.

39. Howie, A. (1991) Blazing the trail from metals to nuclei, *Faraday Discuss.* **92**, 1-11.

40. Zhu. Y. and Tafto J. (1996) Direct imaging of charge modulation, *Phys. Rev. Letts.* **76**, 443-446.

41. Vos,M., Fang, Z., Canney, S., Kheifets, A. and McCarthy, I.E. (1997) Energy-momentum density of graphite by (e,2e) spectroscopy, *Phys. Rev. B* **56**, 963-966.

42. Harada, K., Kasai, H., Matsuda, T., Tonomura, A. and Moshchalkov, V. V. (1996) Direct observation of vortex dynamics in superconducting films with regular arrays of defects, *Science* **274**, 1167-1170.

43. Batson, P.E. (1993) Distortion of the core exciton by the swift electron and plasmon wake in spatially resolved electron energy loss spectroscopy, *Phys. Rev. B* **47**, 6898-6910.

332

44. Morar, J.F., Batson, P.E. and Tersoff, J. (1993) Heterojunction band lineups in Si-Ge alloys using spatially resolved electron-energy-loss spectroscopy, *Phys. Rev.B* **47**, 4107-4110.

45. Rafferty, B. and Brown, L.M. (1997) EELS in the vicinity of the fundamental band gap, in J. Rodenburg (ed.) *Electron microscopy and analysis 1997* Inst. of Physics, Bristol (in press).

46. Mullejans, H. and French, R.H. (1996) Interband electronic structure of a near-$\Sigma 11$ grain boundary in α-alumina determined by spatially resolved valence electron energy loss spectroscopy, *J. Phys. D.* **29**, 1751-1760.

47. Howie, A. (1995) Recent developments in secondary electron imaging, *J. Microsc.* **180**, 192-203.

48. Shih, A., Yater, J., Pehrsson, P., Butler, J., Hor, C.and Abrams, R. (1997) Secondary electron emission from diamond surfaces, *J. Appl. Phys.* **82**, 1860-1867.

49. Perovic, D.D., Castell, M.R., Howie, A., Lavoie, C., Tiedje, T. and Cole J.S.W. (1995) Field emission SEM imaging of compositional and doping layer semiconductor superlattices, *Ultramicrosc.* **58**, 104-113.

50. Pennycook, S.J., Brown, L.M. and Craven, A.J. (1980) Observation of cathodoluminescence at single dislocations by STEM, *Phil. Mag. A* **41**, 589-600.

51. Williamson, J.C., Cao, J., Ihee, H., Frey, H. and Zewail, A.H. (1997) Clocking transient chemical changes by ultrafast electron diffraction, *Nature,* **386**, 159-162.

THE THREE FAMILIES OF ELEMENTARY PARTICLES

E. FERNÁNDEZ
IFAE/U.A. Barcelona
Campus U.A.B., Edifici Cn
E-08193 Bellaterra, Barcelona
Spain

1. Introduction

One of the talks of this conference will be about the discovery of the electron one hundred years ago at the Cavendish Laboratory of Cambridge University. This event marks the beginning of the current understanding of the structure of matter and it is often considered as the starting point of particle physics. The electron is in fact the first elementary particle, in the present sense of the word, which was discovered.

Since 1897 many more particles were identified. In fact, for some time, since the extensive use of particle accelerators in the fifties up to the early seventies, one could argue, with good reason, that there were too many elementary particles, or more precisely, too many particles to be elementary. The situation has changed dramatically since then. Now, in 1997, we have a reasonably small number of elementary particles, the electron still being one of them.

As it will be explained in the next sections, elementary particles can be classified into families, with a remarkable repetition, for each family, of their dynamical properties. The origin of this symmetry is not known, but it remains as one of the more basic problems in particle physics.

2. Two families of elementary particles

From the particle physics point of view matter is indeed very simple. All ordinary matter is made up of 2 leptons and 2 quarks: e (electron), ν_e (elec-

N. García et al. (eds.), Nanoscale Science and Technology, 333–347.

tron neutrino), u (up quark) and d (down quark). All these particles are fermions of spin 1/2. Up and down quarks make up the nucleons (protons and neutrons) which, together with the electrons, make up the atoms. The electron neutrino intervenes for example in beta radioactivity. Unless we look more closely, or more carefully, that is all there is.

For every particle there is an antiparticle. Actually, the neutrino of β^- radioactivity is in fact an anti-neutrino, while in β^+ radioactivity we have positrons (the antiparticles of electrons) and neutrinos. Positrons are also produced by cosmic rays (that is where they were discovered). Nowadays it has even been possible to form atoms of anti-hydrogen in the laboratory, namely at CERN.

In ordinary matter we also have the photon, which is another kind of particle, a boson. The boson is the carrier of the electromagnetic force, very successfully described by Quantum Electro Dynamics (QED), the first quantum field theory to be formulated.

If one looks more carefully things are not so simple. For example cosmic rays, at the earth surface, consists mostly of another charged particle, the muon, of which there are two kinds: μ^+ and μ^-. The muon was observed for the first time in the thirties, and it was taken at first as the carrier of the nuclear force. It is quite interesting that already then, the discoverers noticed something peculiar, e.g. Neddermayer and Anderson (in 1937) wrote: "The experimental fact that penetrating particles [that is muons] occur with positive and negative charges, suggests that they be created in pairs by photons, and that they might be represented as higher mass states of ordinary electrons" [1]. It so happens that the muons are not higher mass states of ordinary electrons, and, in cosmic rays, are not produced in pairs by photons (they are mainly produced in the decay of pions). But, for 1937, that was a very remarkable guess, for they pointed to one of the most fundamental problems still present in elementary particle physics, why the electron and the muon, except for the mass, behave identically. What do we mean by that?

What Neddermayer and Anderson were referring to is that the photon can "materialise" producing an electron-positron pair. This can be represented by the diagram

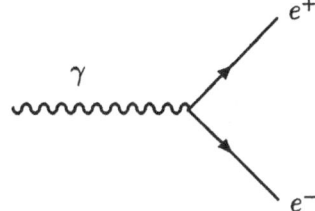

NOTE. (This Feynman diagram, and others in this talk, are only taken here as a convenient pictorial description of the reaction, and not in a technical sense. In the diagram time is running from left to right, and what the diagram says is that, at a given moment, the photon materialises as an electron-positron pair. The arrows represent that the particles emerge from the interaction. In a real Feynman diagram the arrow of the antiparticles should point "backward" in time, a technicality which is also ignored here. It should also be pointed out that the materialization of a photon into an electron-positron pair is not possible in vacuum, because of energy-momentum conservation, but it happens routinely in the presence of an external field. For all these reasons, the Feynman graphs describing this process are necessarily more complicated, but the essentials are well represented by the above diagram). Likewise, muons are also produced in pairs by photons,

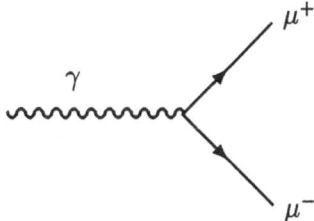

and that is what Neddermayer and Anderson were referring to.

Both reactions are entirely similar. Except for their masses, electrons and muons behave *identically* in the reaction, that is, if we disregard the kinematical effects due to the mass, the two reactions have exactly the same properties, such as rate, angular distribution of the particles, and so on. But the muons are heavier than electrons by a factor of roughly 200, and they can decay, with a lifetime of 2.197×10^{-6}s (the electrons cannot decay, they are absolutely stable as far as we know). Today we know that the decay is into an electron and two neutrinos, and that the two neutrinos are *different*:

$$\mu^- \rightarrow e^- \bar{\nu}_e \nu_\mu$$

We know there is an electron because we see it and measure its properties. We know that there are two neutral particles because we measure the spectrum of the electron, and we know that they are neutrinos because we never (or almost) see them!.

How do we know that the neutrinos are different? Historically, the different nature of the ν_e and ν_μ neutrinos was made manifest by an experiment in which accelerator-produced neutrinos were made to interact with a target [2]. The sequence of steps was the following. First, protons from an accelerator were made to collide with a target. In the proton-nucleon (neutron or proton, which we generically call N) reaction several secondary particles are typically produced, consisting mainly of pions. The pions then decay, and they produce (almost always) a muon an a neutrino:

$$pN \rightarrow ...\pi...$$

$$\pi \rightarrow \mu\nu$$

The muons are then filtered out (by a very large quantity of absorber) in such a way that downstream of the absorber only neutrinos remain. If enough of these reactions take place, an intense flux of neutrinos can be produced and occasionally some of them do interact with a nucleon of the material used as target. The idea of producing such a beam and study the interactions of the resulting neutrinos was proposed independently in 1960 by M. Schwartz [3] and B. Pontecorvo [4]. What was found in the experiment mentioned above [2] is that, in their interactions, the neutrinos so produced (in association with muons), always produce muons and other particles, but <u>never electrons</u>.

This is in marked contrast with what happens in beta-decay, in which a nucleus of a given (A,Z) changes to another nucleus with the same number of nucleons but with ± 1 electrons, that is,

$$(A, Z) \rightarrow (A, Z \pm 1)e\nu$$

This type of reaction takes place copiously, for example, in nuclear reactors, in such a way that an intense ν flux is produced. When these neutrinos interact they always produce electrons (that is how neutrinos were first discovered by Reines and Cowan in 1953 [5]), but <u>never muons</u>.

That is, the neutrinos which are produced together with μ's in π decay produce muons in their interaction with matter, they have a muon-like nature, and that is why they are called *muon* neutrinos and denoted by ν_μ. Likewise, the neutrinos produced together with electrons in β decay have an electron-like nature, they always produce electrons in their interaction with matter and that is why they are called electron neutrinos, and are denoted by ν_e.

The use of diagrams will help in clarifying this matter. For the electron neutrinos

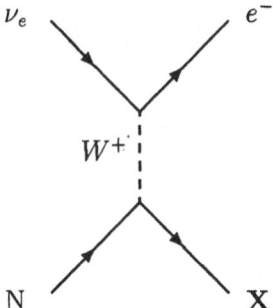

Here the symbol **X** stands for several particles which are the "debris" of the struck nucleon N, and W^+ is the boson carrier of the weak interaction (analogous to the photon for electromagnetic interactions). If the incoming particle were an electron antineutrino, $\bar{\nu}_e$, the final state charged lepton would be a positron, e^+, and the weak boson will be a W^-. In what follows we do not write the charge, with the understanding that the diagram represents both possibilities.

For muon neutrinos the diagram is

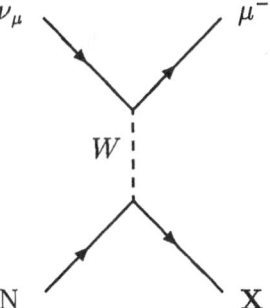

The analogy of the two diagrams is evident. The top vertex describes the weak interaction for leptons. It simply consists in that a neutrino (electron or muon type), undergoes a transition into the corresponding charged lepton (electron or muon), emitting a W particle, the carrier of the weak interaction. (As we have pointed out, the W is a charged particle and this type of interaction is called "charged-current weak interaction". There is another type of weak interaction, called "neutral-current weak interaction" where the exchanged particle is neutral, a Z, and the neutrinos do not change into charged leptons, but remain as neutrinos after the interaction. We will encounter it later). What we want to remark is that electron neutrinos never become muons, or muon neutrinos never become electrons. To describe this fact one introduces two separate "lepton numbers" L_e and L_μ. The e^- and the ν_e have $L_e = +1$ and $L_\mu = 0$ (for the antiparticles the values of the lepton numbers are the opposite, that is e^+ and $\bar\nu_e$ have $L_e = -1$ and $L_\mu = 0$). Analogously the μ^- and ν_μ have $L_e = 0$ and $L_\mu = +1$ (for μ^+ and $\bar\nu_\mu$, $L_e = 0$ and $L_\mu = -1$). In the weak interactions the two *leptonic numbers* are conserved separately.

It seems therefore that, for what concerns the weak interaction, there are two families of leptons, one formed by the electron and the electron-neutrino and another formed by the muon and the muon-neutrino.

But, how do we know that the weak interaction is the same in both cases? After all, electrons and muons are different, and so the upper vertex in the diagrams above, which represents the coupling (interaction) of the leptons and the W, needs not to be the same for both families. One way of checking if this is the case would be to compare the reactions pictured in both diagrams, and this has in fact been done. However the comparison is complicated by the fact that in the lower vertices the W interacts with the quarks, which are not free but bound inside the nucleon.

At present the best comparison between the charged current weak interaction for the electron and the muon families comes from the study of pion decay. The pion (a meson, made of a quark and an antiquark) decays in two different ways: $\pi \to e\nu_e$ and $\pi \to \mu\nu_\mu$. In terms of diagrams we have

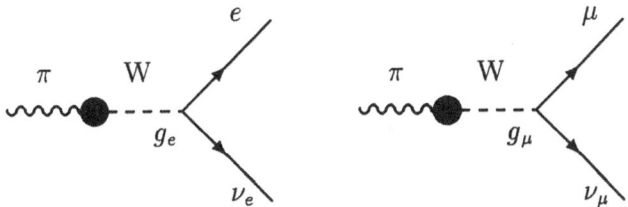

The blob represents the interactions of the quarks inside the pion, from which a W emerges. Since the quarks are not free this interaction is also complicated, and thus the representation by means of a blob, but in any case, the blob is identical for both decay modes and it cancels if one takes the ratio of their rates. From this ratio, it has been possible to infer that the strength of the charged current weak interaction for electrons and muons, represented by the "coupling constants" g_e and g_μ is the same, to 1.5 parts in a thousand, namely [6]

$$\left| \frac{g_\mu}{g_e} \right| = 1.0005 \pm 0.0015$$

(It should be pointed out that the rates for both decay modes are very different, the muonic channel being much more likely. This is due to the large dependence of the weak interaction on the spin orientation of the particles, which, for kinematical reasons it is not the same for electrons and muons produced in the above decays. But this effect is understood and can be taken into account when making the comparison).

3. A third family of leptons

In 1975, the first member of a third family of leptons was discovered by M. Perl and collaborators at SLAC [7]. What was done there was to collide electrons and positrons. The electron-positron system is very clean, since the electrons are point particles, and furthermore only electromagnetic and weak interactions take place. What was seen there was that in some cases the result of the reaction was

$$e^+ e^- \rightarrow e + \mu + (neutral\ particles)$$

The presence of electrons and muons was very striking, remember that the electron and muon families do not mix. Some people even thought that the leptonic lepton numbers were not conserved separately, after all.

But remember that in muon decay, as it was mentioned before we have electrons and neutrinos. The diagram for μ decay is

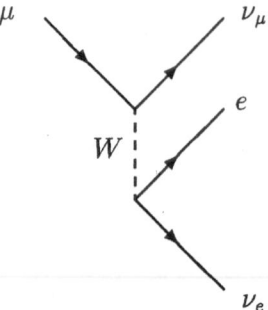

In this case both vertices represent weak interactions of leptons, that at the top for the muon family, the one at the bottom for the electron family.

In the e^+e^- interactions it was concluded, after some effort, that the reaction was,

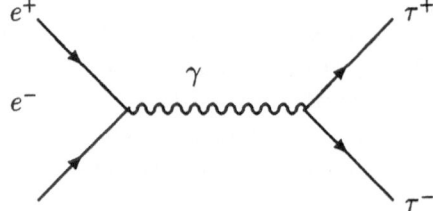

where the τ was a new lepton. What was happening is that, in some cases, one of the taus decayed as

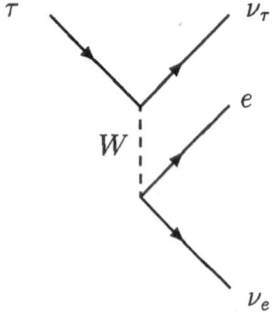

while the other decayed as

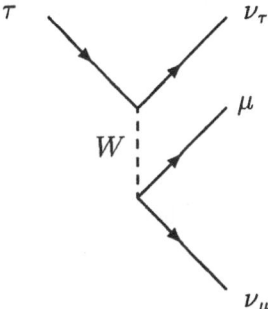

Both decays happen almost immediately after the taus are produced, in such a way that the only particles seen in the detector are an electron and a muon, the four neutrinos escaping undetected. These two decays are very similar to mu decay. The tau, being heavier than both the electron and the muon can decay in the two ways shown above. (The tau, since it is very heavy, 1.776 GeV, can also decay into hadrons, the only lepton for which that is possible).

Much more recently, at the LEP accelerator at CERN, it has been possible to isolate taus in a very clean way and thus to check their decays with great precision. At LEP the production of the taus occurs mainly in the way depicted by the diagram

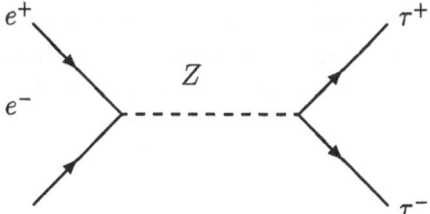

This is also a weak interaction, mediated by a neutral particle, the Z boson. It is the so called *neutral current weak interaction*.

When the energy of the e^+ and the e^- beams is such that their sum is equal to the mass of the Z the reaction is resonant, and the cross section becomes very large. This is precisely what has been done at the first phase of the LEP accelerator at CERN.

Clearly, the ratio of the rates of the two decay modes of the tau shown above will also give us the relative strength of the weak interaction for electrons and for muons. The result is that they are equal, the present precision of the measurement being of 3 parts in a thousand, only slightly worse of what was obtained from pion decay [8].

From the study of *tau* decays it is also possible to test the relative strength of the weak interaction for taus, g_τ (the τ-W-ν_τ vertex), with that of electrons and muons, a comparison which slightly more complicated than the $e - \mu$ comparison explained above. The best number, which is obtained at LEP, gives [8]

$$\left|\frac{g_\tau}{g_\mu}\right| = 1.0024 \pm 0.0045$$

From what it has been said it is clear that there are three families of leptons and that the weak interaction is universal, exactly the same inside each family.

4. Families of Quarks

So far I have avoided to talk about quarks. The study of quarks is more complicated since the quarks also interact strongly, and this interaction often dominates over the weak interaction. Furthermore the quarks are always bound inside hadrons, and therefore the study of their interactions, strong or weak, is more difficult. I will not try to explain here the story of the quarks or of the strong interactions, which are very fascinating by themselves, but I will simply mention some of the characteristics of the weak interaction of quarks, which are relevant for their classification into families.

Today we know that, as in the case of leptons, there are also three families of quarks: (up, down), (charm, strange), (top, bottom). But there are some characteristics which make quark and lepton families very different. For one the mass differences between quarks are much larger than between leptons. Moreover the quark families are slightly "mixed" with respect to the week interactions. What does "mixed" mean?

As for leptons, a charged-current weak interaction for quarks consists on a transition of a member of one family into the other member, emitting a W. In the prototypical weak interaction, β^--decay, a *down* quark inside the proton of a nucleus changes into an *up* quark, emitting a W^-, which in turn "decays" into an e^- and an $\bar{\nu}_e$:

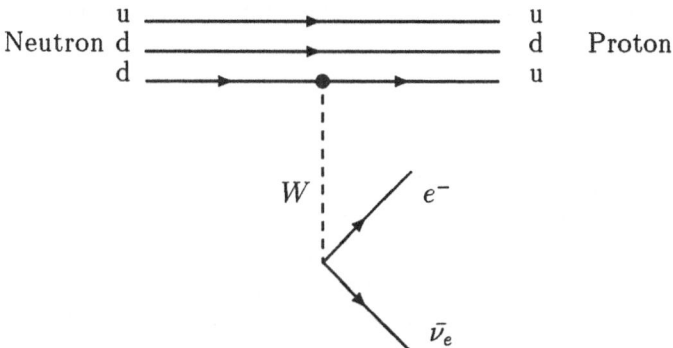

As a consequence a neutron becomes a proton and the resulting nucleus changes the atomic number, $Z \to Z + 1$.

The above diagram is very similar to those previously encountered. At the bottom vertex we have the familiar weak interaction for leptons, while at the top vertex we have a weak interaction for quarks in which the d quark changes into the other member of its family, the u quark, emitting a W. The other two quarks in the neutron, u and d, do not participate in the interaction (they are actually called "spectators"). Up to now everything is very similar to the leptonic weak interaction. However there are cases where a quark, of the down-type say, emits a W^- becoming a up-type quark of a <u>different</u> family. An example is that of the Λ particle, which is a bound state of an u quark, a d quark and a s quark. The Λ decays predominantly (64% of the times) into a p (proton) and a π^-. This is a weak decay in which the s quark, which belongs to the second family, changes into an *up* quark, which belongs the the first family!:

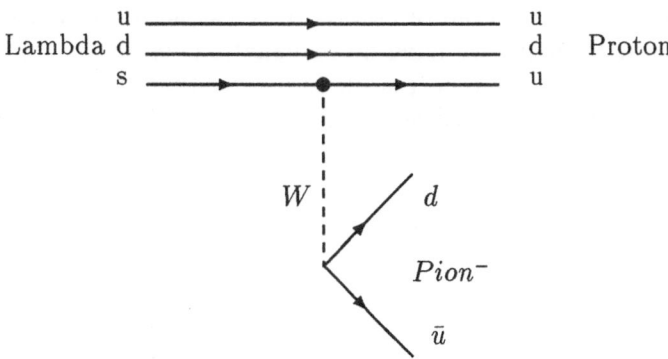

(The $\bar{u}$ and the d of the final state form a bound state, which is the π^- particle). The lifetime of the Λ particle is very long and this reflects the fact that the transition probability of a quark of one family to another is very small. This phenomenon, called mixing, has been observed between the three generations.

One intriguing fact is that there seem to be a connection between masses and mixing. The first evidence for such a possible connection comes from a measurement we did at SLAC in 1983 in which the lifetime of particles containing b quarks was measured for the first time. The b quark, which belongs to the third family, can only undergo weak transitions to quarks of the first or second family. The reason is that the other member of the third family, the top, has a much larger mass. Therefore weak decays of b particles can only occur via mixing with another family. Assuming that the mixing from the third to the second family was equal to the mixing between the second and the first, one could roughly calculate the lifetime, which gave a number of the order of between 10^{-13} to 10^{-15}. This such a small lifetime was almost without reach of the experiments at the time. However, when actually doing the experiment, we found a lifetime of [9]

$$[1.8 \pm 0.6 \ (stat.) \pm 0.4 \ (syst.)] \times 10^{-12} s.$$

a surprisingly long lifetime, which was soon confirmed by another experiment, also at SLAC [10]. The importance of this long lifetime (which is now much better measured, at LEP [11], $(1.6 \pm 0.04) \times 10^{-12} s$.) is that the mixing of the third and the second families is much smaller that that of the second and first. There seems to be some hierarchy in the mixing in parallel to that of the masses. Are these two facts related? At present the question has no answer.

5. Open Questions

One could wonder if the repetition of the structure of matter into families will continue. Since we have no explanation for the pattern into families, to start with, we have no reason to believe that there should not be more families, to be discovered as we increase the reach of our experiments. One of the great achievements of the LEP and SLC experiments is that, under quite general assumptions, there are only three families of particles (see [12] and references therein). The measurement, which I will not attempt to describe in such a short time, concerns specifically the number of neutrino families, with "standard" neutrinos, "standard" meaning here that they have a small mass and that they are subject to the know weak interaction, conditions that are very well satisfied by the three known neutrinos.

A family with a very heavy neutrino cannot be excluded from the LEP experiments, but if would certainly be a very different family from those known. The number of families is also limited to less or equal than 4 from the measured fraction of Helium to Hydrogen in the universe and also very general assumptions about the evolution of the universe.

To finish I will mention briefly one current open issue in particle physics which is related to masses and mixings, that of neutrino oscillations.

Neutrinos are leptons and, from what we have said, no mixing has been observed among lepton families. However, there is no reason a priori for this fact, as there is no reason for the quark mixing. If neutrinos had mass, it could happen that they also "mix" with respect to the weak interaction, as quarks do. It may be that we have not been able yet to observe the effect, as neutrino interactions are very difficult to detect. There are in fact some experimental "anomalies" with neutrino interactions which are intriguing and may point in the direction of mixing.

One anomaly is the so-called "solar neutrino deficit" (see [13] and references therein). Neutrinos are produced copiously in the nuclear reactions and nuclear decays that take place in the Sun. Their interactions have been observed and measured in a variety of experiments for almost 20 years and the flux has been calculated by several authors. There are several models but they give fluxes which are higher, from 30 to 50 per cent, with respect to the observation of several experiments. The discrepancies constitute the "solar neutrino deficit".

Neutrinos are also produced by interaction of the so called primary cosmic rays, mostly protons, which constantly bombard the upper atmosphere. As in accelerators, the primary protons produce $\pi^{\pm}$. The π^- decays mainly into a μ^- and a $\bar{\nu}_{\mu}$ followed by the μ^- decay into an e^-, a ν_{μ} and a $\bar{\nu}_e$ (the same for π^+, with all the particles replaced by the corresponding antiparticles). The neutrinos produced in this chain of reactions are the so called atmospheric neutrinos and they have much larger energies than solar neutrinos. One would expect that the ratio of $\nu_{\mu} + \bar{\nu}_{\mu}$ to $\nu_e + \bar{\nu}_e$ would be approximately 2, while the experimental value is more close to 1. This is know at the "atmospheric neutrino anomaly" (see [13] and references therein).

These facts have led to several interpretation of the data in terms of mixing of the neutrino families, in which a neutrino of a given family changes into a neutrino of a different family, or, more precisely, the quantum state of a a given neutrino becomes a superposition of states consisting of neutrinos of different families, the fraction of each species being a function of time. When the neutrino interacts weakly it does so as one of the three species. The probability of observing a given species will vary with time and thus with distance from the source, and hence the name neutrino oscillation. For the oscillation to take place two conditions are necessary. One is that neutrinos have non-zero mass, the other is that, if neutrinos are "standard", there are leptonic weak transitions between families.

For example one explanation for the solar neutrino deficit would be that the solar neutrinos, which are produced as electron neutrinos in the Sun, would become a mixture of electron and muon neutrinos when they fly away. Since the experiments being done to detect them at the Earth (which are very difficult experiments indeed) assume that they produce electrons, they will fail to see the muon neutrino component and thus see a deficit. Furthermore, an experiment at Los Alamos has published results which would imply electron to muon neutrino oscillations [14].

This is a very active field of research both experimentally and phenomenologically and, for obvious reasons, cannot be treated in this brief article. The results, if interpreted as oscillations, imply most likely very small mass differences and relatively large mixings. Several experiments are being contemplated, in Japan, the US and Europe to study this problem. Some of them consist of sending a beam of neutrinos, produced at an accelerator, to neutrino detectors situated hundreds of kilometers away, to which they would arrive after travelling through the Earth.

6. Conclusions

The classification of elementary particles into families, as explained, for both quarks and leptons, shows a clear regularity. The structure of each family is identical, only masses are different. Is there something behind this regularity? As yet there is no answer, but I hope to have convinced you of the importance of the question.

7. Acknowledgments

I would like to thank Nicolás García for the invitation to give the talk (and for the 28 years of friendship that precede it) and Antonio Correia for his help on many practical matters related to the Conference.

8. References

1. Neddermeyer, S.H. and Anderson, C.D. (1937), *Phys. Rev.*51, 884.

2. Danby, G. et al. (1962), it Phys. Rev. Lett. **9**, 36.

3. Schwartz, M. (1960), *Phys. Rev. Lett.* **4**, 306.

4. Pontecorvo, B. (1960), *Soviet Physics JETP*, **37**, 1236.

5. Reines, F. and Cowan, C.L. (1953), *Phys. Rev.* **92**, 830.

6. Bretton, D.I. et al. (1992), *Phys. Rev. Lett.* **68**, 3000; Czapek, C. et al. (1993), *Phys. Rev. Lett.* **70**, 17.

7. Perl, M.L. et al. (1975), *Phys. Rev. Lett.* **35**, 1489.

8. Rolandi, G. (1997) Highlights of Tau96, *Nuc. Phys. B (Proc. Suppl.)* **55**, 461.

9. Fernández, E. et al. (1983), *Phys. Rev. Lett.* **51**, 1022.

10. Lockyer, N.S. et al. (1983), *Phys. Rev. Lett.* **51**, 1316.

11. Richman, J.D. (1996), in *Proceedings of the 28th International Conference on High Energy Physics*, World Scientific Pub. Co., Singapore, pag. 143.

12. Smith, T.J. (1995), in *Proceedings of the International Europhysics Conference on High Energy Physics*, World Scientific Pub. Co., Singapore, pag. 23.

13. Nussinov, S. (1995), Summary Talk on Neutrino-94, *Nuc. Phys. B (Proc. Suppl.)* **38**. 497.

14. Athanassopoulos, C. et al. (1996), *Phys. Rev. Lett.* **77**, 3082.

Author Index

Index